Microbial Strategies for Vegetable Production

Almas Zaidi • Mohammad Saghir Khan
Editors

Microbial Strategies for Vegetable Production

 Springer

Editors
Almas Zaidi
Department of Agricultural Microbiology
Aligarh Muslim University
Aligarh
India

Mohammad Saghir Khan
Department of Agricultural Microbiology
Aligarh Muslim University
Aligarh
India

ISBN 978-3-319-54400-7   ISBN 978-3-319-54401-4   (eBook)
DOI 10.1007/978-3-319-54401-4

Library of Congress Control Number: 2017942958

Printed on acid-free paper

This Springer imprint is published by Springer Nature
The registered company is Springer International Publishing AG
The registered company address is: Gewerbestrasse 11, 6330 Cham, Switzerland

Preface

Vegetables are one of the most essential components of human dietary systems due to their high nutritional value which provides carbohydrates, proteins, vitamins, and several other useful food elements. Due to these, the consumption of vegetables among health-conscious consumers has increased considerably. In order to fulfill the growing demands of vegetarians, there is an urgent need to enhance the production of fresh and quality vegetables. To optimize production, growers often use energy-intensive fertilizers and pesticides in vegetable farming. The excessive and abrupt application of such agrochemicals for longer duration has, however, been found to disrupt soil fertility and consequently the production of quality vegetables. In addition, soil destruction through abiotic stress and loss of soil fertility following various soil management practices has also compounded vegetable farming in recent times. Therefore, to avoid/minimize the consistent application of expensive and disruptive chemicals in vegetable production practices, viable and practically applicable alternative strategies need to be developed. In this regard, the advent of microbial preparation often called biofertilizers involving many useful soil microbiotas has provided an effective solution to high-input agrochemicals. And hence the use of nonpathogenic rhizosphere microbes to enhance vegetable production is currently considered as a safe, viable, and low-cost alternative to chemicals. Since soil microorganisms are inexpensive and do not cause any pollution, they have been used repeatedly for maximizing the production of many crop plants across different agronomic practices. Also, even though there are no direct connections between many rhizosphere microfloras and vegetables, yet several plant growth-promoting rhizobacteria (PGPR) spanning different genera have been used to facilitate the growth and yield of numerous vegetables through different mechanisms.

Recently, the interest in using renewable resources like beneficial soil microbes for producing fresh and high-quality vegetables has grown substantially. However, most of the farmers engaged in growing organic vegetables, even though they adopt such microbial strategies, do not have correct understanding of such bioformulations and do not know how to apply them properly so that maximum benefits are achieved. Additionally, soil microflora has become important due to its role in disease management and reclamation of derelict soils (salinized/polluted soils). The success of microbes, however, depends largely on their inherent ability and the acceptance and adoption by the vegetable growers. A considerable amount of research work has been conducted to explain the impact of rhizosphere microbes in the enhancement of vegetable crops, but very few efforts have been made to systematize such information that could benefit students/teachers/horticulturists and progressive vegetable-growing farming communities. Considering the importance of beneficial soil microbes and success achieved so far, efforts herein have been directed to highlight the impact of microbiota on the quality and yield of vegetables grown in different agronomic regions of the world. Furthermore, efforts in this book will also be made to identify most suitable organisms which could effectively be applied for optimizing vegetable production.

Microbial Strategies for Vegetable Production edited by experts focuses on the fundamental and practical aspects of beneficial soil microbes employed commonly in the sustainable production of vegetables. This book further presents exceptional, simplified, and wide-ranging information on important soil microbiota which could be used to enhance the production of vegetables in different regions. The book deals with the application of microbial inoculants and many plant growth-promoting rhizobacteria (PGPR) including nitrogen-fixing and phosphate-solubilizing organisms in vegetable production. Even though there is no direct connection between nitrogen-fixing organisms and vegetables as reported for rhizobia and legumes, yet the recent developments in the use of nitrogen-fixing, plant growth-promoting rhizobacteria in sustainable production of vegetables have been sufficiently discussed. The application of PGPR in the improvement of vegetable crop production under stress conditions like salinized soils, drought, high and low temperature, and nutrient stress and heavy metal-stressed conditions is broadly covered in the book. The role of PGPR in growth and yield promotion of tomato is dealt separately. This book also provides information on sources of heavy metal pollution, metal toxicity to vegetables, and bioremediation strategies adopted to clean up metal-contaminated soils. Furthermore, the role of microbes in enhancing the quality and production of vegetables grown under metal-polluted soil is discussed separately. Recent advances in effective disease management by PGPR to control phytopathogens causing diseases on onion (*Allium cepa*), cucumber (*Cucumis sativus* L.), lettuce (*Lactuca sativa*), spinach (*Spinacia oleracea*), and broccoli (*Brassica oleracea*) are discussed paving the way for exploration of microbes for other vegetable crops as well. This book, therefore, can be used as a reference which is likely to be a very useful resource for vegetable growers.

We are highly thankful to our learned colleagues who from different countries contributed their recent and updated chapters in this informative and most demanding book. All chapters presented in this book are written superbly and give elaborate and meaningful information. We would also like to thank our research scholars who were easily available at all times during the preparation/compilation of this book and made this book a reality. *Microbial Strategies for Vegetable Production* provides enough information especially to farmers engaged in vegetable production. The facts and data together with various methodologies presented here may be an imperative source material. This book will practically be valuable for a wide range of people including students/researchers/vegetable growers.

The support and patience of our family members especially our two adorable daughters Zainab and Butool during the entire period of preparation and compilation of this book were commendable for which we are extremely thankful to them. We are also very grateful to the publisher of this book in responding to all our queries very promptly and urgently. Finally, we will be extremely happy and obliged if someone identifies some conceptual or printing mistakes and inform us. We will try to resolve them in our next edition.

Almas Zaidi
Mohammad Saghir Khan

Contents

List of Contributors

Rai Abdelwahab FSNV, Laboratoire de Maitrise des Energies Renouvelables (LMER), Equipe de Biomasse et Environnement, Université de Béjaïa, Targa Ouzemmour, Béjaïa 06000, Algérie

Bilal Ahmad Faculty of Agricultural Sciences, Department of Agricultural Microbiology, Aligarh Muslim University, Aligarh 202002, Uttar Pradesh, India

Sasirekha Bhaktavatchalu Department of Microbiology, Acharya Bangalore B SchoolOff Magadi Road, Bangalore 560 091, India

Ranjit Chatterjee Department of Vegetable and Spice Crops, Uttar Banga Krishi Viswavidyalaya, Pundibari, Cooch Behar, West Bengal, India

Nabti Elhafid FSNV, Laboratoire de Maitrise des Energies Renouvelables (LMER), Equipe de Biomasse et Environnement, Université de Béjaïa, Targa Ouzemmour, Béjaïa 06000, Algérie

Adem Güneş Agricultural Faculty, Soil Science and Plant Nutrition Department, Erciyes University, Kayseri, Turkey

Mohammad Saghir Khan Faculty of Agricultural Sciences, Department of Agricultural Microbiology, Aligarh Muslim University, Aligarh 202002, Uttar Pradesh, India

Nurgul Kitir Engineering Faculty, Genetics and Bioengineering Department, Yeditepe University, Istanbul, Turkey

Mokhtari N.E.P Organic Farming Department, Islahiye Vocational School, Gaziantep University, Gaziantep, Turkey

Emrah Nikerel Engineering Faculty, Genetics and Bioengineering Department, Yeditepe University, Istanbul, Turkey

Bahar Sogutmaz Ozdemir Engineering Faculty, Genetics and Bioengineering Department, Yeditepe University, Istanbul, Turkey

Asfa Rizvi Faculty of Agricultural Sciences, Department of Agricultural Microbiology, Aligarh Muslim University, Aligarh 202002, Uttar Pradesh, India

Ayon Roy Department of Plant Pathology, Uttar Banga Krishi Viswavidyalaya, Pundibari, Cooch Behar, West Bengal, India

Saima Saif Faculty of Agricultural Sciences, Department of Agricultural Microbiology, Aligarh Muslim University, Aligarh 202002, Uttar Pradesh, India

Mohammad Shahid Faculty of Agricultural Sciences, Department of Agricultural Microbiology, Aligarh Muslim University, Aligarh 202002, Uttar Pradesh, India

Srividya Shivakumar Department of Microbiology, Center for Post Graduate Studies, Jain University, Bangalore 560 011, India

Ravi Kiran Thirumdasu Department of Vegetable and Spice Crops, Uttar Banga Krishi Viswavidyalaya, Pundibari, Cooch Behar, West Bengal, India

Metin Turan Engineering Faculty, Genetics and Bioengineering Department, Yeditepe University, Istanbul, Turkey

Ceren Unek Engineering Faculty, Genetics and Bioengineering Department, Yeditepe University, Istanbul, Turkey

Ertan Yildirim Horticulture and Viticulture Department, Ataturk University Agricultural Faculty, Erzurum, Turkey

Almas Zaidi Faculty of Agricultural Sciences, Department of Agricultural Microbiology, Aligarh Muslim University, Aligarh 202002, Uttar Pradesh, India

The Editors

Almas Zaidi received her M.Sc. and Ph.D. (agricultural microbiology) from Aligarh Muslim University, Aligarh, India, and is currently serving as guest teacher/assistant professor at the Department of Agricultural Microbiology, Aligarh Muslim University, Aligarh, India. Dr. Zaidi has been teaching microbiology at postgraduate level for the last 12 years and has research experience of 16 years. She has published above 50 research papers, book chapters, and review articles in journals of national and international repute. Dr. Zaidi has edited seven books published by leading publishers. Her main focus of research is to address problems related to rhizo-microbiology, microbiology, environmental microbiology, and biofertilizer technology.

Mohammad Saghir Khan Ph.D., is a professor at the Department of Agricultural Microbiology, Aligarh Muslim University, Aligarh, India. Dr. Khan received his M.Sc. from Aligarh Muslim University, Aligarh, India, and his Ph.D. (microbiology) from Govind Ballabh Pant University of Agriculture and Technology, Pantnagar, India. He has been teaching microbiology to postgraduate students for the last 20 years and has research experience of 24 years. In addition to his teaching, Dr. Khan is engaged in guiding students for their doctoral degree in microbiology. He has published over 100 scientific papers including original research articles, review articles, and book chapters in various national and international publication media. Dr. Khan has also edited nine books published by leading publishers. Dr. Khan is deeply involved in research activities focusing mainly on rhizobiology, microbiology, environmental microbiology especially heavy metals-microbes-legume interaction, bioremediation, pesticide-PGPR-plant interaction, biofertilizers, and rhizo-immunology.

Microbial Inoculants in Organic Vegetable Production: Current Perspective

1

Ranjit Chatterjee, Ayon Roy, and Ravi Kiran Thirumdasu

Abstract

Vegetable crops provide food and nutritional security to millions of people. They are rich in moisture and essential nutrients that make them susceptible to diseases and pests. To increase the productivity and to prevent disease and pest attack, a wide range of agrochemicals are applied to the crop which leave harmful residues in vegetables and consequently pollute the soil and groundwater. In the present situation, the growing awareness on consumption of contaminated food products and the ill effects of chemical farming on environment make people more concern for food quality and safety leading to more focus on organic vegetable production. Generally, organic farming avoids or largely excludes the use of synthetic fertilisers, pesticides, plant growth regulators, etc. but primarily rely upon biological cycle within the farming system. As a component of organic farming, microbial inoculants performs pivotal role in crop production through decomposition of organic residues, improving nutrient uptake and availability, mineralization, nutrient recycling, detoxification of organic and inorganic substances, supply of plant growth-promoting compound and suppression of disease and pest. Due to constantly diminishing biological wealth, utilisation of bioinoculants will be one of the promising alternatives as renewable resource for promoting organic vegetables. Here, an attempt has been made to highlight the potential microbial inoculants and their benefits in sustainable cultivation of organic vegetables.

R. Chatterjee (✉) • R.K. Thirumdasu
Department of Vegetable and Spice Crops, Uttar Banga Krishi Viswavidyalaya,
Pundibari, Cooch Behar, West Bengal, India
e-mail: ranchat22@rediffmail.com

A. Roy
Department of Plant Pathology, Uttar Banga Krishi Viswavidyalaya,
Pundibari, Cooch Behar, West Bengal, India

© Springer International Publishing AG 2017
A. Zaidi, M.S. Khan (eds.), *Microbial Strategies for Vegetable Production*,
DOI 10.1007/978-3-319-54401-4_1

1.1   Introduction

The modern-day conventional agricultural practices demand different agrochemicals, namely, fertilisers, pesticides, growth regulators, etc., for optimum crop yield. Excessive application of fertilisers, pesticides and synthetic hormones are, however, causing severe damage to the soil fertility and environment as well as harvested produce (Chatterjee 2009). The residues of pesticides and fertilisers that persist in soil invariably destroy the beneficial microorganisms, earthworms and other soil habitants and contaminate the water and water bodies that ultimately become unfit for human consumption (Bishnu et al. 2009). For example, common, annually used pesticides such as thiram, lindane, dimethoate, linuron, maleic hydrazide and glyphosate had variable impacts on numbers and activities of different soil microorganisms and pesticide residues in soil and on yield of carrot (*Daucus carota*), grown in Central Finland. The pesticide treatments in general reduced the growth of soil algae but increased the total number of microorganisms and the number of aerobic spore-forming bacteria. Furthermore, the carrot yield in pesticide-treated plots was only 20–60% of the yield in the hand-weeded plots (Heinonen-Tanski et al. 1986). Realising the threats of agrochemicals in general, people are gradually realising the danger of modern-day production system and are therefore asking for fertilisers and pesticide residue-free food items. In order to circumvent the toxic impact of chemicals, organic farming is being considered as a sound and viable alternative for sustainability of the production system, soil health and environment across the world (Lauridsen et al. 2005).

Organic farming is a system of agriculture which avoids or largely excludes the use of off-farm inputs such as synthetic fertilisers, pesticides, plant growth regulators, antibiotics and livestock feed additives. Moreover, organic farming primarily relies upon natural organic inputs like animal manures, organic residues, crop rotations, mineral-grade rock additives and biological plant protection measures mainly for diverse population of soil organisms (Lampin 1990). The aim of organic farming is essentially to promote and enhance the health of varying agroecosystems, including biodiversity, biological cycles and soil biological activities. Organic farming builds a healthy soil that is alive with beneficial organisms that release, transform and transfer nutrients. The rich soil organic matter contributes to good soil structure and water-holding capacity. Properly managed organic farming reduces or eliminates pesticide residues in food product and soil and water pollution and helps to conserve agro-biodiversity and sustainability of the production system (Palaniappan and Annadurai 1999). Organic farming systems have a strong potential to withstand climate variability, including erratic rainfall and temperature variations and other unexpected events (Chatterjee and Thirumdasu 2015). Food production under long-term organic management is more resistant to the production threats like drought and sudden change of climatic parameters, and, hence, the production system becomes more stable and sustainable.

The relatively high success of organic farming in western countries is due to growing public concern about safety and quality of food and the high awareness of the health problems caused by the consumption of contaminated food products.

The adverse impacts of chemical-based agriculture practices on the environment have also motivated farmers to adopt organic farming. The organic market is growing very fast throughout the world. India has vast potential to compete with the emerging intentional market because most of the cultivated area still remains free of contamination from chemicals, spread over distinctly varying agro-climatic conditions that can be easily converted to organic farming. There is an emerging need to create public awareness regarding ill effect of chemical farming and educating about benefits of organic products and production system on soil, environment and human/animal health.

1.2   Need of Organic Vegetables

Vegetable crops are group of herbaceous plants, in which different plant parts like root, stem, leaf, flower, fruits, etc. are consumed as raw or after cooking. They are the cheaper source of natural protective nutrients, namely, carbohydrates (cassava, sweet potato, potato, *Colocasia*), protein (peas, beans, drumstick, agathi flower, fenugreek leaves), fat (*Colocasia* leaves, drumstick leaves), minerals (spinach, amaranth, fenugreek, coriander) and vitamins (carrot, beet, cabbage, tomato, chilli). Vegetables are good source of essential amino acids that are lacking in cereals and pulses. A crop like *Amaranthus*, okra, ridge gourd and sponge gourd is valuable source of roughages in the form of dietary fibre and neutralises the acid formed during the digestion of protein and fatty foods. Coloured vegetables, for example, orange carrot, purple cabbage/cauliflower, yellow pepper and red tomato, are rich source of antioxidant and anticancerous molecules. The Indian Council of Medical Research (ICMR) recommended that an adult human should consume at least 300 g of vegetable per day of which 125 g should be leafy vegetables, 100 g roots and tuber vegetables and 75 g fruit vegetables (Anonymous 2009).

India is the land of vegetable crops and ranks second in production just after China in the world, and during 2013–2014, India produced 162.90 million tons of vegetables grown in 9.40 million hectare area and accounting for nearly 14% of the world production (Anonymous 2014). To obtain higher vegetable yields, farmers are indiscriminately using different inorganic fertilisers, synthetic hormones and pesticide that make the plant susceptible to pests and diseases besides deteriorating the quality of end produce (Rembialkowska 2003). Also, the rich moisture and essential nutrient contents make vegetable more susceptible to diseases and pest attack which causes significant yield losses. To prevent disease and pest attack, a wide range of pesticides including fungicides, insecticides, acaricides, herbicide, etc. are sprayed to the crops which leave harmful residues in/on the surface of the vegetables. Regular consumption of fertilisers and pesticide residue-rich vegetable, however, may disrupt the functions of central nervous system, cardiovascular system and respiratory system (Gilden et al. 2010). After uptake, pesticide residues may concentrate in the liver and kidney and can damage these organs and in later stages can lead to cancer (Bassil et al. 2007). Beside these, the nitrite and nitrate form of nitrogenous fertiliser pollute the groundwater resulting in serious environmental threats.

A good number of vegetables are consumed in raw form as salad like tomato, cucumber, carrot, bell pepper, radish, onion, pea, etc. Again several vegetables are consumed after partial boiling or after light frying like okra, bottle gourd, bitter gourd, pumpkin, pea, French bean, palak, spinach, fenugreek, etc. The fertiliser and pesticide residue of these vegetables is emerging as major concern for the health and safety of the common people. In the present scenario, the growing awareness about health and environment makes people more worried for food quality and safety leading to more focus on organic vegetable production. Therefore, the demand for organically grown vegetables is increasing rapidly throughout the globe. The increasing emphasis on quality, particularly taste, has also encouraged growers to produce organic vegetables. The demand for organic vegetables is increasing sharply both in domestic and international market. In contrast, several evidences have shown that the nutritional value of important commercial vegetables is declining when grown under conventional farming practices (Woese et al. 1995; Worthington 2001). However, there is evidence that suggest that organic vegetables are significantly higher in phosphorus, iron, magnesium and vitamin C content and lower in nitrate content (Worthington 2001). In a study, Magkos et al. (2006) reported that vitamin C in organic leafy vegetables and potato was comparatively higher than normal crops. Lower nitrate in organically grown vegetable could possibly be due to relatively lower availability of N in organic farming system (Bourn and Prescott 2002). Brand and Molgaard (2001) estimated that organic vegetable may contain 10–50% higher defence-related secondary metabolites than conventionally grown vegetables. Ren et al. (2001) reported that organically grown onion, green pepper and leafy vegetables contain 1.3–10.4 times higher quercetin concentration and possessed higher antioxidant and antimutagenic activities compared to conventionally grown vegetables.

1.3 Essential Characteristics of Organic Farming Systems

Organic farming encourages and enhances the biological cycle within the farming system by involving the soil microbes, soil flora and fauna and plant and animal residues. To adopt a complete organic vegetable production system, the growers have to follow certain guidelines in the farming practices. First, a suitable organic resource-based production system "organic system plan" has to be designed that will describe the practices used in producing crops and livestock products. Emphasis should be given on crop cultivation with the use of on-farm resources or locally available renewable resources such as animal manures (farmyard manure, compost, vermicompost, etc.), organic waste recycling, crop rotations, mineral-grade rock additives and biological inputs (biofertilisers). The use of purchased inputs is discouraged for crop cultivation. As part of plant protection strategy, the use of resistant varieties, summer ploughing, soil solarisation, diverse bioinoculants, natural predators, beneficial insects and birds, etc. is promoted to reduce the pressure of harmful pesticides. In organic farming system, the organic fields need to be separated from the conventional field through a divider or buffer zone. It will prevent the inadvertent contamination by synthetic farm chemicals from adjacent conventional fields. More emphasis is given on conservation of soil, environment, wildlife and natural habitats to enhance agro-biodiversity and ecological balance of the farm. In case of urgency or export purpose, only permitted and allowable chemical substances may be used for crop production. In any case, the prohibited and restricted substances should not be used in the field. A detailed record-keeping system covering the entire production system, input details and dose and time of application is an essential part of organic cultivation.

1.4   Microbial Inoculants as Components of Organic Production

Microbial inoculants are the products containing living cells of efficient strains of different types of microorganisms which are capable of mobilising nutritive elements from insoluble to soluble form through biological processes. They accelerate different microbial processes in soil and enhance the availability of nutrients to growing plants. When applied as seed or soil inoculants, they multiply and participate in nutrient cycling and benefit crop productivity (Singh et al. 2011). They play active role in transforming atmospheric nitrogen into usable N form (Lhuissier et al. 2001); solubilisation or mobilisation of important plant nutrients such as P, K, Zn, S and Fe (Chen et al. 2006; Coyne and Mikkelsen 2015); release of plant growth-promoting hormones like indoleacetic acid, gibberellic acid, cytokinins and ethylene (Youssef and Eissa 2014); and biodegradation of organic matter in the soil (Sinha et al. 2014).

The beneficial soil microbiota can be classified as (1) biofertilisers (N_2 fixers, P solubiliser and siderophore producer), (2) general microbial plant growth promoter (AM – fungi, fungal biocontrol agents like *Trichoderma*, *Gliocladium*, etc.), (3) microbial plant growth regulator (which is able to produce different types of phytohormones, namely, auxins, brassinosteroids, cytokinins, gibberellins, abscisic acid, ethylene and the recently discovered strigolactones) and (4) phytoremediator (microorganism having the potential to degrade pesticides). Thus, the use of plant growth-promoting microorganisms (PGPMs) offers an attractive way to replace chemical fertilisers, pesticides and supplements. The PGPMs due to their ability to colonise and establish on the crop roots are also known as plant growth-promoting rhizobacteria (PGPR). The major rhizobacterial genera include species belonging to genera *Acetobacter*, *Arthrobacter*, *Azospirillum*, *Azotobacter*, *Bacillus*, *Flavobacterium*, *Pseudomonas*, *Proteus*, *Rhizobium*, *Serratia*, *Xanthomnonas* and others (Rodriguez and Fraga 1999; Sturz and Nowak 2000; Sudhakar et al. 2000; Esitken et al. 2006; Bhattacharjee and Dey 2014). Some PGPMs like *Trichoderma*, *Gliocladium* and fluorescent pseudomonads also help in the disease control in plants (Glick 1995). *Rhizobia* have a good potential to be used as biological control agents against some plant pathogens. Strains of *Sinorhizobium meliloti* are antagonistic to *Fusarium oxysporum* (Antoun and Kloepper 2001), and *Rhizobia* antagonistic to *F. solani f. sp. phaseoli* isolated from commercial snap bean appeared to have a good potential for controlling *Fusarium* rot (Buonassisi et al. 1986). Recently, the use of mixture of microorganisms with diverse activities especially in the management of plant diseases has received greater attention. In this context, PGPR strains have been tested individually and in combinations (two/more strains) as biological control agents against multiple plant pathogens (Raupach and Kloepper 1998; Harish et al. 2009).

1.4.1 Microbes as Biofertilisers in Organic Cultivation

Biofertilisers are the biologically active product containing carrier-base (solid or liquid) active strain of selective microbial inoculants like bacteria, fungi, algae or their combination. According to Bhattacharjee and Dey (2014), "Biofertilizers are those substances that contain living microorganisms and they colonize the rhizosphere of the plant and increase the supply or availability of primary nutrient and/or growth stimulus to the target crop". The different benefits of biofertilisers in the improvement of vegetable production (Table 1.1) and enhancing the quality (Table 1.2) are discussed in the following section.

Table 1.1 Effect of biofertilisers on yields of some commonly grown vegetables

Biofertilisers	Vegetable crops	Yield		% increase in yield	Reference
		B⁻	B⁺		
Rhizobium + PSB	Garden pea (t/ha)	3.71	6.61	78.17	Jaipaul et al. (2011)
Potassium-mobilising bacteria	Garden pea (g/plant)	107.33	164.33	53.11	Pawar et al. (2014)
Rhizobium + PSB	French bean (t/ha)	12.72	13.86	8.96	Zahida et al. (2016)
AMF	French bean (t/ha)	5.55	6.66	20.00	Ramana et al. (2010)
Rhizobium + PSB	Cowpea (t/ha)	2.91	3.74	28.52	Khan et al. (2015)
Rhizobium + *Azotobacter* + PSB + AMF	Cluster bean (t/ha)	4.28	4.99	16.59	Deshmukh et al. (2014)
Azospirillum + *Azotobacter* + *PSB*	Potato (t/ha)	10.88	17.66	62.32	El-Sayed et al. (2015)
Azotobacter + PSB	Brinjal (g/plant)	2039.91	2554.70	25.24	Doifode and Nandkar (2014)
Azotobacter + PSB	Capsicum (t/ha)	7.13	9.27	30.01	Jaipaul et al. (2011)
Azospirillum + AMF + *Frateuria*	Okra (t/ha)	1.55	1.63	5.16	Anisa et al. (2016)
Azospirillum	Okra (t/ha)	1.72	4.59	166.86	Shaheen et al. 2007

Table 1.1 (continued)

Biofertilisers	Vegetable crops	Yield		% increase in yield	Reference
		B⁻	B⁺		
Azotobacter + Azospirillum + PSB	Okra (g/plant)	387.38	548.74	41.65	Mal et al. (2013)
Azotobacter	Cucumber (t/ha)	4.34	4.39	1.15	Saeed et al. (2015)
Azotobacter	Cabbage (t/ha)	31.77	35.62	12.12	Sarkar et al. (2010)
Azotobacter + PSB	Broccoli (kg/plant)	1.10	1.29	17.27	Singh et al. (2014b)
Azotobacter + Azospirillum	Onion (t/ha)	18.17	22.24	22.40	Ghanti and Sharangi (2009)
Azotobacter + Bacillus circulans + Mycorrhiza	Carrot (t/ha)	22.48	23.18	3.11	Naby et al. (2013)
Azospirillum	Lettuce (g/plant)	690.00	993.3	43.96	Chamangasht et al. (2012)

B^- without biofertiliser, B^+ with biofertiliser, *PSB* phosphate-solubilising bacteria, *AMF* arbuscular mycorrhizal fungi

Table 1.2 Quality improvement of different vegetable crops applying biofertilisers

Biofertilisers	Crop	Quality attributes		% increase	Reference
		B⁻	B⁺		
Azotobacter + PSB	Tomato (TSS—°Brix)	4.33	4.80	10.85	Singh et al. (2015)
Azotobacter + Azospirillum	Tomato (protein—mg/g)	0.39	0.43	10.26	Ramakrishnan and Selvakumar (2012)
Azospirillum + PSB	Chilli (vitamin C mg/100 g)	1.14	1.62	42.11	Singh et al. (2014a)
Azotobacter + PSB	Capsicum (vitamin C mg/100 g)	19.26	25.23	31.00	Jaipaul et al. (2011)
Rhizobium + PSB	French bean (seed protein—%)	19.70	21.20	7.61	Zahida et al. (2016)
Rhizobium + PSB	Cowpea (seed protein—%)	21.56	24.44	13.36	Khan et al. (2015)
Rhizobium + Azospirillum + PSB	Cowpea (vitamin C mg/100 g)	0.85	2.56	201.18	Sivakumar et al. (2013)
Azotobacter + PSB	Cabbage (vitamin C mg/100 g)	15.53	32.68	110.43	Kumar et al. (2015)
Azotobacter + PSB	Carrot (TSS—°Brix)	10.30	12.30	19.42	Sarma et al. (2015)
AMF + *PSB*	Onion (reducing sugar—%)	0.94	0.97	3.19	Ghanti and Sharangi (2009)

B^- without biofertiliser, B^+ with biofertiliser

1.4.1.1 Nitrogen-Fixing Biofertilisers

Nitrogen Fixation Through Nodule Formation in Leguminous Vegetable Crops

Rhizobia are soil bacteria that fix nitrogen inside root nodules of legume crops. However, not all *Rhizobia* nodulate all legumes. The symbiosis between *Rhizobia* and legumes appears to be precisely matched, although in some cases a certain level of mismatching is tolerated. To obtain the full benefits, it is extremely important to provide farmers with the correct *Rhizobia* for their legume crop. Table 1.3 provides the information of cross inoculation groups of *Rhizobium* species. However, apart from their role in legume improvement exclusively through N supply, *Rhizobia* have also been used to enhance the nonlegume production. For example, García-Fraile et al. (2012) in a study showed that *Rhizobium* strains colonise the roots of tomato and pepper plants promoting their growth in different production stages increasing yield and quality of seedlings and fruits.

Azotobacter as Free-Living Nitrogen Fixer

Azotobacter species are free-living, aerobic soil dwelling, oval or spherical bacteria that form thick-walled cysts and are capable of fixing an average of 20 kg N/ha/year. These bacteria utilise atmospheric nitrogen for their cell protein synthesis (Jnwali et al. 2015). This cell protein is then mineralized in soil after the death of *Azotobacter* cells thereby contributing towards the N availability to the crop plants. The agronomically important species of *Azotobacter* are *A. chroococcum*, *A. vinelandii*, *A. salinestris*, *A. beijerinckii*, etc. Like other PGPRs, *Azotobacter* has also been reported to enhance germination, growth and yield of vegetable crops (Dhumal 1992; Verma and Shende 1993). For instance, application of *Azotobacter* was found to increase the yield of vegetable crops such as tomato, brinjal, chilli, potato, cabbage, cauliflower, broccoli and cucumber by 15–50% over control and was also able of producing antibacterial and antifungal compounds, hormones and siderophores (Dahama 1997; Ramakrishnan and Selvakumar 2012; Anburani and Manivannan 2002). In a similar investigation, the influence of *Azotobacter chroococcum* both in isolation and in combination with phosphate-solubilising bacteria (*Bacillus polymyxa*)

Table 1.3 Cross inoculation group of *Rhizobia* bacteria

Hos Cross inoculation group	Host legumes	*Rhizobium* species
Pea group	Garden pea, lentil, field pea	*Rhizobium leguminosarum*
Bean group	French bean	*Rhizobium phaseoli*
Soybean group	Soybean	*Rhizobium japonicum*
Clover group	Red clover	*Rhizobium trifolii*
Alfa-alfa group	Alfa-alfa	*Rhizobium meliloti*
Lupine group	Lupine	*Rhizobium lupine*
Cicer group	Bengal gram	*Rhizobium* species
Cowpea group	Cowpea, groundnut	*Rhizobium* species

Source: Morel et al. (2012)

was evaluated in the presence of different doses of chemical fertiliser (NPK) on brinjal (*Solanum melongena* L.) crop grown during the Kharif season to explore the possibility of reducing doses of chemical fertilisers and for better soil health (Doifode and Nandkar 2014). The results revealed significant improvement in growth characters such as height of plant (11.03–37.54%), stem diameter (6.38–23.79%), length of root (5.56–36.93%), number of functional leaves (5.67–51.51%), weight of fresh shoot (7.90–35.91%) and weight of dry shoot (7.14–46.94%) of inoculated brinjal over control. Similarly, the number of fruits picked per plant (11.30–52.81%) and yield of fruits (11.89–54.61%) was more in inoculated plants. The attack of shoot-root borer, fruit borer and little leaf infestation was less (26.71–50.14%) as compared to uninoculated plants. It also helps to sustain the plant growth and yield even in case of low phosphate content in soil, as well as helps in uptake of macro- and certain micronutrients which facilitates better utilisation of plant root exudates (Revillas et al. 2000).

Azospirillum as Nitrogen Fixer

Bacteria of the genus *Azospirillum* are known for many years as plant growth-promoting rhizobacteria (PGPR). They were isolated from the rhizosphere of many grasses and cereals all over the world, in tropical as well as in temperate climates. Both in greenhouse and in field trials, *Azospirillum* was shown to exert beneficial effects on plant growth and crop yields (Singh 2014; Singh et al. 2014b). At present, five species have been described: *A. lipoferum*, *A. brasilense*, *A. amazonense*, *A. halopraeferens* and *A. irakense*. *Azospirillum* bacteria are gram-negative free-living nitrogen-fixing rhizosphere bacteria. Ammonium, nitrite, amino acids and molecular nitrogen can serve as N sources. In unfavourable conditions, such as stress and nutrient limitation, *Azospirillum* can convert into enlarged cyst-like forms, accompanied by an outer coat of polysaccharides and accumulation of abundant poly-β-hydroxybutyrate granules, which can serve as energy source under condition of stress and starvation. Motility of the bacteria offers the advantage of moving towards favourable nutrient conditions.

1.4.1.2 Phosphorus-Solubilising Microorganisms

Release of phosphorus by phosphate-solubilising bacteria (PSB) from insoluble and fixed/adsorbed forms is an important aspect regarding phosphorus availability in soils. Microbial biomass assimilates soluble phosphorus and prevents it from adsorption or fixation (Khan and Joergensen 2009). Microorganisms enhance the phosphorus availability to plants by mineralising organic phosphorus in soil and by solubilising precipitated phosphates (Pradhan and Sukla 2005).

These bacteria in the presence of labile carbon serve as a sink for phosphorus by rapidly immobilising it even in low phosphorus soils (Bunemann et al. 2004). Subsequently, PSB become a source of phosphorus to plants upon its release from their cells. The PSB and plant growth-promoting rhizobacteria together could reduce phosphorus fertiliser application by 50% without any significant reduction of crop yield (Yazdani et al. 2009). Bacteria are more effective in phosphorus solubilisation than fungi (Alam et al. 2002). Among the whole microbial population in soil, PSB constitute 1–50%, while phosphorus solubilising fungi (PSF) are only 0.1–0.5%.

Strains from bacterial genera *Pseudomonas*, *Bacillus*, *Rhizobium* and *Enterobacter* along with *Penicillium* and *Aspergillus* fungi are the most powerful phosphorus solubilises (Whitelaw 2000). A nematode fungus *Arthrobotrys oligospora* also has the ability to solubilise the phosphate rocks (Duponnois et al. 2006).

1.4.1.3 Potassium-Solubilising Bacteria

Potassium is a major nutrient required in large quantities to run various metabolic reactions of crop plants. Potassium promotes root growth, enhances stem strength, promotes quality of crop, increases resistance to cold as well as water stress and reduces pest and disease incidence (Parmar and Sindhu 2013; Shanware et al. 2014; Prajapati and Modi 2016). A wide range of bacteria, namely, *Pseudomonas*, *Burkholderia*, *Acidithiobacillus ferrooxidans*, *Bacillus mucilaginosus*, *B. extorquens*, *B. edaphicus*, *B. circulans*, *Paenibacillus* sp. and *Clostridium pasteurianum*, have been reported to release potassium in accessible form from potassium-bearing minerals such as mica, illite, muscovite, biotite and orthoclases from soil (Bennett et al. 2001) and increased the potassium availability up to 15% (Supanjani et al. 2006). These potassium-solubilising bacteria (KSB) were found to dissolve potassium, silicon and aluminium from insoluble K-bearing minerals such as micas, illite and orthoclases, by excreting organic acids which either directly dissolved rock potassium or chelated silicon ions to bring potassium into the solution. *B. mucilaginosus*, *A. chroococcum* and *Rhizobium* resulted in significant higher mobilisation of potassium from waste mica, which in turn acted as a source of potassium for plant growth. Therefore, potassium-solubilising bacteria are extensively used as biofertilisers in significant areas of cultivated soils of Korea and China because soil in these countries is deficient in soil-available potassium. Thus, application of potassium-solubilising bacteria as biofertiliser for agriculture improvement can reduce the use of agrochemicals and support sustainable crop production.

1.4.1.4 Mycorrhiza as Biofertiliser

A mycorrhiza is a symbiotic association between a fungus and the roots of a vascular plant. In this association, the fungus colonises the host plant's root cortex and develops an extramatrical mycelium either intracellularly (endomycorrhiza) as in arbuscular mycorrhizal fungi or extracellularly (ectomycorrhizal fungi) that helps plant to acquire mineral nutrients from soil. Endomycorrhizae are variable and are further classified as arbuscular, ericoid, arbutoid, monotropoid and orchid mycorhizae. Arbuscular mycorrhiza (AM)-forming genera of the family include *Acaulospora*, *Entrophospora*, *Gigaspora*, *Glomus*, *Sclerocystis* and *Scutellospora*. The mycorrhizal efficacy varies in soil, mycorrhizosphere, inoculation sequences and modification of cultural practices (Sharma et al. 2004). The main advantage of mycorrhiza is its greater soil exploration and increasing uptake of N, P, K, Zn, Cu, S, Fe, Ca, Mg and Mn and supply of these nutrients to the host roots. AM fungi may increase the effectiveness of absorbing capability of surface host root as much as ten times. Ions such as P, Zn and Cu do not diffuse readily through soil. Because of this poor diffusion, roots deplete these immobile soil nutrients from the zone immediately surrounding the root (Smith and Smith 2011). The increase in plant growth resulting from AM symbiosis is usually associated with increased nutrient uptake by the hyphae from the soil.

1.5 Microbes as Biocontrol Agents for Suppression of Plant Diseases

Plant disease management is indeed a challenge due to the involvement of multiple pathogens. And for any individual crop, the grower deals with variety of phytopathogens including fungi, bacteria, viruses and nematodes. This situation becomes even more complicated and challenging for organic vegetable growers because they are expected to produce a wide variety of vegetables without applying conventional synthetic pesticides/ fungicides. Moreover, the world market continues to be extremely competitive and consistently requires that growers supply high-quality and disease-free produce with an acceptable shelf life. Disease management is, therefore, a critical consideration in organic vegetable production. In an organic system, the growth and multiplication of diversified soil inhabiting and epiphytic microorganisms are encouraged to exert beneficial and pathogen-antagonistic influences. Utilisation of different microbial inoculants for controlling various diseases of vegetable crops is presented in Table 1.4.

Table 1.4 Examples of vegetable disease management using microbial inoculants

Biofertilisers	Vegetables	Disease/pathogen	Reference
P. fluorescens	French bean	*Colletotrichum lindemuthianum*	Ravi et al. (1999)
P. fluorescens	Garden pea	*P. ultimum*	Naseby et al. (2001)
P. fluorescens	Tomato	Fruit rot	Hegde and Anahosur (2001)
P. fluorescens	Cauliflower	*F. moniliforme*	Rajappan and Ramaraj (1999)
Bacillus subtilis	Garden pea	Powdery mildew	Villanueva et al. (2014)
Trichoderma sp.	Sweet potato	*Lasiodiplodia theobromae*	Palomar and Palermo (2004)
Trichoderma sp.	Cabbage	*Plasmodiophora brassicae*	Cuevas and Kebasen (2005)
T. hamatum, T. viride, T. harzianum, T. koningii	Tomato	*Alternaria solani*	Selim (2015)
Trichoderma spp.	Tomato	*F. oxysporum f.sp. lycopersici*	Someshwar et al. (2013)
Trichoderma sp.	Brinjal	*Macrophomina phaseolina*	Singh and Singh (2014)
T. harzianum + *Pseudomonas* sp.	Tomato	*Sclerotium rolfsii*	Singh et al. (2014c)
T. viride + *T. harzianum* + *P. Fluorescens* + *Azotobacter* + *Azospirillum* + PSB	Tomato	*Pythium aphanidermatum, Ralstonia solanacearum, Fusarium oxysporum* f.sp. *lycopersici*	Thakur and Tripathi (2015)
Pseudomonas fluorescens	Sugar beet	Damping off	Kumar et al. (2002)
T. harzianum, T. viride	Cabbage, Chinese cabbage	*Sclerotinia sclerotiorum*	Ha (2010)

The mechanisms by which antagonistic microorganisms protect plants are generally attributed to parasitism, competition and the production of secondary metabolites with toxic effects on pathogens (Harman et al. 2012). Among naturally occurring diversified microflora, many fungi such as *Trichoderma* sp., *Chaetomium* sp., *Coniothyrium minitans* and *Gliocladium* sp., bacteria including *Pseudomonas* sp. and *Bacillus* sp. and actinomycetes, for example, *Streptomyces* sp., have been screened and developed to control the soilborne pathogens like, *Pythium* sp., *Phytophthora* sp., *Fusarium* sp., *Rhizoctonia* sp. and *Sclerotium* sp. (Pan et al. 2001; Pal and Gardener 2006; Whipps et al. 2008). Some antagonistic microorganisms, such as *Trichoderma* sp., *Ampelomyces quisqualis*, *Bacillus* sp. and *Ulocladium* sp., have also been developed for foliar disease control of powdery mildew and Botrytis rot (Paulitz and Belanger 2001). *Pseudomonas floculosa* was developed for the control of powdery mildew (Paulitz and Belanger 2001). The *B. amyloliquefaciens* has been developed to control soilborne and foliar diseases, anthracnose and bacterial pustule of soybean (Prathuangwong and Kasem 2003). The PGPR microorganism can also produce bacteriocin that inhibits closely related bacterial species and induce systemic resistance (ISR) in plants. The ISR mechanism of PGPR can inhibit pathogen infection or induce special structures creating a physical barrier and/or biochemical barriers that induce defensive plant mechanism and systemic resistance against severe strains (Khan et al. 2006).

1.5.1 *Trichoderma* in Organic Disease Management

Trichoderma is considered a potent fungal biocontrol agent against a range of plant pathogen and, hence, has attracted considerable scientific attention (Rini and Sulochana 2007) due to its variable antifungal activities (Zaidi and Singh 2004). Antagonistic efficiency of any antagonist can be considered as a function of different attributes of an antagonist, the acceptance of the host, and the pathogen concern and its armoury of attack. The probable mechanisms involved in biological control of plant pathogens may include mycoparasitism/hyperparasitism, antibiosis/toxins, competition/rhizosphere competence, lytic enzymes, induction of resistance and plant growth promotion (Howell 2003; Vinale et al. 2008). Competition is considered as a classical mechanism of biological control. The competition between antagonist and plant pathogens occurs mainly for space and nutrient (Harman 2006). For example, competition for carbon and nitrogen by *T. harzianum* suppressed the infection of *F. oxysporum* f. sp. *melonis* and *F. oxysporum* f.sp. *vasinfectum* (Sivan and Chet 1989). More recently, involvement of lytic enzymes causing destruction of fungal cell wall made up of chitin or β-glucans has been found effective in biological management of diseases (Strakowska et al. 2014). Many research works with *Trichoderma* recorded the phenomenon of hyperparasitism in which the hyphae of the antagonist parasitized hyphae of other fungi in vitro and caused several morphological changes (coiling, haustoria production, disorganisation of host cell contents, penetration of the host, etc.) during destruction (Anitha and Murugesan 2001). Biocontrol potential of *Trichoderma* against pathogens also depends on the secretion of antibiotic substances like trichodermin, dermadin, trichoviridin and sesquiterpene heptalic acid (Godwin and Arnize 2000) that are produced during nutrient-limiting condition. The nonvolatile and volatile compounds secreted by *Trichoderma* inhibited fungal growth at very low concentration (Jelen et al. 2013). And there are evidence which suggest that volatile compounds produced by *Trichoderma* inhibited the colony growth of *Macrophomina phaseolina* (Maheshwari et al. 2001), *Rhizoctonia solani* (Bunker et al. 2001), *Fusarium* (Pandey and Upadhyay 1997) and *Sclerotium rolfsii* (Aparecido and Figueiredo 1999). However, the inhibitory effect of volatiles decreased with increase in incubation. In addition to the ability of *Trichoderma* species to attack or inhibit the growth of the pathogen directly, recent discoveries indicate that they can also induce systemic and localised resistance to a variety of plant pathogens. Several studies revealed that some BCAs including *Trichoderma* spp. are also able to reduce disease through plant-mediated systematically activated resistance mechanisms referred to as induced systemic resistance (ISR) (Thrane et al. 1997). Treatment of seeds, roots, cuttings, soil or artificial growing media with *Trichoderma* induces plant growth promotion in terms of increased germination, early emergence, fresh and dry weight of roots or shoots, root length, yield and flowering (Joshi et al. 2007; Bhagat and Pan 2010).

1.5.2 Fluorescent Pseudomonads as Biocontrol Agent in Organic Cultivation

Numerous fluorescent pseudomonads including *Pseudomonas fluorescens*, *P. aureofaciens*, *P. putida* and *P. syringae* have been reported to exhibit biocontrol activity against a wide range of phytopathogens (Thomashow and Weller 1996; Jayraj et al. 2007; Mansoor et al. 2007; Sen et al. 2009). *Pseudomonas* is gram-negative, strictly aerobic, polarly flagellated rods belongs to gammaproteobacteria of order *Pseudomonadales*. They aggressively colonise the rhizosphere of various crop plants and have a broad spectrum antagonistic activity against plant pathogens. The antagonistic activity results from antibiosis (Cartwright et al. 1995; Rosales et al. 1995; Maurhofer et al. 1995), siderophores production (Winkelmann and Drechsel 1997) and nutrition or site competition (Bull et al. 1991). Several antibiotic-like substances like bacteriocins and phenazine antibiotics produced by pseudomonads have been identified (Hamdan et al. 1991). *Pseudomonas* spp. is known to further produce one or an array of antifungal metabolites and lytic enzymes such as chitinase and glucanase in culture (Viswanathan and Samiyappan 2001) that inhibit the mycelial growth of certain fungi (Radhajeyalakshmi et al. 2009). Of the various metabolites secreted by pseudomonads, siderophores have been found to efficiently sequester iron, thereby depriving the pathogen from this essential element during its deleterious activities in the rhizosphere (Kloepper et al. 1980). Production of HCN is yet another mechanism by which certain strains of fluorescent pseudomonads suppress the soilborne pathogens (Voisard et al. 1989) by inhibiting the electron transport and the energy supply to the cell leading to the death of the organisms (Corbett 1974). Certain fluorescent pseudomonads have been reported to induce defence gene products that include peroxidase (PO) and polyphenol oxidase (PPO) that catalyse the formation of lignin and phenylalanine ammonia lyase (PAL) that are involved in phytoalexins and phenolic synthesis. Other defence enzymes include pathogenesis-related proteins (PRs) such as β-1,3-glucanases (PR-2 family) and chitinases (PR-3 family) which degrade the fungal cell wall and cause lysis of fungal cell (Friedlender et al. 1993). Siderophore producing *Pseudomonas* species thus plays a vital role in stimulating plant growth and in controlling several plant diseases (Lemanceau and Albouvette 1993).

1.6   Microbes for Decomposition of Organic Residues

In India, enormous amount of crop residues is produced from crop cultivation. A considerable part of which remains unutilised and is either burnt or dumped in nearby sites that creates pollution, harbours pathogens for diseases and causes severe problems of disposal. These residues are valuable sources of organic carbon as well as several essential nutrients. Therefore, instead of disposing, it can be used as source of organic residues and can effectively be recycled for compost production. Composting is a microbiological process that depends on the growth and activity of mixed populations of bacteria, actinomycetes and fungi that are indigenous to the wastes being composted. Different microbes involved in composting process are presented in Table 1.5. During the composting process, organic wastes are decomposed, plant nutrients are mineralized into plant-available forms, pathogens are destroyed and malodours are decreased (Parr et al. 1992). Compostable waste materials normally contain a large number of different types of bacteria, fungi and actinomycetes. During the decomposition, change in the nature and number of these microorganisms takes place. Fungi and actinomycetes play an important role in the decomposition of cellulose, lignin and other materials.

Table 1.5 Representative organisms involved in composting process

Bacteria	Fungi	Actinomycetes
Mesophilic	**Mesophilic**	**Thermo-tolerant and thermophilic**
Cellulomonas folia	*Fusarium roseum*	*Micromonospora vulgaris*
Chondrococcus exiguus	*F. culmorum*	*Nocardia brasiliensis*
Myxococcus virescens	*Coprinus cinereus*	*Streptomyces rectus*
Thermophilic	*C. lagopus*	*S. thermofuscus*
Bacillus stearothermophilus	*Trichoderma viride*	*S. thermovulgaris*
	T. lignorum	
	Rhizopus nigricans	

Source: Tauro et al. (1986)

1.7   Factors Affecting the Efficacy of Microbial Inoculants

Since the performance of the microbial inoculants generally is host plant specific and strain specific, the improper selection of inoculants may hamper the favourable response. Also, variation in plant genotypes, age of plants and soil chemical, physical and biological properties largely influence the performance of the inoculants. Microbial inoculants when introduced into the new soil, it faces stiff competition from the inherent microbes which severely reduce their beneficial effects. However, there are several reports that a single species with a particular strain is highly effective under diverse plant types in different environments. Therefore, identification and exploitation of such strain for commercial formulation are likely to be more useful in different agro-climatic regions. A longer shelf life of the inoculants responds favourably and improves the efficiency of the strains. To improve the strain efficiency, different carriers can be used in the formulation process. Granular inoculants perform better under poor soil conditions. Liquid inoculants on the contrary have limited use due to shorter shelf life. The microbial inoculants must also be compatible with the applied agrochemicals in the crop production. Application of soil amendments to maintain the pH balance or supplying micronutrients stimulates the activities of the microbial inoculants. Adoption of proper application procedure in the presence of sufficient organic manures with optimum inoculum density and within the limit of expiry period will lead to a better performance under field condition.

Conclusions

In the present scenario, over dependence and non-judicious use of chemical fertilisers and pesticides are posing serious threats to ecological balance. With the introduction of new hybrids/high-yielding varieties of different vegetable crops in different countries, new pathogens have emerged in the soil system. Also, the use of modern pesticides which are highly target specific or site specific in action is encouraging the development of resistance among pathogens. The adverse effect of global climate change has also transformed the soil pathogen interactions, and the earlier minor soil pathogens are emerging as new threat for different vegetable crops. Maintenance of soil health is therefore a bigger challenge for organic cultivation practices. So to counteract such challenges, there is a greater need to produce and use diverse microbial inoculants so that the use of chemical fertilisers and pesticides in the crop production could be minimised, if not completely eliminated. Identification of potential local indigenous soil microbes and exploiting their beneficial activities are likely to reduce dependence on the use of chemicals in vegetable production. Large-scale use of microbial inoculants is likely to make organic vegetable cultivation more effective, economical and sustainable.

References

Alam S, Khalil S, Ayub N, Rashid M (2002) *In vitro* solubilization of inorganic phosphate by phosphate solubilizing microorganism (PSM) from maize rhizosphere. Int J Agric Biol 4:454–458

Anburani A, Manivannan K (2002) Effect of integrated nutrient management on growth in brinjal (*Solanum melongena* L.) cv. Annamalai. South Indian Hortic 5:377–386

Anisa NA, Markose BL, Joseph S (2016) Effect of biofertilizers on yield attributing characters and yield of okra (*Abelmoschus esculentus* (L.) Moench). Int J Appl Pure Sci Agric 2: 59–62

Anitha R, Murugesan K (2001) Mechanism of action of *Gliocladium virens* on *Alternaria helianthi*. Indian Physician 54:449–452

Anonymous (2009) Nutrient requirements and recommended dietary allowances for Indians. National Institute of Nutrition, Indian Council of Medical Research, Hyderabad, India

Anonymous (2014) Indian Horticulture Database—2014. National Horticultural Board, Govt. of India, Gurgaon, India

Antoun H, Kloepper JW (2001) Plant growth-promoting rhizobacteria (PGPR). In: Brenner S, Miller JH (eds) Encyclopedia of genetics. Academic Press, New York, pp 1477–1480

Aparecido CC, Figueiredo MB (1999) Antagonism of *Trichoderma viride* against two different bean soil borne pathogenic fungi. O-Biologico 61:17–21

Bassil KL, Vakil C, Sanborn M, Cole DC, Kaur JS, Kerr KJ (2007) Cancer health effects of pesticides: systematic review. Can Farm Physician 53:1704–1711

Bennett CB, Lewis LK, Karthikeyan G (2001) Genes required for ionizing radiation resistance in yeast. Nat Genet 29:26–34

Bhagat S, Pan S (2010) Biopriming of seeds for improving germination behavior of chilli, tomato and brinjal. J Mycol Plant Pathol 40:375–379

Bhattacharjee R, Dey U (2014) Biofertilizers, a way towards organic agriculture: a review. Afr J Microbiol Res 8:2332–2342

Bishnu A, Saha T, Ghosh PB, Mazumdar D, Chakraborty A, Chakrabarti K (2009) Effect of pesticide residues on microbiological and biochemical soil indicators in tea gardens of Darjeeling Hills, India. World J Agric Sci 5:690–697

Bourn D, Prescott J (2002) A comparison of the nutritional value, sensory qualities and food safety of organically and conventionally produced foods. Crit Rev Food Sci Nutr 42:1–34

Brand K, Molgaard JP (2001) Organic agriculture: does it enhance or reduce the nutritional value of plant foods? J Sci Food Agric 81:924–931

Bull CT, Weller DM, Thomashow LS (1991) Relationship between root colonization and suppression of *Gaeumannomyces graminis* var. *tritici* by *Pseudomonas fluorescens* strain 2-79. Phytopathology 81:950–959

Bunemann EK, Steinebrunner F, Smithson PC, Frossard E, Oberson A (2004) Phosphorus dynamics in a highly weathered soil as revealed by isotopic labelling techniques. Soil Sci Soc Am J 68:1645–1655

Bunker RN, Mathur K, Mathur K (2001) Antagonism of local biocontrol agents to *Rhizoctonia solani* inciting dry root rot of chilli. J Mycol Plant Pathol 31:50–53

Buonassisi AJ, Copeman RJ, Pepin HS, Eaton GW (1986) Effect of *Rhizobium* spp. on *Fusarium solani* f.sp. *phaseoli*. Can J Phytopathol 8:140–146

Cartwright DK, Chilton WS, Benson DM (1995) Pyrrolnitrin and phenazine production by *Pseudomonas cepacia*, strain 5.5 B, a biological agent of *Rhizoctonia solani*. Appl Microbiol Biotechnol 43:211–221

Chamangasht S, Ardakani MR, Khavazi K, Abbaszadeh B, Mafakheri S (2012) Improving lettuce (*Lactuca sativa* L.) growth and yield by the application of biofertilizers. Ann Biol Res 3:1876–1879

Chatterjee R (2009) Production of vermicompost from vegetable wastes and its effect on integrated nutrient management for vegetable production. Ph.D. thesis, UBKV, Pundibari, West Bengal

Chatterjee R, Thirumdasu RK (2015) Climate change mitigation through organic farming in vegetable production. Agric Biol Sci J 1:76–82

Chen YP, Rekha PD, Arunshen AB, La WA, Young CC (2006) Phosphate solubilizing bacteria from subtropical soil and their tricalcium phosphate solubilizing abilities. Appl Soil Ecol 34:33–41

Corbett JR (1974) The biochemical mode of action of pesticides. Academic Press, Inc., London, pp 44–86

Coyne MS, Mikkelsen R (2015) Soil microorganisms contribute to plant nutrition and root health. Better Crops 99:18–20

Cuevas VC, Kebasen SB (2005) Ecological approach in the control of club root disease of cabbage. In: 7th annual scientific meeting and symposium, Mycological Society of the Philippines, ERDB, College, Laguna, 8 Apr 2005

Dahama AK (1997) Organic farming for sustainable agriculture. Ashila Offset Printers, Daruagung, New Delhi, India

Deshmukh RP, Nagre PK, Wagh AP, Dod VN (2014) Effect of different bio-fertilizers on growth, yield and quality of cluster bean. Indian J Adv Plant Res 1:39–42

Dhumal KN (1992) Effect of *Azotobacter* on germination, growth and yield of some vegetables. J Maharashtra Agric Univ 17:500

Doifode VD, Nandkar PB (2014) Influence of biofertilizers on the growth, yield and quality of brinjal crop. Int J Life Sci A2:17–20

Duponnois R, Kisa M, Plenchette C (2006) Phosphate solubilising potential of the nemato-fungus *Arthrobotrys oligospora*. J Plant Nutr Soil Sci 169:280–282

El-Sayed SF, Hassan AH, El-Mogy MM (2015) Impact of bio- and organic fertilizers on potato yield, quality and tuber weight loss after harvest. Potato Res 58:67–81

Esitken A, Pirlak L, Turan M, Sahin F (2006) Effects of floral and foliar application of plant growth promoting rhizobacteria (PGPR) on yield, growth and nutrition of sweet cherry. Sci Hortic 110:324–327

Friedlender M, Inbar J, Chet I (1993) Biological control of soil borne plant pathogens by a β-1,3-glucanase producing *Pseudomonas cepacia*. Soil Biol Biochem 25:1211–1221

García-Fraile P, Carro L, Robledo M et al (2012) *Rhizobium* promotes non-legumes growth and quality in several production steps: towards a biofertilization of edible raw vegetables healthy for humans. PLoS ONE 7:e38122

Ghanti S, Sharangi AB (2009) Effect of bio-fertilizers on growth, yield and quality of onion cv. Sukhsagar. J Crop Weed 5:120–123

Gilden RC, Huffling K, Sattler B (2010) Pesticides and health risk. J Obstet Gynecol Neonatal Nurs 39(1):103–110

Glick BR (1995) The enhancement of plant growth by free-living bacteria. Can J Microbiol 41:109–117

Godwin EMI, Arnize AE (2000) The induction of some hydrolytic enzymes and antibiotics in *Trichoderma harzianum* and *Fusarium oxysporum* using some food wastes. Global J Pure Appl Sci 6(1):31–36

Ha NT (2010) Using *Trichoderma* species for biological control of plant pathogens in Vietnam. J ISSAAS 16:17–21

Hamdan H, Weller DM, Thomashow LS (1991) Relative importance of fluorescent siderophores and other factors in biological control of *Gaeumannomyces graminis* var. *tritici* by *Pseudomonas fluorescens* 2–79 and M4-80R. Appl Environ Microbiol 57:3270–3277

Harish S, Kavino M, Kumar N, Balasubramanian P, Samiyappan R (2009) Induction of defense-related proteins by mixtures of plant growth promoting endophytic bacteria against Banana bunchy top virus. Biol Control 51:16–25

Harman GE (2006) Overview of mechanisms and uses of *Trichoderma* spp. Phytopathology 96:190–194

Harman GE, Herrera-Estrella AH, Horwitz BA, Lorito M (2012) *Trichoderma*—from basic biology to biotechnology. Microbiology 158:1–2

Hegde GM, Anahosur KH (2001) Evaluation of fungi toxicants against fruit rot of chilli and their effect on biochemical constituents. Karnataka J Agric Sci 14(3):836–838

Heinonen-Tanski H, Siltanen H, Kilpi S, Simojoki P, Rosenberg C, Mäkinen S (1986) The effect of the annual use of some pesticides on soil microorganisms, pesticide residues in soil and carrot yields. Pest Manage Sci 17:135–142

Howell CR (2003) Mechanisms employed by *Trichoderma* species in the biological control of plant disease: the history and evolution of current concepts. Plant Dis 87:4–10

Jaipaul SS, Dixit AK, Sharma AK (2011) Growth and yield of capsicum (*Capsicum annum*) and garden pea (*Pisum sativum*) as influenced by organic manures and biofertilizers. Indian J Agric Sci 81(7):637–642

Jayraj J, Parthasarathi T, Radhakrishnan NV (2007) Characterization of a *Pseudomonas fluorescens* strain from tomato rhizosphere and its use for integrated management of tomato damping off. Biocontrol 52:683–702

Jelen H, Błaszczyk L, Chełkowski J, Rogowicz K, Strakowska J (2013) Formation of 6-n-pentyl-2H-pyran-2-one (6-PAP) and other volatiles by different *Trichoderma* species. Mycol Prog 13(3):589–600

Jnwali AD, Ojha RB, Marahatta S (2015) Role of *Azotobacter* in soil fertility and sustainability—a review. Adv Plant Agric Res 2(6):64–69

Joshi N, Brar KS, Pannu PPS, Singh P (2007) Field efficacy of fungal and bacterial antagonists against brown spot of rice. J Biol Control 21(1):159–162

Khan KS, Joergensen RG (2009) Changes in microbial biomass and P fractions in biogenic household waste compost amended with inorganic P fertilizers. Bioresour Technol 100:303–309

Khan MR, Fischer S, Egan D, Doohan FM (2006) Biological control of fusarium seedling blight disease of wheat and barley. Phytopathology 96(4):386–394

Khan VM, Manohar KS, Verma HP (2015) Effect of vermicompost and biofertilizer on yield, quality and economics of cowpea. Ann Agric Res 36(3):309–311

Kloepper JW, Leong J, Teintze M, Schroth MN (1980) *Pseudomonas* siderophores: a mechanism explaining disease suppressive soils. Curr Microbiol 4:317–320

Kumar NR, Arasu VT, Gunasekaran P (2002) Genotyping of antifungal compounds producing plant growth-promoting rhizobacteria, *Pseudomonas fluorescens*. Curr Sci 82:1465–1466

Kumar J, Phookan DB, Lal N, Kumar H, Sinha K, Hazarika M (2015) Effect of organic manures and biofertilizers on nutritional quality of cabbage (*Brassica oleracea* var. capitata). J Eco-friendly Agric 10(2):114–119

Lampin N (1990) Organic farming. Farming Press Books, Ipswick, UK

Lauridsen C, Jorgensen H, Halekoh U, Christensen L P (2005) Organic food and health—status and future perspectives. In: Paper presented at Researching Sustainable Systems, international scientific conference on organic agriculture, Adelaide, Australia, pp 21–23

Lemanceau P, Albouvette C (1993) Suppression of fusarium wilts by fluorescent pseudomonads; mechanism and applications. Biocontrol Sci Tech 3:219–234

Lhuissier FGP, de Ruijter NCA, Sieberer BJ, Esseling JJ, Emons AMC (2001) Time course of cell biological events evoked in legume root hairs by *Rhizobium* nod factors: state of the art. Ann Bot 87:289–302

Magkos F, Arvaniti F, Zampelas A (2006) Organic food: buying more safety or just peace of mind? A critical review of the literature. Crit Rev Food Sci Nutr 46:23–56

Maheshwari DK, Dubey RC, Sharma VK (2001) Biocontrol effects of *Trichoderma virens* on *Macrophomina phaseolina* causing Indian. J Microbiol 41(4):251–256

Mal B, Mahapatra P, Mohanty S, Mishra HN (2013) Growth and yield parameters of okra (*Abelmoschus esculentus*) influenced by diazotrophs and chemical fertilizers. J Crop Weed 9(2):109–112

Mansoor FK, Sultana V, Haque SE (2007) Enhancement of biocontrol of *Pseudomonas aeruginosa* and *Paecilomyces lilacinus* against root rot of mung- bean by a medicinal plant *Launaea nudicaulis* L. Pak J Bot 39:2113–2119

Maurhofer M, Keel C, Haas D, Defago G (1995) Influence of plant species on disease suppression by *Pseudomonas fluorescens* strain CHAO with enhanced antibiotic production. Plant Pathol 44:40–50

Morel MA, Brana V, Castro-Sowinski S (2012) Legume crops, importance and use of bacterial inoculation to increase production. In: Goyal A (ed) Crop plant. InTech, Rijeka, pp 217–240

Naby HMEA, Dawa KK, El-Gamily EE, El-Hameed SMA (2013) Effect of organic, bio and mineral fertilization on yield and quality of carrot plants. J Plant Prod 4(2):335–349

Naseby DC, Way JA, Bainton NJ, Lynch JM (2001) Biocontrol of *Pythium* in the pea rhizosphere by antifungal metabolite producing and non-producing *Pseudomonas* strains. J Appl Microbiol 90:421–429

Pal KK, Gardener BM (2006) Biological control of plant pathogens. Plant Health Instructor. doi:10.1094/PHI-A-2006-1117-02

Palaniappan SP, Annadurai K (1999) Organic farming: theory and practice. Scientific Publishers, Jodhpur, India, p 257

Palomar MK, Palermo VG (2004) Microbial control of sweet potato to tuber rot caused by *Lasiodiplodia theobromae* using *Trichoderma* F17c. In: Proceedings of the 35th anniversary and annual scientific conference of PMCP, Amigo Terrace Hotel, Iloilo City, pp 102–103

Pan S, Roy A, Hazra S (2001) *In vitro* variability of biocontrol potential among some isolates of *Gliocladium virens*. Adv Plant Sci 14:301–303

Pandey KK, Upadhyay JP (1997) Selection of potential biocontrol agents based on production of volatile and non volatile antibiotics. Veg Sci 24(2):144–146

Parmar P, Sindhu SS (2013) Potassium solubilisation by rhizosphere bacteria: influence of nutritional and environmental conditions. J Microbiol Res 3:25–31

Parr JF, Papendick RI, Hornick SB, Meyer RE (1992) Soil quality: attributes and relationship to alternative and sustainable agriculture. Am J Altern Agric 7:5–11

Paulitz TM, Belanger RR (2001) Biological control in greenhouse system. Ann Rev Phytopathol 39:103–133

Pawar YD, Varma LR, Joshi HN, Verma P (2014) Growth, flowering and yield parameters of garden pea (*Pisum sativum* L.) as influenced by different biofertilizers. In: Mishra GC (ed) Agriculture: towards a new paradigm of sustainability. Excellent Publishing House, New Delhi, pp 290–292

Pradhan N, Sukla LB (2005) Solubilization of inorganic phosphates by fungi isolated from agriculture soil. Afr J Biotechnol 5:850–854

Prajapati K, Modi HA (2016) Growth promoting effect of potassium solubilizing *Enterobacter hormaechei* (KSB-8) on cucumber (*Cucumis sativus*) under hydroponic conditions. Int J Adv Res Biol Sci 3:168–173

Prathuangwong S, Kasem S (2003) Potential of new antagonists for controlling soybean bacterial pustule and reducing bactericide application. In: Proceedings of the sum of the 7th international conference of plant pathology, 31 Jan–6 Feb 2003, Christchurch, New Zealand, pp 2–11

Radhajeyalakshmi R, Velazhahan R, Samiyappan R, Doraiswamy S (2009) Systemic induction of pathogenesis related proteins (PRs) in *Alternaria solani* elicitor sensitized tomato cells as resistance response. Sci Res Essays 4:685–689

Rajappan K, Ramaraj B (1999) Evaluation of fungal and bacterial antagonists against *Fusarium moniliforme* causing wilt of cauliflower. Ann Plant Protect Sci 7:205–207

Ramakrishnan K, Selvakumar G (2012) Effect of biofertilizers on enhancement of growth and yield on tomato (*Lycopersicum esculentum* Mill.) Int J Res Botany 2(4):20–23

Ramana V, Ramakrishna M, Purushotham K, Reddy KB (2010) Effect of bio-fertilizers on growth, yield attributes and yield of French bean (*Phaseolus vulgaris* L.) Legume Res 33:178–183

Raupach GS, Kloepper JW (1998) Mixtures of plant growth promoting rhizobacteria enhance biological control of multiple cucumber pathogens. Phytopathology 88:1158–1164

Ravi S, Doraiswamy S, Valluvaparidasan V, Jeyalakshmi C, Doraiswany S (1999) Effect of iocontrol agents on seed-borne *Colletotrichum* in French bean. Plant Dis Res 14:146–151

Rembialkowska E (2003) Organic farming as a system to provide better vegetable quality. Acta Hortic 604:473–479

Ren H, Endo H, Hayashi T (2001) Antioxidative and antimutagenic activities and polyphenol content of pesticide-free and organically cultivated green vegetables using water-soluble chitosan as a soil modifier and leaf surface spray J Sci Food Agic 81:1426–1432

Revillas JJ, Rodelas BC, Pozo C, Martonez-Toledo MV, Gonzalez-Lopez J (2000) Production of B-group vitamins by two *Azotobacter* strains with phenolic compounds as sole carbon source under diazotrophic and adiazotrophic conditions. J Appl Microbiol 89:486–493

Rini CR, Sulochana KK (2007) Substrate evaluation for multiplication of *Trichoderma* spp. J Trop Agric 45(1–2):58–60

Rodriguez H, Fraga R (1999) Phosphate solubilizing bacteria and their role in plant growth promotion. Biotechnol Adv 17:319–339

Rosales AM, Thomashow L, Cook RJ, Mew TW (1995) Isolation and identification of antifungal metabolites produced by rice-associated antagonistic *Pseudomonas* spp. Phytopathology 85:1028–1032

Saeed KS, Ahmed SA, Hassan IA, Ahmed PH (2015) Effect of bio-fertilizer and chemical fertilizer on growth and yield in cucumber (*Cucumis sativus*) in green house condition. Pak J Biol Sci 18(3):129–134

Sarkar A, Mandal AR, Prasad PH, Maity TK (2010) Influence of nitrogen and biofertilizer on growth and yield of cabbage. J Crop Weed 6(2):72–73

Sarma I, Phookan DB, Boruah S (2015) Influence of manures and biofertilizers on carrot (*Daucus carota* L.) cv. Early Nantes growth, yield and quality. J Eco-friendly Agric 10:25–27

Selim ME (2015) Effectiveness of *Trichoderma* biotic applications in regulating the related defense genes affecting tomato early blight disease. J Plant Pathol Microbiol 6:311

Sen S, Rai M, Acharya R, Dasgupta S, Saha A, Acharya K (2009) Biological control of pathogens causing the *Cymbidium* Pseudobulb rot complex using *Pseudomonas fluorescent* strain BRL-1. J Plant Pathol 91:617–621

Shaheen AM, Rizk FA, Sawan OM, Ghoname AA (2007) The integrated use of bio-inoculants and chemical nitrogen fertilizer on growth, yield and nutritive value of two okra (*Abelmoschus Esculentus*, L.) cultivars. Australian J. Basic Appl Sci 1(3):307–312

Shanware AS, Kalkar SA, Trivedi MM (2014) Potassium solubilizers: occurrence, mechanism and their role as competent biofertilizers. Int J Curr Microbiol Appl Sci 3:622–629

Sharma MP, Gaur A, Tanu U, Sharma OP (2004) Prospects of *Arbuscular mycorrhiza* in sustainable management of root and soil borne diseases of vegetable crops. In: Mukerji KG (ed) Diseases management of fruits and vegetable. Kluwer Academic Publishers, Netherlands, pp 501–539

Singh SP (2014) Effect of bio-fertilizer azospirillum on growth and yield parameters of coriander (*Coriandrum sativum* L.) cv. Pant Haritima. Int J Seed Spices 4:73–76

Singh SP, Singh HB (2014) Effect of mixture of *Trichoderma* isolates on biochemical parameters in leaf of *Macrophomina phaseolina* infected brinjal. J Environ Biol 35:871–876

Singh JS, Pandey VC, Singh DP (2011) Efficient soil microorganisms: a new dimension for sustainable agriculture and environmental development. Agric Ecosyst Environ 140: 339–353

Singh CK, John AS, Jaiswal D (2014a) Effect of organics on growth, yield and biochemical parameters of chilli (*Capsicum annum* L.) IOSR J Agric Vet Sci 7:27–32

Singh A, Maji S, Kumar S (2014b) Effect of biofertilizers on yield and biomolecules of anticancerous vegetable broccoli. Int J Bio-resource Stress Manage 5:262–268

Singh SB, Singh HB, Singh DK (2014c) Biocontrol potential of mixture of *Trichoderma* isolates on damping-off and collar rot of tomato. The Bioscan 9(3):1301–1304

Singh SK, Sharma HR, Shukla A, Singh U, Thakur A (2015) Effect of biofertilizers and mulch on growth, yield and quality of tomato in mid-hills of Himachal Pradesh. Int J Farm Sci 5(3):98–110

Sinha RK, Valani D, Chauhan K, Agarwal S (2014) Embarking on a second green revolution for sustainable agriculture by vermiculture biotechnology using earthworms: reviving the dreams of Sir Charles Darwin. Int J Agric Health Saf 1:50–64

Sivakumar T, Ravikumar M, Prakash M, Thamizhmani R (2013) Comparative effect on bacterial biofertilizers on growth and yield of green gram (*Phaseolus radiata* L.) and cow pea (*Vigna sinensis* Edhl.) Int J Curr Res Aca Rev 1(2):20–28

Sivan A, Chet I (1989) The possible role of competition between *Trichoderma harzianum* and *Fusarium oxysporum* on rhizosphere colonization. Phytopathology 79:198–203

Smith SE, Smith FA (2011) Roles of arbuscular mycorrhizas in plant nutrition and growth: new paradigms from cellular to ecosystem scales. Ann Rev Plant Biol 62:227–250

Someshwar B, Bambawale OM, Tripathi AK, Ahmad I, Srivastava RC (2013) Biological management of fusarial wilt of tomato by *Trichoderma* spp. in Andamans. Indian J Hortic 70:397–403

Strakowska J, Błaszczyk L, Chełkowski J (2014) The significance of cellulolytic enzymes produced by *Trichoderma* in opportunistic lifestyle of this fungus. J Basic Microbiol 54(Suppl 1):S2–13

Sturz AV, Nowak J (2000) Endophytic communities of rhizobacteria and the strategies required to create yield enhancing associations with crops. Appl Soil Ecol 15:183–190

Sudhakar P, Chattopadhyay GN, Gangwar SK, Ghosh JK (2000) Effect of foliar application of *Azotobacter, Azospirillum* and *Beijerinckia* on leaf yield and quality of mulberry (*Morus alba*). J Agric Sci 134:227–234

Supanjani S, Habiba A, Mabooda F, Leea KD, Donnellya D, Smith DL (2006) Nod factor enhances calcium uptake by soybean. Plant Physiol Biochem 44:866–872

Tauro P, Kapoor KK, Yadav KS (1986) An introduction to microbiology. New Age International (P) Limited Publishers, New Delhi, India, p 412

Thakur N, Tripathi A (2015) Biological management of damping-off, buckeye rot and fusarial wilt of tomato (cv. Solan Lalima) under mid-hill conditions of Himachal Pradesh. Agric Sci 6:535–544

Thomashow LS, Weller DM (1996) Current concepts in the use of introduced bacteria for biological disease control: mechanisms and antifungal metabolites. In: Stacey G, Keen NT (eds) Plant-microbe interactions, vol 1. Chapman & Hall, New York, pp 187–236

Thrane C, Tronsmo A, Jenson DF (1997) Endo β-1,3 glucanase and cellulose from *Trichoderma harzianum*: biological activity against plant pathogenic spp. Eur J Plant Pathol 103:331–344

Verma OP, Shende ST (1993) *Azotobacter* a biofertilizer for vegetable crops. Biofert Newsletter 1:6–10

Villanueva LM, Ibis LM, Dayao AS (2014) Potential of *Bacillus subtilis* against powdery mildew of garden pea. In: Reddy MS, Ilao RI, Faylon PS, Dar WD, Batchelor WD, Sudini RSH, Kumar KVK, Armanda A, Gopalkrishnan S (eds) Recent advances in biofertilizers and biofungicides (PGPR) for sustainable agriculture, Cambridge Scholars Publishing, United Kingdom, pp 31–42

Vinale F, Sivasithamparam K, Ghisalberti EL, Marra R, Woo SL, Lorito M (2008) *Trichoderma–*plant–pathogen interactions. Soil Biol Biochem 40:1–10

Viswanathan R, Samiyappan R (2001) Antifungal activity of chitinase produced by fluorescent pseudomonads against *Colletotrichum falcatum* Went. causing red rot disease in sugarcane. Microbiol Res 155:305–314

Voisard C, Keel C, Haas D, Defago G (1989) Cyanide production by *Pseudomonas fluorescens* helps suppress black root rot of tobacco under gnotobiotic conditions. EMBO J 8:351–358

Whipps JM, Sreenivasaprasad S, Muthumeenakshi S, Rogers CW, Challen MP (2008) Use of *Coniothyrium minitans* as a biocontrol agent and some molecular aspects of sclerotial mycoparasitism. Euro J. Plant Pathol 121:323–330

Whitelaw MA (2000) Growth promotion of plants inoculated with phosphate solubilizing fungi Adv Agron 69:99–151

Winkelmann G, Drechsel H (1997) Microbial siderophores. In: Rehm HJ, Reed G (eds) Biotechnology, vol 7, 2nd edn. VCH, Weinheim, pp 199–246

Woese K, Lange D, Boess C, Bogl KW (1995) A comparison of organically and conventionally grown foods—results of a review of the relevant literature. J Sci Food Agric 74:281–293

Worthington V (2001) Nutritional quality of organic versus conventional fruits, vegetables and grains. J Altern Complement Med 7:161–173

Yazdani M, Bahmanyar MA, Pirdashti H, Esmaili MA (2009) Effect of phosphate solubilization microorganisms (PSM) and plant growth promoting rhizobacteria (PGPR) on yield and yield components of corn (*Zea mays* L.) Proc World Acad Sci Eng Technol 37:90–92

Youssef MMA, Eissa MFM (2014) Biofertilizers and their role in management of plant parasitic nematodes. A review. E3 J Biotechnol Pharm Res 5:1–6

Zahida R, Dar SB, Mudasir R, Inamullah S (2016) Productivity and quality of French bean (*Phaseolus vulgaris* L.) as influenced by integrating various sources of nutrients under temperate conditions of Kashmir. Int J Food Agric Vet Sci 6:15–20

Zaidi NW, Singh US (2004) Development of improved technology for mass multiplication and delivery of fungal (*Trichoderma*) and bacterial (*Pseudomonas*) biocontrol agents. Indian J Mycol Plant Pathol 34:732–741

Plant Growth-Promoting Bacteria: Importance in Vegetable Production

2

Abdelwahab Rai and Elhafid Nabti

Abstract

A large number of soil bacteria are able to colonize the surface/interior of root system and stimulate plant growth and health. This group of bacteria, generally referred to as plant growth-promoting rhizobacteria (PGPR), enhances the growth of plants including vegetables in both conventional and stressed soil. In addition, many PGPR facilitate crop production indirectly by inhibiting various phytopathogens. Conclusively, PGPR affects plant growth via nitrogen fixation, phosphate solubilization and mineral uptake, siderophore production, antibiosis, and hydrolytic enzymes synthesis. Some of the notable PGPR capable of facilitating the growth of a varied range of vegetables such as potato, carrot, onion, etc. belong to genera *Azotobacter*, *Azospirillum*, *Pseudomonas*, and *Bacillus*. Vegetables play a major role in providing essential minerals, vitamins, and fiber, which are not present in significant quantities in staple starchy foods. Hence, to optimize vegetable production without chemical inputs, the use of PGPR in vegetable cultivation is recommended. Here, an attempt is made to highlight the role of PGPR in vegetable production under both normal and derelict soils.

2.1 Introduction

Human population is growing very rapidly, and according to the United Nations estimate, it is expected to be 8.9 billion by the end of 2050 (UN 2004, 2015; Ashraf et al. 2012). In order to feed the growing populations, there is an increasing food demand whose production needs to be augmented alarmingly in the next few years.

A. Rai • E. Nabti (✉)
FSNV, Equipe de Biomasse et Environnement, Laboratoire de Maitrise des Energies Renouvelables (LMER), Université de Béjaïa, Targa Ouzemmour, Béjaïa 06000, Algeria
e-mail: elhnabti1977@yahoo.fr

© Springer International Publishing AG 2017   23
A. Zaidi, M.S. Khan (eds.), *Microbial Strategies for Vegetable Production*,
DOI 10.1007/978-3-319-54401-4_2

In this regard, the Center for Study of Carbon Dioxide and Global Change indicated that 70–100% increase in agricultural production is required to feed the ever-increasing human populations. It also published a presumptive model estimating that only 34.5–51.5% increase will be achieved between 2009 and 2050. Of the various food items, vegetables play an important role in human dietary systems. And hence, among vegetables, total potato production is estimated to raise from 329 to 416 million tons between 2009 and 2050 due to advancements in agricultural technology and scientific research (techno-intel effect) and to 466 million tons due to the combined consequences of techno-intel effect and CO_2 aerial fertilization effect. Also, total bean production is estimated to increase from about 21 to 26 and 32 million tons between 2009 and 2050 due to techno-intel effect alone or due to the combined techno-intel effect and CO_2 aerial fertilization (Idso 2011). However, the average vegetable supply available per person in the world was about 102 kg per person by the year 2000. In addition, between 1979 and 2000, it augmented from 45.4 to 52 kg in Africa and from 43.2 to 47.8 kg in South America, while the highest improvement was found in Asia (from 56.6 to 116.2 kg per person per year), noting that global vegetable production jumped from 326.616 to 691.894 million tons (Fresco and Baudoin 2002). However, due to environment degradation, biodiversity destruction, and soil fertility loses, considerable reduction in agricultural production including those of vegetable production leading to inadequate food supply to human populations has been recorded (Shahbaz and Ashraf 2013).

2.2 Place of PGPR in Food Safety and Agricultural Challenges

Because of different factors threatening agriculture, scientists are searching for alternatives involving natural and eco-friendly solutions. Among these options, microbe-based (bacteria, fungi) ecological engineering strategies have been developed for ecological conservation and to improve agronomic practices for enhancing food production (Ashraf et al. 2012). Among soil microflora, the use of plant growth-promoting rhizobacteria (PGPR) began about 100 years ago where some countries like China, European countries, the former Soviet Union, and the United States started practical programs to develop PGPR inoculants at a larger scale for the use in agriculture. However, the term "rhizobacteria" was introduced first by Kloepper and Schroth (1978) to qualify bacterial community that aggressively colonize roots and improve plant growth. The PGPR application is considered one of the most viable and inexpensive methods for increasing agricultural productivity through plant growth stimulation, plant pathogens control, and pollutant biodegradation, bioremediation (Bhattacharyya and Jha 2012; Landa et al. 2013). In this chapter, different mechanisms by which beneficial soil bacteria improve plant growth, plant defenses against phytopathogens, and soil health and how they participate in the interactive plant-soil-bacteria system are discussed. Furthermore, the importance of PGPR in vegetable production under different agroclimatic conditions is highlighted. It is important to mention that vegetables play a major role in providing essential minerals, vitamins, and fiber, which are not present in significant quantities in starchy foods, and represent an important supply of proteins and carbohydrates (Nichols and Hilmi 2009).

2.3 Mechanism of Growth Promotion by PGPR: A General Perspective

2.3.1 Nitrogen Fixation

Nitrogen fixation, one of the most important means of adding N to soil nutrient pool (Reddy 2014), is mediated both by symbiotic prokaryotic microorganisms like *Rhizobium*, *Mesorhizobium*, *Bradyrhizobium*, *Azorhizobium*, *Allorhizobium*, and *Sinorhizobium* and asymbiotic/free-living organisms such as *Azoarcus*, *Azospirillum*, *Burkholderia*, *Gluconacetobacter*, *Pseudomonas*, *Azotobacter*, *Arthrobacter*, *Acinetobacter*, *Bacillus*, *Enterobacter*, *Erwinia*, *Flavobacterium*, *Klebsiella*, and *Acetobacter*. These bacterial genera and some others have been described as nitrogen-fixing PGPR with substantial ability to promote plant growth and yield (Gupta et al. 2015; Miao et al. 2014; Sivasakthi et al. 2014; Verma et al. 2013). Nitrogen fixation is carried out by a highly conserved and energetically expensive enzyme called nitrogenase. The conventional nitrogenase is composed of two metalloprotein subunits. The first one is composed of two heterodimers (250 kDa) and encoded by *nifD* and *nifK* genes; it contains the active site for nitrogen reduction. The second one (two identical subunits/70 kDa, encoded by *nifH* gene) ensures ATP hydrolysis and electron transfer between subunits that are coordinated by Fe-S containing Mo. Mo is replaced by V (*vnfH*) in "alternative nitrogenase" and by Fe (*anfH*) in "second alternative nitrogenase" (Zehr et al. 2003). Of the various nitrogen fixers, bacteria belonging to group "rhizobia" are known to establish symbiotic relations with host-specific legumes and to provide a major plant nutrient N to plants. The species *R. meliloti*, *R. trifolii*, *R. leguminosarum*, *R. phaseoli*, *R. japonicum*, etc. can supply N to plants such as lucerne, sweet clover, pea, lentil, bean, cowpea, etc. (Yamaguchi 1983). In addition, some other associative nitrogen fixers, for example, *Azospirillum* inoculation, have been reported to enhance growth and yield of several winter legumes such as pea and chickpea (Sarig et al. 1986). The role of two PGPR strains (*Serratia liquefaciens* 2-68 or *S. proteamaculans* 1-102) in increasing nodulation, nitrogen fixation, and total nitrogen yield of two soybean cultivars in a short season area was reported (Dashti et al. 1998). Strains increased soybean nodulation and accelerated nitrogen fixation onset. Fixed N, expressed as a percentage of total plant N, and protein and N yield were increased by PGPR inoculation. Pishchik et al. (1998) on the other hand reported the inoculation effect of nitrogen-fixing *Klebsiella* on yield of nonlegumes such as potato. A significant increase in potato yield and N content was obtained after inoculation with *K. mobilis* strains CIAM880 and CIAM853 when low doses of nitrogenous fertilizer were used. Recently, Naqqash et al. (2016) observed that inoculation of nitrogen-fixing bacteria, namely, *Azospirillum*, *Enterobacter*, and *Rhizobium*, under axenic conditions resulted in differential growth responses of potato. Of these, *Azospirillum* sp. TN10 showed the highest increase in fresh and dry weight of potato over control plants. In addition, a significant augmentation in N contents of shoot and roots of *Azospirillum* sp.-inoculated potato plants was observed.

2.3.2 Nitrification

Bacterial nitrification is a biological process in which energy is extracted by sequential oxidation of nitrogen that occurs as ammonia. Complete oxidation of nitrate is carried out by two metabolically distinct groups of bacteria: (i) ammonia-oxidizing bacteria, for example, *Nitrosomonas, Nitrosospira, Nitrosovibrio, Nitrosolobus*, and *Nitrosococcus*, transform ammonia to nitrite, and (ii) nitrite is transformed to nitrate by nitrifying bacteria like *Nitrobacter, Nitrococcus, Nitrospira*, and *Nitrospina*. Nitrification is important for soil and ecosystem health because it completes the mineralization of organic nitrogen started with ammonification process (nitrogen fixation) (Ardisson et al. 2014; Cohen and Mazzola 2006; Cohen et al. 2010). Among others, nitrification is considered as an important trait to select beneficial bacteria able to improve plant growth and crop yield (Prasad et al. 2015). It is believed that nitrification is the principal source of nitric oxide (NO) emitted from the soil. However, recent works have described NO as a signal molecule in plant-PGPR interaction. For example, *Azospirillum* strains produced tenfold of NO than the amount found in plant. Nevertheless, when bacterial nitric oxide was sequestered with specific scavenger (cPTIO), results clearly showed that the ability of *Azospirillum* inoculation to induce lateral root development in tomato was lost suggesting the involvement of NO in the *Azospirillum*-plant root association (Cohen et al. 2010; Skiba et al. 1993).

2.3.3 Denitrification

The first description of soil organic matter degradation that resulted in release of nitrogen gas into atmosphere was realized by Reyest in 1856. Later on, Gayon and Dupetit were the first to describe denitrification in 1886 (Elmerich 2007). Denitrification is defined as a microbial respiratory process during which soluble N oxides are used as alternative electron acceptor when O_2 is not available for aerobic respiration. It involves sequential reduction of NO^{3-} into dinitrogen in four steps coupled with energy conservation (NO to NO_2, NO_2 to NO, NO to N_2O, and N_2O to N_2). Denitrification completes the N cycle and usually balances the total biological N fixation in the global N cycle (Hofstra and Bouwman 2005; Philippot et al. 2007). Among denitrifying bacteria, *Agrobacterium, Aquaspirillum, Azoarcus, Azospirillum, Bradyrhizobium, Hyphomicrobium, Magnetospirillum, Paracoccus, Rhodobacter, Rhodopseudomonas, Cytophaga, Sinorhizobium, Flexibacter, Alcaligenes, Neisseria, Nitrosomonas*, and *Thiobacillus* are the most commonly found in nature, especially in soil (Knowles 2004).

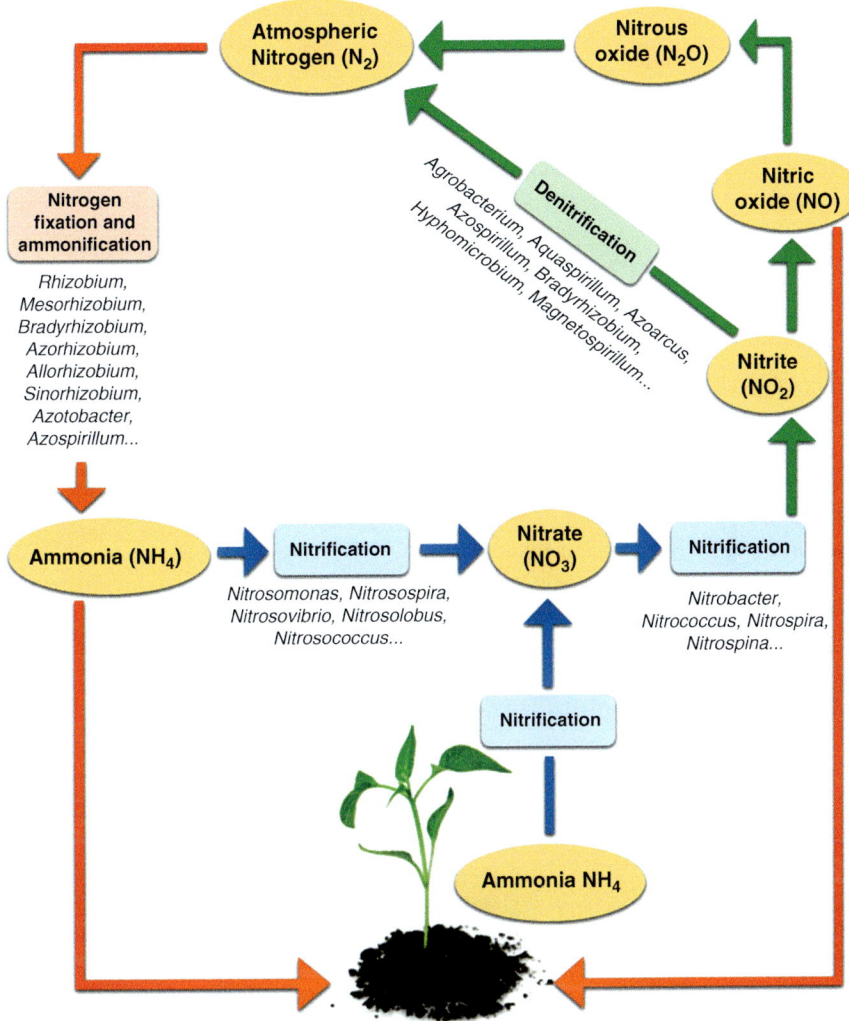

Fig. 2.1 Role of PGPR in nitrogen recycling and plant growth stimulation (modified from Cohen et al. 2010; Reddy 2014)

Ecologically, denitrification is a key mechanism for biological elimination of N. In fact, 15–70% of ammonium derived from organic matter mineralization is reported to be eliminated through nitrification and denitrification process (Bertrand et al. 2015). In rhizosphere, oxygen concentration could be lowered because of root and microorganism's respiration. In addition, organic compounds released by plants' roots can be used as electron donors in denitrification process, suggesting that denitrifiers could constitute highly competitive microorganisms in rhizosphere (Fig. 2.1). Denitrifying bacteria may prevent nitrogen accumulation to toxic levels, reduce nitrate contents in groundwater, and maintain a balance between soil and atmospheric nitrogen avoiding serious problems that could occur if no alternative mechanism is available to return nitrogen to atmosphere (Antoun and Prévost 2005; Gupta et al. 2000; Philippot et al. 2007). Due to these and in addition to the presence of positive correlation between bacterial denitrification ability and rhizosphere colonization, Kumar et al. (2014) considered nitrification as an important trait to isolate and select fluorescent PGP *Pseudomonas*. Furthermore, in a recent work conducted by Muriel et al. (2015), denitrification was regarded as an important plant growth trait in PGP *Pseudomonas fluorescens* F113. Otherwise, denitrification in legumes may be a species-dependent mechanism to maintain optimum rates of N_2 fixation within root nodule; hence, NO has been reported as inhibitor of nitrogenase activity (Williams et al. 2014). Denitrification in nodules could also ensure detoxification of cytotoxic compounds produced as intermediates during denitrification reactions or emerging from host plant such as nitrite and NO (O'Hara and Daniel 1985; Sánchez et al. 2011). In addition, Lombardo et al. (2006) reported that when lettuce plants were grown hydroponically, root epidermis did not form root hairs. The addition of 10 µM sodium nitroprusside (a nitric oxide (NO) donor) resulted in almost all rhizodermal cells differentiated into root hairs. They also found that treatment with synthetic auxin 1-naphthyl acetic acid exhibited a significant increase of root hair formation that was prevented by the specific NO scavenger carboxy-PTIO.

2.3.4 Phosphate Solubilization

After nitrogen, phosphorus (P) is the most important macronutrient for biological processes, for example, cell division and development, energy transport, signal transduction, macromolecular biosynthesis, photosynthesis, and plant respiration. Phosphorus is present at levels of 400–1200 mg/kg of soil. However, only a very small amount (1 mg or less) of P is in soluble forms, while the rest is insoluble and, hence, not available for plant uptake (Khan et al. 2009). It is important to mention that a big part of P applied to agricultural fields as fertilizer is rapidly immobilized and, hence, becomes inaccessible for plants (Oteino et al. 2015). In addition, the process of traditional phosphorus fertilizer production is environmentally undesirable because of contaminants release into the main product, gas stream and by-products, and accumulation of Cd or other heavy metals in soil and crops because of repetitive use of phosphatic fertilizers (Sharma et al. 2013; Song et al. 2008). To avoid these problems, a group of soil microorganisms, called phosphate-solubilizing microorganisms (PSM), is considered as one of the best eco-friendly options for providing inexpensive P to plants. Through their activities, insoluble forms of P are hydrolyzed to soluble forms through solubilization (inorganic P) and mineralization (organic p) processes. On the contrary, immobilization is the reverse reaction of mineralization, during which, microorganisms convert inorganic forms to organic phosphate (Sharma et al. 2013; Khan et al. 2014). Some of the notable PGPR possessing P-solubilizing activity are *Achromobacter xylosoxidans* (Ma et al. 2009), *Bacillus polymyxa* (Nautiyal 1999), *Pseudomonas putida* (Malboobi et al. 2009),

Acetobacter diazotrophicus (Sashidhar and Podile 2010), *Agrobacterium radiobacter* (Leyval and Berthelin 1989), *Bradyrhizobium mediterranium* (Peix et al. 2001), *Enterobacter aerogenes*, *Pantoea agglomerans* (Chung et al. 2005), *Gluconacetobacter diazotrophicus* (Crespo et al. 2011), and *Rhizobium meliloti* (Krishnaraj and Dahale 2014). Among non-symbiotic bacteria, *Azotobacter* has also been found as phosphate solubilizer and plant growth-enhancing bacterium (Nosrati et al. 2014). Malboobi et al. (2009) evaluated the performance of three PSB *P. agglomerans* strain P5, *Microbacterium laevaniformans* strain P7, and *P. putida* strain P13 in potato's rhizosphere. All experiments proved that these isolates compete well with naturally occurring soil microorganisms in potato's rhizosphere. The combinations of strains P5 + P13 and P7 + P13 led to higher biomass and potato tuber in greenhouse and in field trials. The effect of other phosphate solubilizers such as *B. megaterium* var. phosphaticum, *P. agglomerans*, *M. laevaniformans*, *P. putida*, *P. cepacia*, *P. fluorescens*, *Xanthomonas maltophilia*, *Enterobacter cloacae*, *Acidovorans delafieldii*, *Rhizobium* sp., *A. chroococcum*, and *Burkholderia anthina* on some of the widely grown and consumed vegetables such as potato, tomato, pepper, cucumber, pea, brinjal, etc. has been reported by others (Bahena et al. 2015; Pastor et al. 2014; Rizvi et al. 2014 and Walpola and Yoon 2013).

2.3.5 Siderophores, a Powerful Tool for Antagonism and Competition

Iron is a central element for life on earth, especially for plant growth and development. It participates in formation of several types of vegetable proteins such as ferredoxin, cytochrome, and leghemoglobin (Fukuyama 2004; Liu et al. 2014). This element is relatively insoluble in soil solution. So why plants secrete soluble organic compounds (binders) which bind to ferric ion (Fe^{3+}) to form the chelator-Fe^{3+} complex (Tokala et al. 2002; Vessey 2003)? Several studies on iron utilization by plants allowed scientists to distinguish two strategies used by plants for iron acquisition from soil (Bar-Ness et al. 1992). In the first one, iron chelators (siderophores: from the Greek "iron carriers") secreted by plants are immediately absorbed with Fe^{3+} through the plasma lemma. In the second one, formed complex (chelator-Fe^{3+}) helps to keep ferric ions in solution, then exposes to root surface where they are reduced to ferrous ions (Fe^{2+}) and immediately absorbed (Neilands 1995; Vessey 2003). In addition to these two strategies, plants can also use microbial siderophores (fungi and bacteria) which are synthesized under iron-starved conditions. Broadly, siderophores are defined as low-molecular-weight compounds (500–1500 daltons) possessing high affinity for ferric iron. They are mainly produced by bacteria (Kümmerli et al. 2014), fungi (Renshaw et al. 2002), and graminaceous plants (Hider and Kong 2010) to scavenge iron from environment.

According to the chemical nature, siderophores are divided into five classes, (1) catecholates, (2) phenolates, (3) hydroxamates, (4) carboxylates, and (5) mixed siderophores, which contain at least two of the abovementioned classes. In agriculture, the secretion of bacterial siderophores is important for two reasons: (1) it provides iron to plants, and (2) it limits the availability of iron to plant pathogens (Miethke and Marahiel 2007; Tailor and Joshi 2012). Additionally, siderophores may stimulate biosynthesis of other antimicrobial compounds (Beneduzi et al. 2012; Laslo et al. 2011). Impressively, it has been reported that some nodule bacteria, for example, *Rhizobium*, can require an intact siderophore system to express some vital activities such as nitrogenase (Neilands 1995).

Until 2014, more than 500 siderophore-type molecules have been identified (Kannahi and Senbagam 2014). Genera like *Azotobacter* (Fekete et al. 1983), *Azospirillum* (Tortora et al. 2011), *Pseudomonas* (Tailor and Joshi 2012), *Agrobacterium* (Rondon et al. 2014), *Alcaligenes* (Sayyed and Chincholkar 2010), *Serratia* (Seyedsayamdost et al. 2012), *Enterobacter* and *Achromobacter* (Tian et al. 2009), *Rhizobium* (Datta and Chakrabartty 2014), *Bradyrhizobium* (Abd-Alla 1998), etc. are known to promote growth of many crops through siderophore production. Therefore, siderophores secreted by many PGPR are used as a specific trait for selection and application of effective bacteria in crop production. For example, the indigenous isolate *B. subtilis* CTS-G24 producing a hydroxamate type of siderophore was found to be efficient in inhibiting wilt and dry root rot disease caused by both *Fusarium oxysporum* f. sp. ciceri and *Macrophomina phaseolina* in chickpea (Patil et al. 2014). In other study, a yellow-green pigment (pseudobactin) exhibiting properties typical of a siderophore was isolated from broth cultures of fluorescent *Pseudomonas* strain B10, grown in iron-deficient medium (Kloepper et al. 1980). The application of B10 as inoculant and pure pseudobactin significantly improved potato growth in greenhouse assay compared to water-treated controls. In addition, strain B10 and pseudobactin significantly reduced fungal population in potato's rhizoplane (control, 5.5; B10, 2.3; pseudobactin, 1.4 CFU per 10 cm roots) suggesting that bacterial siderophores play a crucial role in enhancing plant growth by sequestering iron in root zone and by antagonism to potentially deleterious phytopathogens. The role of siderophore-producing bacteria in enhancing potato growth has also been reported by others (Bakker et al. 1986; Weisbeek et al. 1987). Moreover, in a hydroponic culture experiment, siderophores from bacterial strain *Chryseobacterium* C138 were found effective in supplying Fe to iron-starved tomato plants by roots inoculated with or without bacteria (Radzki et al. 2013). Similarly, the role of fluorescent siderophore (pyoverdin) in suppression of *Pythium*-induced damping-off in tomato by *Pseudomonas aeruginosa* RBL 101 has been reported by Jagadeesh et al. (2001). Thus, hyperactive mutants (Flu++ Sid++) (RBL 1015 and 1011) with higher siderophore production suppressed wilt disease more efficiently (75 and 37%, respectively) than the wild type (12.5%). In a follow-up study, Valencia-Cantero et al. (2007) observed a significant increase Fe content and growth of bean plants inoculated with *B. megaterium* UMCV1, *Arthrobacter* spp. UMCV2, *S. maltophilia* UMCV3, and *S. maltophilia* UMCV4, compared to uninoculated plants grown in sterilized soil. Similarly, the role of bacteria such as *Pseudomonas aeruginosa*, *P. fluorescens*, *P. putida*, and *S. marcescens* in inducing siderophore-dependent resistance in vegetables such as bean, tomato, radish, and cucumber against plant pathogens like *Colletotrichum lindemuthianum*, *C. orbiculare*, *Botrytis cinerea*, and *Fusarium* was also reported (Höfte and Bakker 2007).

2.3.6   Bacterial Phytohormones and Plant Growth Regulation

Phytohormones or "plant growth hormones" are naturally occurring organic substances that exert, at low concentrations, a major influence on plant growth and upregulation of physiological process. Among phytohormones, auxin, the term derived from Greek word αυξειν (auxein means "grow or increase"), was the first plant hormone discovered by Kende and Zeevaart (1997). Auxin remained the only synonym of phytohormone until 1973, when Went and Thimann published their book *Phytohormones*. Since then, other phytohormones such as gibberellin, ethylene, cytokinin, and abscisic acid have been discovered (Tran and Pal 2014). Phytohormones are produced by plants (Bari and Jones 2009), by microorganisms (Narayanasamy 2013), and even by algae (Kiseleva et al. 2012). Among microbes, PGPR can also modulate phytohormone levels in plant tissues affecting hormonal balance of host plant (Figueiredo et al. 2016). Some of the most common phytohormones affecting plant growth are discussed in the following section.

2.3.6.1 Auxins: Biosynthesis and Their Place in the Plant-PGPR Interaction

Among phytohormones, auxins have the ability to affect, practically, all plant physiological aspects from promotion of cell enlargement and division, apical dominance, root initiation, and differentiation of vascular tissue to modulation of reactive oxygen species (Tomić et al. 1998). Recently, it has been reviewed that auxins affect other plant hormone activities, such as cytokinin, abscisic acid, ethylene, jasmonate, and salicylic acid, and modulates various plant defense-signaling pathways (Vidhyasekaran 2015). Indole acetic acid (IAA) is the major naturally occurring phytohormone which is also produced by bacteria involved in plant growth and health enhancement (Gao and Zhao 2014; Etesami et al. 2015; Spaepen and Vanderleyden 2010). In most cases, tryptophan (Trp) serves as physiological precursor in IAA synthesis (Spaepen et al. 2007a). IAA biosynthesis in bacteria involves five Trp-dependent pathways: indole-3-acetamide pathway, indole-3-pyruvic acid pathway, tryptamine pathway, indole-3-acetonitrile pathway and Trp side chain oxidase pathway, and one Trp-independent pathway (Spaepen et al. 2007b; Di et al. 2016).

Beyeler et al. (1999) reported that a genetically modified strain of *P. fluorescens* CHA0, which overproduced IAA, was more effective for cucumber growth improvement than the wild strain. Accordingly, mutant strain CHA0/pME3468 increased fresh root weight of cucumber by 17–36%, compared to the effect of wild CHA0 strain; Gravel et al. (2007) found that IAA (10 μg/ml) application by drenching to the growing medium or by spraying on shoots reduced symptoms caused by *P. ultimum* on tomato plants. Furthermore, Khan et al. (2016) reported that among other tested strains, endophyte *B. subtilis* LK14 produced the highest (8.7 μM) amount of IAA on the fourteenth day of growth and significantly increased shoot and root biomass and chlorophyll (a and b) contents in tomato as compared to control plants.

2.3.6.2 Gibberellins: Miraculous Molecules for Plant Growth Regulation

Gibberellins were first isolated in 1962 from fungus *Fusarium moniliforme* (*Gibberella fujikuroi* in sexual form) by Kurosawa (Japan). In 1938, two other Japanese workers (Yakutat and Sumiki) isolated active principles as crystals from culture medium and named them gibberellins A and B (Takahashi et al. 1991). Macmillan and Suter (1958) identified the first plant gibberellin (GA1) from *Phaseolus coccineus* seeds. However, gibberellins are synthesized not only by plants and fungi but also by bacteria (Morrone et al. 2009). In this context, Maheshwari et al. (2015) mentioned that the bacterial gibberellins were reported first time in 1988 in *R. meliloti*. Later on, based on gibberellins pathways synthesis occurring in plant and fungi, it was suggested that its synthesis in bacteria started with geranyl-PP conversion into ent-kaurene via ent-copalyl diphosphate. After this, ent-kaurene is converted into GA12-aldehyde through ent-kaurene oxidase and ent-kaurenoic acid oxidase synthesis. GA12-aldehyde is then oxidized into GA_{12} and metabolized into other GA (Kang et al. 2014). Morrone et al. (2009) described an operon in *Bradyrhizobium japonicum* genome, whose enzymatic composition indicates that gibberellin biosynthesis in bacteria represents a third independently assembled pathway relative to plants and fungi.

Currently, gibberellins include a wide range of tetracyclic diterpene acids that regulate, in combination with other phytohormones, diverse processes in plant growth such as germination, stem elongation, flowering, fruiting, root growth promotion, root hair abundance, vegetative/reproductive bud dormancy, and delay of senescence in many plant organs (Cassán et al. 2014; Kang et al. 2012; Niranjana and Hariprasad 2014). Bacteria such as *Acetobacter diazotrophicus* (Bastian et al. 1998), *Azospirillum lipoferum* (Bottini et al. 1989), *A. brasilense* (Janzen et al. 1992), *Bacillus pumilus* (Joo et al. 2005), *B. cereus* (Joo et al. 2005), *B. macroides* (Joo et al. 2005), *Herbaspirillum seropedicae* (Kang et al. 2014), *Acinetobacter calcoaceticus* (Kang et al. 2009), *Burkholderia cepacia* (Joo et al. 2009), and *Promicromonospora* sp. (Kang et al. 2012) have been reported as gibberellin producers. In addition, Kang et al. (2012) described the role of gibberellin-producing *Promicromonospora* sp. SE188 in *Solanum lycopersicum* plant growth improvement. *Promicromonospora* sp. produced physiologically active (GA1 and GA4) and inactive (GA9, GA12, GA19, GA20, GA24, GA34, and GA53) gibberellins. In addition to plant growth improvement, tomato inoculated with this bacterium resulted in a downregulation of the stress hormone abscisic acid, while salicylic acid was significantly higher compared to control plants. Joo et al. (2004, 2005) reported the positive effect of gibberellin-producing bacteria (*B. cereus* MJ-1, *B. macroides* CJ-29, and *B. pumilus* CJ-69) on red pepper growth and its endogenous gibberellins content. Inoculation with *B. cereus* MJ-1 improved shoots and roots fresh weight of red pepper by 1.38- and 1.28-fold, respectively. Among 864 bacterial isolates tested on cucumber and crown daisy for growth promotion, the most efficient strain for plant growth enhancement, *Burkholderia* sp. KCTC 11096BP, was found to produce physiologically active gibberellins (GA_1, 0.23; GA_3, 5.11; and GA_4 2.65 ng/100 ml) and inactive gibberellins (GA_{12}, GA_{15}, GA_{20}, and GA_{24}) (Joo et al.

2009). Moreover, Khan et al. (2014) reported tomato growth-promoting activity of IAA and gibberellin-producing bacteria *Sphingomonas* sp. LK11 isolated from leaves of *Tephrosia apollinea*. In culture broth, the strain LK11 released active (GA4, 2.97 ng/ml) and inactive gibberellins (GA9, 0.98 and GA20, 2.41 ng/ml). Tomato plants inoculated with endophytic *Sphingomonas* sp. LK11 had significantly higher shoot length, chlorophyll contents, and dry matter accumulation in shoot and root compared to control suggesting the potential role of phytohormones in crop growth improvement.

2.3.6.3 Cytokinins and Plant Growth Regulation

Cytokinins are N6-substituted aminopurines or adenine compounds with an isoprene, modified isoprene, aromatic side chain attached to the N6-amino group, or zeatin and trans-zeatin. These molecules have the ability to influence physiological and developmental processes of plants. Cytokinins affect cell division, cell cycle, leaf senescence, nutrient mobilization, apical dominance, shoot apical meristems formation and activity, floral development, breaking of bud dormancy and seed germination, chloroplast differentiation, autotrophic metabolism, and leaf and cotyledon expansion (Maheshwari et al. 2015; Wong et al. 2015). Apart from plant roots, cytokinins can also be derived from microalgae, bacteria, mycorrhizal fungi, and nematodes in rhizosphere (Reddy 2014). For a long time, cytokinins have been considered as an important plant growth regulator. Hence, several works reported the role of cytokinin-producing bacteria like *Azotobacter* (Taller and Wong 1989), *Azospirillum* (Conard et al. 1992), *Agrobacterium* (Akiyoshi et al. 1987), *Pseudomonas* (Akiyoshi et al. 1987), *Paenibacillus* (Timmusk et al. 1999), *Bacillus* (Ortíz Castro et al. 2008), *Achromobacter* (Donderski and Głuchowska 2000), *Enterobacter* (Kämpfer et al. 2005), and *Klebsiella* (Conard et al. 1992) in plant growth regulation.

The impact of cytokinins produced by some bacterial strains isolated from rhizosphere on growth and cell division in cucumber cotyledons have been reported (Hussain and Hasnain 2009). Chlorophyll contents, cell division, and fresh weight were increased in cucumber cotyledons placed at 2 mm distance from cytokinin-producing *B. licheniformis* Am2, *B. subtilis* BC1, and *P. aeruginosa* E2 cultures under green light. Major cytokinin species detected were zeatin and zeatin riboside. Arkhipova et al. (2007) followed the consequences of inoculating growing medium with cytokinin-producing *Bacillus* (strain IB-22) under conditions of water sufficiency and deficit on 12-day-old lettuce seedlings. Inoculation increased shoot cytokinins, shoot abscisic acid, accumulation of shoot mass, and shortened roots, while it showed a smaller effect on root mass and root/shoot ratios by stimulating shoot growth, but did not raise stomatal conductance. Likewise, Arkhipova et al. (2005) evaluated the ability of cytokinin-producing *B. subtilis* in influencing growth and endogenous hormone content of lettuce plants. Recently, the osmotolerant cytokinin-producing *Citricoccus zhacaiensis* and *B. amyloliquefaciens* were found to enhance tomato growth under irrigation deficit conditions (Selvakumar et al. 2016). They observed that microbial inoculation significantly enhanced stomatal conductivity, transpiration rates, photosynthesis, and relative water contents of tomato plants across stress levels. Moreover, *C. zhacaiensis* enhanced the yield by 24 and 9%, while *B. amyloliquefaciens* increased the yield by 42 and 12.7%, at 50 and 25% water holding capacity, respectively. Ortiz Castro et al. (2008) described the important role played by cytokinin receptors in plant growth promotion by *B. megaterium*, initially isolated from bean plants rhizosphere. Inoculation with *B. megaterium* promoted biomass production of bean plants. This effect is related to altered root system architecture in inoculated plants (inhibition in primary root growth followed by an increase in lateral root formation and root hair length). These promoting effects on plant development were found to be independent of auxin and ethylene signaling.

2.3.6.4 Ethylene

Ethylene is a gaseous hormone produced by plants and plays an important role in various developmental processes, such as leaf senescence, leaf abscission, epinasty, and fruit ripening (Gray and Smith 2004; Vogel et al. 1998). Ethylene is synthesized from methionine in three steps that starts with methionine activation to S-adenosyl-L-methionine by the enzyme SAM synthetase. The second step consists to convert S-adenosyl-L-methionine to 1-aminocyclopropane-1-carboxylic acid (ACC), which is catalyzed by ACC synthase. After that, the enzyme ACC oxidase ensures ACC conversion to ethylene via an oxygenation reaction (Ma et al. 2014). At the beginning, ethylene was considered as a stress hormone because under stress conditions (salinity, drought, water logging, heavy metals, and pathogenicity), plants synthesize high amount of ethylene, leading to the alteration of their physiological performance and, consequently, to the reductions in root and shoot growth. Later, other vital functions such as seed germination, root hair development, adventitious root formation, nodulation, leaf and fruit abscission, and flower and leaf senescence have been found to be influenced by ethylene (Bakshi et al. 2015; Shrivastava and Kumar 2015).

2.3.6.5 Abscisic Acid

Abscisic acid (ABA) is a sesquiterpene phytohormone, synthesized by plants, bacteria, fungi, algae, and animals (Gomez-Cadenas et al. 2015; Karadeniz et al. 2006; Tuomi and Rosenquist 1995). ABA affects many physiological processes of plants including vegetables (Porcel et al. 2014). For example, ABA regulates several events during late seed development and plays an important role in circumventing environmental stresses such as desiccation, salt, and cold. Abscisic acid also controls plant growth and inhibits root elongation (Pilet and Chanson 1981) suggesting that a negative correlation exists between growth and the endogenous ABA plants content (Pilet and Saugy 1987). The prokaryotic pathway for abscisic acid biosynthesis originates from isoprene known as isopentenyl pyrophosphate that is synthesized from mevalonate pathway (Endo et al. 2014). Abscisic acid is the main hormone that balances many plant physiological responses to abiotic stress. However, its signaling pathways act in a complex interconnection with other hormone signal (Gomez-Cadenas et al. 2015).

2.3.6.6 Bacterial ACC Deaminase: A Hormone Balancing Signal Molecule

The enzyme 1-aminocyclopropane-1-carboxylate (ACC) deaminase synthesized by a wide range of rhizospheric bacteria (Glick et al. 2007) decreases the deleterious ethylene amounts and balances ABA levels in stressed plants. Enzyme ACC deaminase degrades ACC into α-ketobutyrate and ammonia to supply N and energy and, hence, lowers the ethylene levels in plant (Glick et al. 2007; Penrose and Glick 2003). It has been reviewed that many biotic (viruses, bacteria, fungi, and insects) and abiotic (salt, heavy metals, drought, radiation, etc.) stresses could be relieved by ACC deaminase-producing bacteria (Lugtenberg and Kamilova 2009; Shaharoona et al. 2012). Among microorganisms, soil bacteria belonging to genera *Agrobacterium*, *Azospirillum*, *Alcaligenes*, *Bacillus*, *Burkholderia*, *Enterobacter*, *Methylobacterium*, *Pseudomonas*, *Ralstonia*, *Rhizobium*, *Rhodococcus*, *Sinorhizobium*, *Kluyvera*, *Variovorax*, and *Paradoxus* have been reported to produce ACC deaminase (Barnawal et al. 2012; Glick 2014; Hao et al. 2010; Saleem et al. 2007; Toklikishvili et al. 2010).

The bacterial strain *M. ciceri* LMS-1 was transformed by triparental mating with plasmid pRKACC containing ACC deaminase gene (*acdS*) of *P. putida* UW4 cloned in pRK415. By expressing ACC deaminase under free-living conditions, ACC deaminase-producing mutant *Mesorhizobium* LMS-1 (pRKACC) increased chickpea nodulation performance and plant total biomass compared to LMS-1 wild-type strain (127 and 125%, respectively). These results suggest that the use of bacteria with improved ACC deaminase activity might be very important to develop microbial inocula for agricultural purposes (Nascimento et al. 2012). Like other crops, the role of ACC deaminase positive bacteria in vegetable growth is reported. As an example, Mayak et al. (2004) described the role of ACC deaminase-producing *Achromobacter piechaudii* in conferring resistance in tomato plants to salt stress. This bacterium significantly reduced ethylene levels in seedlings and increased fresh and dry weights of tomato grown in presence of up to 172 mM NaCl. Under salt stress, the bacterium also increased water use efficiency by plants compared to the control, suggesting the usefulness of such ACC deaminase-producing bacteria in alleviating salt stress. Similarly, ACC deaminase-producing and halotolerant *Brevibacterium iodinum*, *B. licheniformis*, and *Zhihengliuela alba* were found to regulate ethylene levels and consequently enhanced growth and salt tolerance of red pepper, grown in salt-stressed conditions (Siddikee et al. 2011). The inoculation with *B. licheniformis* RS656, *Z. alba* RS111, and *B. iodinum* RS16 reduced ethylene production by 44, 53 and 57%, respectively. In addition, when red pepper was grown in salt-stressed condition, salt stress caused 1.3-fold reduction in root/shoot dry weight ratio, while bacterial inoculation on the contrary relieved the stress, and the red pepper plants grew normally similar to those of control plants. Numerous other studies have also been conducted to validate the role of PGPR in vegetable improvement across many production systems (Ali et al. 2014; Belimov et al. 2015; Husen et al. 2011).

2.4   PGPR Hydrolytic Enzymes

Bacterial lytic enzymes such as urease, esterase, lipase, protease, chitinase, amylase, and cellulase are key protagonists in the biological transformation processes of N, H, and C (Rana et al. 2012; Reddy 2013; Xun et al. 2015). Enzymes like chitinase and cellulase play a major role as biocontrol agents by degrading fungal cell walls (Sindhu and Dadarwal 2001). Kathiresan et al. (2011) reported that an *Azotobacter* sp. produced high amounts of amylase, cellulase, lipase, chitinase, and protease and participated in biodegradation process of soil organic matter. Bacteria belonging to *Bacillus* and *Pseudomonas* sp. reduced growth of filamentous fungi by secreting lytic enzymes such as chitinases and glucanase. The application of such bacteria for biological protection of crops from pathogens, especially those that contain chitin and glucans within their cell wall structure, is widely assumed (Prasad et al. 2015). Kohler et al. (2007) observed that inoculation of lettuce plants with *B. subtilis* increased significantly urease, protease, and phosphatase activity in rhizosphere, hence participated in plant growth enhancement and potassium/calcium uptake. A bacterial isolate (MIC 3) produced lytic enzymes (protease, amylase, cellulase, chitinase, and pectinase) and exhibited high in vitro antagonistic activity against *F. oxysporum* and *Phoma* sp. (Avinash and Rai 2014). Recently, the role of chitinolytic *Streptomyces vinaceusdrappus* S5MW2 in enhancing tomato plant growth and biocontrol efficacy through chitin supplementation against *Rhizoctonia solani* is reported (Yandigeri et al. 2015). Under greenhouse experiment, chitin supplementation with S5MW2 showed a significant growth of tomato plants and superior disease reduction as compared to untreated control and without CC-treated plants. The role of chitinase-producing *S. maltophilia* and *Chromobacterium* sp. in inhibiting egg hatch of potato cyst nematode *Globodera rostochiensis* was reported by Cronin et al. (1997). Xu and Kim (2016) evaluated the role of cellulase-/protease-producing *Paenibacillus polymyxa* strain SC09-21 as biocontrol agent of *Phytophthora* blight and growth stimulation in pepper plants. Strain SC09-21 significantly reduced *Phytophthora* blight severity and increased phenylalanine ammonia-lyase, peroxidase, polyphenol oxidase, and superoxide dismutase activities. In addition, SC09-21 boosted pathogenesis-related protein gene expression in pepper plants. Singh et al. (1999) observed that two chitinolytic bacterial strains, *Paenibacillus* sp. 300 and *Streptomyces* sp. 385, suppressed *Fusarium* wilt of cucumber caused by *F. oxysporum* f. sp. *cucumerinum* in non-sterile, soilless potting medium.

2.5   Systemic Tolerance and Systemic Resistance Induction by PGPR

Apart from extreme temperatures, salinity, drought, unfavorable pH, heavy metals, and organic pollutants that hit the vegetable production hardest around the world, losses due to phytopathogens are equally substantial in many countries. As an example, about 28–40% of potatoes, cotton, wheat, rice, and maize yields loss are reported due to biotic factors, where the highest loss (40%) was observed in potato due to pathogen diseases (Ashraf et al. 2012; Schwarz et al. 2010). Recently, several works have been published highlighting the PGPR role as enhancers of plant tolerance to abiotic stress. PGPR-induced physiological and biochemical changes in plants that result in enhanced tolerance to environmental stress (drought, salinity, heavy metals, etc.) is known as induced systemic tolerance (IST) (Choudhary and Varma 2016; Nadeem et al. 2015). Species belonging to the genera *Bacillus*, *Halomonas*, *Planococcus*, *Azospirillum*, *Azotobacter*, *Rhizobium*, *Achromobacter*, and *Pseudomonas* can promote potato, chickpea, tomato, bean, lettuce, and cucumber growth under high salinities (Egamberdieva and Lugtenberg 2014; Gururani et al. 2013; Qurashi and Sabri 2012). In growth chamber experiment, Barassi et al. (2006) reported that lettuce seeds inoculated with *Azospirillum* had better germination and vegetative growth than non-inoculated plants exposed to varying levels of NaCl. Several other workers have also reported that *Bacillus*, *Pseudomonas*, *Achromobacter*, *Variovorax*, *Citrobacter*, *Bacillus*, and *Mesorhizobium* could be used to improve potato and tomato growth under drought stress (Belimov et al. 2015; Bensalim et al. 1998; Gururani et al. 2013; Ullah et al. 2016). Also, a novel osmotolerant plant growth-promoting *Actinobacterium citricoccus zhacaiensis* B-4 (MTCC 12119) was found to enhance onion seed germination under osmotic stress conditions (Selvakumar et al. 2015). On the other hand, Wang et al. (2015) evaluated the effect of a bacterial consortium (*Bacillus cereus* AR156, *B. subtilis* SM21, and *Serratia* sp. XY21) on alleviating cold stress in tomato seeds after 7 days of chilling treatment (4 °C) and 1 week recovery at normal 28 °C. Treated tomato plants had a survival rate of 93% on average six times more than control plants (16%). The same consortium (*B. cereus* AR156, *B. subtilis* SM21, and *Serratia* sp. XY21) was previously reported to be an efficient eco-friendly tool to induce drought tolerance in cucumber plants (Wang et al. 2012).

There are numerous reports where PGPR have been found to stimulate plant defense by inhibiting phytopathogens. They induce physical or chemical changes in plants and, hence, improve plant resistance, which is designated by induced systemic resistance (ISR) (Nadeem et al. 2015; Niranjana and Hariprasad 2014). For instance, *Bacillus subtilis* B4 and *B. subtilis* B5 when tested in pot trials against *Sclerotium cepivorum*, causing onion white rot, decreased disease incidence by 33.33% and 41.67%, respectively, compared with the control. In contrast, under field conditions, disease incidence was declined by 25% (*B. subtilis* B5) and 16.67% (*B. subtilis* B4) compared with the control. Due to their disease-reducing ability, strains of *Bacillus* were considered suitable for enhancing growth and productivity of onion plants (Shalaby et al. 2013). Furthermore, the ability of endophytic *Pseudomonas* sp. strain to promote growth and resistance of potato plants toward infection by necrotroph *Pectobacterium atrosepticum* is also reported (Pavlo et al. 2011). Apart from its ability to promote potato shoots growth, *Pseudomonas* sp. increased plant resistance toward soft rot disease. Disease inhibition was inversely proportional to the size of inoculated bacterial population. Raupach et al. (1996) studied the effect of two bacterial strains *P. fluorescens* 89B-27 and *S. marcescens* 90–166 to protect cucumber and tomato against cucumber mosaic *Cucumovirus* (CMV). The two strains showed high ability to stimulate tomato and cucumber defenses against phytopathogen virus CMV, and the results suggest that the two strains should be evaluated for their potential to contribute toward management of viral plant diseases. Equally, PGPR such as *Pseudomonas, Alcaligenes, Paenibacillus*, and *Chryseobacterium* have been reported as systemic resistance inducers in potato, tomato, pea, bean, and Chinese cabbage against pathogens like *Bemisia tabaci, Fusarium, Macrophomina phaseolina, Rhizoctonia, Ralstonia solanacearum, C. orbiculare, Botrytis cinerea*, and *Pectobacterium carotovorum* (Ben Abdallah et al. 2016; Lee et al. 2014; Moradi et al. 2012; Murthy et al. 2014; Valenzuela-Soto et al. 2010). Recently, Konappa et al. (2016) reported the role of lactic acid bacterium *Lactobacillus paracasei* in mediating induction of defense enzymes to enhance resistance against *Ralstonia solanacearum* causing bacterial wilt in tomato. Inoculation of tomato seedlings with bacterial isolate induced a significant amount of peroxidase, polyphenol oxidase, phenylalanine ammonia-lyase, total phenolics, and β-1,3-glucanase activities. In field experiment, treatment with lactic acid bacteria increased the yield by 15% (8.2 kg/m^2), and pathogen-infected plants as well as pretreated with bacteria gave an average of 55% yield (28.3 kg/m^2 compared to infected plots). The results indicated that bacterial inoculation reduced the bacterial wilt by 61% in tomato.

Conclusion

Vegetables constitute an important part of human healthy foods. They provide many important nutrient elements such as calcium, magnesium, potassium, iron, beta-carotene, vitamin B complex, vitamin C, vitamin A, vitamin K, and antioxidants. Vegetables also provide soluble as well as insoluble dietary fiber collectively known as non-starch polysaccharides (NSP) such as cellulose, mucilage, hemicellulose, gums, pectin, etc. Like many other crops, vegetables are threatened by biotic and abiotic stresses. Thus, scientists and vegetable growers are working hard to develop different strategies to overcome these problems. Among various strategies, the use of PGPR in agricultural practices has received greater attention. It is clear that until now, there is no clear antithesis about beneficial and eco-friendly effect of PGPR in a sustainable agriculture establishment worldwide. However, there are many challenges that need to be addressed in order to make full use of this technology. Among various reasons, the lack of uniformity and variation in responses are of prime concern. Moreover, the detection of vegetable-specific PGPR and understanding the interactive relationship between PGPR and vegetable require special attention so that vegetable-specific inoculant is developed. In addition to these, the difficulties encountered in inoculum production, storage, delivery, viability, and its competitiveness in the new environment after application are some of the other major challenges that require immediate and considerable attention of both scientists and farmers to make full use of this technology for enhancing the vegetable production in different agroecological niches.

References

Abd-Alla MH (1998) Growth and siderophore production in vitro of *Bradyrhizobium* (Lupin) strains under iron limitation. Eur J Soil Biol 34:99–104

Akiyoshi DA, Regier DA, Gordon MP (1987) Cytokinin production by *Agrobacterium* and *Pseudomonas* spp. J Bacteriol 169:4242–4248

Ali S, Charles TC, Glick BR (2014) Amelioration of high salinity stress damage by plant growth-promoting bacterial endophytes that contain ACC deaminase. Plant Physiol Biochem 80:160–167

Antoun H, Prévost D (2005) Ecology of plant growth promoting rhizobacteria. In: Siddiqui ZA (ed) PGPR: biocontrol and biofertilization. Springer, Netherlands, pp 1–38

Ardisson GB, Tosin M, Barbale M, Degli-Innocenti F (2014) Biodegradation of plastics in soil and effects on nitrification activity. A laboratory approach. Front Microbiol. doi:10.3389/fmicb.2014.00710

Arkhipova TN, Prinsen E, Veselov SU, Martinenko EV, Melentiev AI, Kudoyarova GR (2007) Cytokinin producing bacteria enhance plant growth in drying soil. Plant Soil 292:305–315

Arkhipova TN, Veselov SU, Melentiev AI, Martynenko EV, Kudoyarova GR (2005) Ability of bacterium *Bacillus subtilis* to produce cytokinins and to influence the growth and endogenous hormone content of lettuce plants. Plant Soil 272:201–209

Ashraf M, Ahmad MSA, Öztürk M, Aksoy A (2012) Crop improvement through different means: challenges and prospects. In: Ashraf M et al (eds) Crop production for agricultural improvement. Springer Science + Business Media BV, Dordrecht, Netherlands, pp 1–15

Avinash TS, Rai RV (2014) Antifungal activity of plant growth promoting rhizobacteria against *Fusarium oxysporum* and *Phoma* sp. of cucurbitaceae. In: Kharwar RN et al (eds) Microbial diversity and biotechnology in food security. Springer, India, pp 257–264

Bahena MHR, Salazar S, Velázquez E, Laguerre G, Peix A (2015) Characterization of phosphate solubilizing rhizobacteria associated with pea (*Pisum sativum* L.) isolated from two agricultural soils. Symbiosis 67:33–41

Bakker PAHM, Lamers JG, Bakker AW, Marugg JD, Weisbeek PJ, Schippers B (1986) The role of siderophores in potato tuber yield increase by *Pseudomonas putida* in a short rotation of potato. Neth J Plant Pathol 92:249–256

Bakshi A, Shemansky JM, Chang C, Binder BM (2015) History of research on the plant hormone ethylene. J Plant Growth Regul 34:809–827

Barassi CA, Ayrault G, Creus CM, Sueldo RJ, Sobrero MT (2006) Seed inoculation with *Azospirillum* mitigates NaCl effects on lettuce. Sci Hortic 109:8–14

Bari R, Jones JDG (2009) Role of plant hormones in plant defence responses. Plant Mol Biol 69:473–488

Barnawal D, Bharti N, Maji D, Chanotiya CS, Kalra A (2012) 1-Aminocyclopropane-1-carboxylic acid (ACC) deaminase-containing rhizobacteria protect *Ocimum sanctum* plants during waterlogging stress via reduced ethylene generation. Plant Physiol Biochem 58:227–235

Bar-Ness E, Hadar Y, Chen Y, Shanzer A, Libman J (1992) Iron uptake by plants from microbial siderophores. Plant Physiol 99:1329–1335

Bastian F, Cohen A, Piccoli P, Luna V, Baraldi R, Bottini R (1998) Production of indole-3-acetic acid and gibberellins A1 and A3 by *Acetobacter diazotrophicus* and *Herbaspirillum seropedicae* in chemically defined media. Plant Growth Regul 24:7–11

Belimov AA, Dodd IC, Safronova VI, Shaposhnikov AI, Azarova TS, Makarova NM, Davies WJ, Tikhonovich IA (2015) Rhizobacteria that produce auxins and contain 1 amino-cyclopropane-1-carboxylic acid deaminase decrease amino acid concentrations in the rhizosphere and improve growth and yield of well-watered and water-limited potato (*Solanum tuberosum*). Ann Appl Biol 167:11–25

Ben Abdallah RA, Mokni-Tlili S, Nefzi A, Jabnoun-Khiareddine H, Daami-Remadi M (2016) Biocontrol of Fusarium wilt and growth promotion of tomato plants using endophytic bacteria isolated from *Nicotiana glauca* organs. Biol Control 97:80–88

Beneduzi A, Ambrosini A, Passaglia LMP (2012) Plant growth-promoting rhizobacteria (PGPR): their potential as antagonists and biocontrol agents. Genet Mol Biol 35:1044–1051

Bensalim S, Nowak J, Asiedu SK (1998) A plant growth promoting rhizobacterium and temperature effects on performance of 18 clones of potato. Am J Potato Res 75:145–152

Bertrand JC, Bonin P, Caumette P, Gattuso JP, Grégori G, Guyoneaud R, Le Roux X, Matheron R, Poly F (2015) Biogeochemical cycles. In: Bertrand JC et al (eds) Environmental microbiology: fundamentals and applications: microbial ecology. Springer Science + Business Media, Dordrecht, pp 511–617

Beyeler M, Keel C, Michaux P, Haas D (1999) Enhanced production of indole-3-acetic acid by a genetically modified strain of *Pseudomonas fluorescens* CHA0 affects root growth of cucumber, but does not improve protection of the plant against Pythium root rot. FEMS Microbiol Ecol 28:225–233

Bhattacharyya PN, Jha DK (2012) Plant growth-promoting rhizobacteria (PGPR): emergence in agriculture. World J Microbiol Biotechnol 28:1327–1350

Bottini R, Fulchieri M, Pearce D, Pharis RP (1989) Identification of gibberellins A1, A3, and iso-A3 in cultures of *Azospirillum lipoferum*. Plant Physiol 90:45–47

Cassán F, Vanderleyden J, Spaepen S (2014) Physiological and agronomical aspects of phytohormone production by model plant-growth-promoting rhizobacteria (PGPR) belonging to the genus *Azospirillum*. J Plant Growth Regul 33:440–459

Choudhary DK, Varma A (2016) Microbial-mediated induced systemic resistance in plants. Springer Science +Business Media, Singapore

Chung H, Park M, Madhaiyan M, Seshadri S, Song J, Cho H, Sa T (2005) isolation and characterization of phosphate solubilizing bacteria from the rhizosphere of crop plants of Korea. Soil Biol Biochem 37:1970–1974

Cohen MF, Mazzola M (2006) Resident bacteria, nitric oxide emission and particle size modulate the effect of *Brassica napus* seed meal on disease incited by *Rhizoctonia solani* and *Pythium* spp. Plant Soil 286:75–86

Cohen MF, Lamattina L, Yamasaki H (2010) Nitric oxide signaling by plant-associated bacteria. In: Hayat S et al (eds) Nitric oxide in plant physiology. WILEY-VCH Verlag GmbH and Co. KGaA, Weinheim

Conard K, Bettin D, Neumann S (1992) The cytokinin production of *Azospirillum* and *Klebsiella* and its possible ecological effects. In: Kamínek M et al (eds) Physiology and biochemistry of cytokinins in plants: Proc Intern Symp Physiol Biochem of cytokinins in plants. SPB Academic Publishing BV, The Hague, Netherlands, pp 401–405

Crespo J, Boiardi J, Luna M (2011) Mineral phosphate solubilization activity of *gluconacetobacter diazotrophicus* under P-limitation and plant root environment. Agric Sci 2:16–22

Cronin D, Moenne-Loccoz Y, Dunne C, O'Gara F (1997) Inhibition of egg hatch of the potato cyst nematode *Globodera rostochiensis* by chitinase-producing bacteria. Eur J Plant Pathol 103:433–440

Dashti N, Zhang F, Hynes R, Smith DL (1998) Plant growth promoting rhizobacteria accelerates nodulation and increase nitrogen fixation activity by field grown soybean [*Glycine max* (L.) Merr.] under short season conditions. Plant Soil 200:205–213

Datta B, Chakrabartty PK (2014) Siderophore biosynthesis genes of *Rhizobium* sp. isolated from *Cicer arietinum* L. 3 Biotech 4:391–401

Di DW, Zhang C, Luo P, An CW, Guo GQ (2016) The biosynthesis of auxin: how many paths truly lead to IAA? Plant Growth Regul 78:275–285

Donderski W, Głuchowska M (2000) Production of cytokinin-like substances by planktonic bacteria isolated from lake Jeziorak. Pol J Environ Stud 9:369–376

Egamberdieva D, Lugtenberg B (2014) Use of plant growth-promoting rhizobacteria to alleviate salinity stress in plants. In: Miransari M (ed) Use of microbes for the alleviation of soil stresses, vol 1. Springer Science + Business Media, New York, pp 73–96

Elmerich C (2007) Historical perspective: from bacterization to endophytes. In: Elmerich C, Newton WE (eds) Associative and endophytic nitrogen-fixing bacteria and *Cyanobacterial* associations. Springer, The Netherlands, pp 1–16

Endo A, Okamoto M, Koshiba T (2014) ABA biosynthetic and catabolic pathways. In: Zhang DP (ed) Abscisic acid: metabolism, transport and signaling. Springer Science + Business Media, Dordrecht, pp 21–46

Etesami H, Alikhani HA, Hosseini HM (2015) Indole-3-acetic acid (IAA) production trait, a useful screening to select endophytic and rhizosphere competent bacteria for rice growth promoting agents. MethodsX 2:72–78

Fekete FA, Spence JT, Emery T (1983) Siderophores produced by nitrogen-fixing *Azotobacter vinelandii* OP in iron-limited continuous culture. Appl Environ Microbiol 46:1297–1300

Figueiredo MVB, Bonifacio A, Rodrigues AC, Araujo FF (2016) Plant growth-promoting rhizo-bacteria: key mechanisms of action. In: Choudhary DK, Varma A (eds) Microbial-mediated induced systemic resistance in plants. Springer Science + Business Media, Singapore, pp 23–37

Fresco LO, Baudoin WO (2002) Food and nutrition security towards human security. In: ICV souvenir paper. International Conference on Vegetables, World Food Summit: five years later, 11–13 June 2002, Rome, Italy

Fukuyama K (2004) Structure and function of plant-type ferredoxins. Photosynth Res 81: 289–230

Gao Y, Zhao Y (2014) Auxin biosynthesis and catabolism. In: Zažímalová E et al (eds) Auxin and its role in plant development. Springer-Verlag, Vienna, pp 21–38

Glick BR (2014) Bacteria with ACC deaminase can promote plant growth and help to feed the world. Microbiol Res 169:30–39

Glick BR, Cheng Z, Czarny J, Duan J (2007) Promotion of plant growth by ACC deaminase-producing soil bacteria. Eur J Plant Pathol 119:329–339

Gomez-Cadenas A, Vives V, Zandalinas SI, Manzi M, Sanchez-Perez AM, Perez-Clemente RM, Arbona V (2015) Abscisic acid: a versatile phytohormone in plant signaling and beyond. Curr Protein Pept Sci 16:413–434

Gravel V, Antoun H, Tweddell RJ (2007) Effect of indole-acetic acid (IAA) on the development of symptoms caused by *Pythium ultimum* on tomato plants. Eur J Plant Pathol 119:457–462

Gray EJ, Smith DL (2004) Intracellular and extracellular PGPR: commonalities and distinctions in the plant–bacterium signaling processes. Soil Biol Biochem 37:395–412

Gupta A, Gupal M, Tilak KVBR (2000) Mechanism of plant growth promotion by rhizobacteria. Indian J Exp Biol 38:856–862

Gupta G, Parihar SS, Ahirwar NK, Snehi SK, Singh V (2015) Plant growth promoting rhizobacte-ria (PGPR): current and future prospects for development of sustainable agriculture. J Microb Biochem Technol 7:096–102

Gururani MA, Upadhyaya CP, Baskar V, Venkatesh J, Nookaraju A, Park SW (2013) Plant growth-promoting rhizobacteria enhance abiotic stress tolerance in *Solanum tuberosum* through induc-ing changes in the expression of ros-scavenging enzymes and improved photosynthetic performance. J Plant Growth Regul 32:245–258

Hao Y, Charles TC, Glick BR (2010) ACC deaminase increases the *Agrobacterium tumefaciens*-mediated transformation of commercial canola cultivars. FEMS Microbiol Lett 307:185–190

Hider RC, Kong X (2010) Chemistry and biology of siderophores. Royal Soc Chem 27:637–657

Hofstra N, Bouwman AF (2005) Denitrification in agricultural soils: summarizing published data and estimating global annual rates. Nutr Cycl Agroecosyst 72:267–278

Höfte M, Bakker PAHM (2007) Competition for iron and induced systemic resistance by sidero-phores of plant growth promoting rhizobacteria. In: Varma A, Chincholkar SB (eds) Soil biol-ogy, vol. 12. Microbial siderophore. Springer-Verlag, Berlin Heidelberg, pp 121–133

Husen E, Wahyudi AT, Suwanto A, Giyanto (2011) Growth enhancement and disease reduction of soybean by 1-aminocyclopropane-1-carboxylate deaminase-producing pseudomonas. Am J Appl Sci 8:1073–1080

Hussain A, Hasnain S (2009) Cytokinin production by some bacteria: its impact on cell division in cucumber cotyledons. Afr J Microbiol Res 3:704–712

Idso CD (2011) Estimates of global food production in the year 2050: will we produce enough to adequately feed the world? Center for the Study of Carbon Dioxide and Global Change. www.co2science.org

Jagadeesh KS, Kulkarni JH, Krishnaraj PU (2001) Evaluation of the role of fluorescent siderophore in the biological control of bacterial wilt in tomato using Tn5 mutants of fluorescent *Pseudomonas* sp. Curr Sci 81:882–883

Janzen R, Rood S, Dormar J, McGill W (1992) *Azospirillum brasilense* produces gibberellins in pure culture and chemically-medium and in co-culture on straw. Soil Biol Biochem 24:1061–1064

Joo GJ, Kang SM, Hamayun M, Kim SK, Na CI, Shin DH, Lee IJ (2009) *Burkholderia* sp. KCTC 11096BP as newly isolated gibberellin producing bacterium. J Microbiol 47:167–171

Joo GJ, Kim YM, Kim JT, Rhee IK, Kim JH, Lee IJ (2005) Gibberellins-producing rhizobacteria increase endogenous gibberellins content and promote growth of red peppers. J Microbiol 43:510–515

Joo GJ, Kim YM, Lee IJ, Song KS, Rhee IK (2004) Growth promotion of red pepper plug seedlings and the production of gibberellins by *Bacillus cereus*, *Bacillus macroides* and *Bacillus pumilus*. Biotechnol Lett 26:487–491

Kämpfer P, Ruppel S, Remus R (2005) *Enterobacter radicincitans* sp. nov., a plant growth promoting species of the family Enterobacteriaceae. Syst Appl Microbiol 28:213–221

Kang SM, Joo GJ, Hamayun M, Na CI, Shin DH, Kim HY, Hong JK, Lee IJ (2009) Gibberellin production and phosphate solubilization by newly isolated strain of *Acinetobacter calcoaceticus* and its effect on plant growth. Biotechnol Lett 31:277–281

Kang SM, Khan AL, Hamayun M, Hussain J, Joo GJ, You YH, Kim JG, Lee IJ (2012) Gibberellin-producing *Promicromonospora* sp. SE188 improves *Solanum lycopersicum* plant growth and influences endogenous plant hormones. J Microbiol 50:902–909

Kang SM, Waqas M, Khan AL, Lee IJ (2014) Plant-growth-promoting rhizobacteria: potential candidates for gibberellins production and crop growth promotion. In: Miransari M et al (eds) Use of microbes for the alleviation of soil stresses, vol 1. Springer Science + Business Media, New York, pp 1–19

Kannahi M, Senbagam N (2014) Studies on siderophore production by microbial isolates obtained from rhizosphere soil and its antibacterial activity. J Chem Pharm Res 6:1142–1145

Karadeniz A, Topcuoğlu SF, İnan S (2006) Auxin, gibberellin, cytokinin and abscisic acid production in some bacteria. World J Microbiol Biotechnol 22:1061–1064

Kathiresan K, Saravanakumar K, Anburaj R, Gomathi V, Abirami G, Sahu SK, Anandhan S (2011) Microbial enzyme activity in decomposing leaves of mangroves. Int J Adv Biotechnol Res 2:382–389

Kende H, Zeevaart JAD (1997) The five "classical" plant hormones. Plant Cell 9:1197–1210

Khan AL, Halo BA, Elyassi A, Ali S, Al-Hosni K, Hussain J, Al-Harrasi A, Lee IJ (2016) Indole acetic acid and ACC deaminase from endophytic bacteria improves the growth of *Solanum lycopersicum*. Electron J Biotechnol 21:58–64

Khan AA, Jilani G, Akhtar MS, Naqvi SMS, Rasheed M (2009) Phosphorus solubilizing bacteria: occurrence, mechanisms and their role in crop production. J Agric Biol Sci 1:48–58

Khan AL, Waqas M, Kang SM, Al-Harrasi A, Hussain J, Al-Rawahi A, Al-Khiziri S, Ullah I, Ali L, Jung HY, Lee IJ (2014) Bacterial endophyte *Sphingomonas* sp. LK11 produces gibberellins and iaa and promotes tomato plant growth. J Microbiol 52:689–695

Khan MS, Zaidi A, Ahmad E (2014) Mechanism of phosphate solubilization and physiological functions of phosphate-solubilizing microorganisms. In: Khan MS et al (eds) Phosphate solubilizing microorganisms. Springer International Publishing, Switzerland, pp 31–62

Kiseleva AA, Tarachovskaya ER, Shishova MF (2012) Biosynthesis of phytohormones in algae. Russ J Plant Physiol 59:595–610

Kloepper JW, Schroth MN (1978) Plant growth-promoting rhizobacteria on radishes. In: Proceedings of the 4th international conference on plant pathogenic bacteria, vol 2. Station de Pathologie Végétale et de Phytobactériologie, INRA, Angers, France, pp 879–882

Kloepper JW, Leong J, Teintze M, Schroth MN (1980) Enhanced plant growth by siderophores produced by plant growth-promoting rhizobacteria. Nature 286:885–886

Knowles R (2004) Nitrogen cycle. In: Schaechter M (ed) The desk encyclopedia of microbiology. Elsevier, China, pp 690–701

Kohler J, Caravaca F, Carrasco L, Roldán A (2007) Interactions between a plant growth-promoting rhizobacterium, an AM fungus and a phosphate-solubilising fungus in the rhizosphere of *Lactuca sativa*. Appl Soil Ecol 35:480–487

Konappa NM, Maria M, Uzma F, Krishnamurthy S, Nayaka SC, Niranjana SR, Chowdappa S (2016) Lactic acid bacteria mediated induction of defense enzymes to enhance the resistance in tomato against *Ralstonia solanacearum* causing bacterial wilt. Sci Hortic 207:183–192

Krishnaraj PU, Dahale S (2014) Mineral phosphate solubilization: concepts and prospects in sustainable agriculture. Proc Indian Natl Sci Acad 80:389–405

Kumar NR, Krishnan M, Kandeepan C, Kayalvizhi N (2014) Molecular and functional diversity of PGPR fluorescent *Pseudomonas* isolated from rhizosphere of rice (*Oryza sativa* L.) Int J Adv Biotechnol Res 5:490–505

Kümmerli R, Schiessl KT, Waldvogel T, McNeill K, Ackermann M (2014) Habitat structure and the evolution of diffusible siderophores in bacteria. Ecol Lett 17:1536–1544

Landa BB, Montes-Borrego M, Navas-Cortés JA (2013) Use of PGPR for controlling soilborne fungal pathogens: assessing the factors influencing its efficacy. In: Maheshwari DK (ed) Bacteria in agrobiology: disease management. Springer-Verlag, Berlin Heidelberg, pp 259–292

Laslo E, György E, Mathé I, Mara G, Tamas E, Abraham B, Lanyi S (2011) Replacement of the traditional fertilizer with microbial technology: isolation and characterization of beneficial nitrogen fixing rhizobacteria. U P B Sci Bull 73:109–114

Lee SW, Lee SH, Balaraju K, Park KS, Nam KW, Park JW, Park K (2014) Growth promotion and induced disease suppression of four vegetable crops by a selected plant growth-promoting rhizobacteria (PGPR) strain *Bacillus subtilis* 21-1 under two different soil conditions. Acta Physiol Plant 36:1353–1362

Leyval C, Berthelin J (1989) Interaction between *Laccaria laccata*, *Agrobacterium agrobacter* and beech roots: influence on P, K, Mg and Fe mobilization from minerals and plant growth. Plant Soil 117:103–110

Liu J, Chakraborty S, Hosseinzadeh P, Yu Y, Tian S, Petrik I, Bhagi A, Lu Y (2014) Metalloproteins containing cytochrome, iron–sulfur, or copper redox centers. Chem Rev 114:4366–4469

Lombardo MC, Graziano M, Polacco JC, Lamattina L (2006) Nitric oxide functions as a positive regulator of root hair development. Plant Signal Behav 1:28–33

Lugtenberg B, Kamilova F (2009) Plant-growth-promoting rhizobacteria. Annu Rev Microbiol 63:541–556

Ma B, Chen H, Chen SY, Zhang JS (2014) Roles of ethylene in plant growth and responses to stresses. In: Tran LSP, Pal S (eds) Phytohormones: a window to metabolism, signaling and biotechnological applications. Springer Science + Business Media, New York, pp 81–118

Ma Y, Rajkumar M, Freitas H (2009) Inoculation of plant growth promoting bacterium *Achromobacter xylosoxidans* strain Ax10 for the improvement of copper phytoextraction by *Brassica juncea*. J Environ Manage 90:831–837

MacMillan J, Suter PJ (1958) The occurrence of gibberellin A1 in higher plants: isolation from the seed of runner bean (*Phaseolus multiflorus*). Naturwissenschaften 45:46–64

Maheshwari DK, Dheeman S, Agarwal M (2015) Phytohormone-producing PGPR for sustainable agriculture. In: Maheshwari DK (ed) Bacterial metabolites in sustainable agroecosystem, sustainable development and biodiversity, vol 12. Springer International Publishing, Switzerland, pp 159–182

Malboobi MA, Behbahani M, Madani H, Owlia P, Deljou A, Yakhchali B, Moradi M, Hassanabadi H (2009) Performance evaluation of potent phosphate solubilizing bacteria in potato rhizosphere. World J Microbiol Biotechnol 25:1479–1484

Mayak S, Tirosh T, Glick BR (2004) Plant growth-promoting bacteria confer resistance in tomato plants to salt stress. Plant Physiol Biochem 42:565–572

Miao G, Jian-jiao Z, En-tao W, Qian C, Jing X, Jian-guang S (2014) Multiphasic characterization of a plant growth promoting bacterial strain, *Burkholderia* sp. 7016 and its effect on tomato growth in the field. J Integr Agric 14:1855–1863

Miethke M, Marahiel MA (2007) Siderophore-based iron acquisition and pathogen control. Microbiol Mol Biol Rev 71:413–451

Moradi H, Bahramnejad B, Amini J, Siosemardeh A, Haji-Allahverdipoor K (2012) Suppression of chickpea (*Cicer arietinum* L.) *Fusarium* wilt by *Bacillus subtilis* and *Trichoderma harzianum*. POJ 5(2):68–74

Morrone D, Chambers J, Lowry L, Kim G, Anterola A, Bender K, Peters RJ (2009) Gibberellin biosynthesis in bacteria: separate ent-copalyl diphosphate and ent-kaurene synthases in *Bradyrhizobium japonicum*. FEBS Lett 583:475–480

Muriel C, Jalvo B, Redondo-Nieto M, Rivilla R, Martín M (2015) Chemotactic motility of *Pseudomonas fluorescens* F113 under aerobic and denitrification conditions. PLoS ONE 10(7):e0132242

Murthy KN, Uzma F, Chitrashree, Srinivas C (2014) Induction of systemic resistance in tomato against *Ralstonia solanacearum* by *Pseudomonas fluorescens*. AJPS 5:1799–1811

Nadeem SM, Naveed M, Ahmad M, Zahir ZA (2015) Rhizosphere bacteria for crop production and improvement of stress tolerance: mechanisms of action, applications, and future prospects. In: Arora NK (ed) Plant microbes symbiosis: applied facets, vol 1. Springer, India, pp 1–36

Naqqash T, Hameed S, Imran A, Hanif MK, Majeed A, van Elsas JD (2016) Differential response of potato toward inoculation with taxonomically diverse plant growth promoting rhizobacteria. Front Plant Sci 7:144

Narayanasamy P (2013) Mechanisms of action of fungal biological control agents. In: Narayanasamy P (ed) Biological management of diseases of crops, progress in biological control. Springer Science + Business Media, Dordrecht, pp 99–200

Nascimento F, Brigido C, Alho L, Glick BR, Oliveira S (2012) Enhanced chickpea growth-promotion ability of a *Mesorhizobium* strain expressing an exogenous ACC deaminase gene. Plant Soil 353:221–230

Nautiyal CS (1999) An efficient microbiological growth medium for screening phosphate solubilizing microorganisms. FEMS Microbiol Lett 170:265–270

Neilands JB (1995) Siderophores: structure and function of microbial iron transport compounds. J Biol Chem 270:26723–26726

Nichols M, Hilmi M (2009) Growing vegetables for home and market. Diversification booklet number 11. FAO, Rome, Italy

Niranjana SR, Hariprasad P (2014) Understanding the mechanism involved in pgpr-mediated growth promotion and suppression of biotic and abiotic stress in plants. In: Goyal A, Manoharachary C (eds) Future challenges in crop protection against fungal pathogens, fungal biology. Springer Science + Business Media, New York, pp 59–108

Nosrati R, Owlia P, Saderi H, Rasooli I, Malboobi MA (2014) Phosphate solubilization characteristics of efficient nitrogen fixing soil *Azotobacter* strains. Iran J Microbiol 6:285–295

O'Hara GW, Daniel RM (1985) Rhizobial denitrification: a review. Soil Biol Biochem 17:1–9

Ortiz Castro R, Valencia-Cantero E, Lopez-Bucio J (2008) Plant growth promotion by *Bacillus megaterium* involves cytokinin signaling. Plant Signal Behav 3:263–265

Oteino N, Lally RD, Kiwanuka S, Lloyd A, Ryan D, Germaine KJ, Dowling DN (2015) Plant growth promotion induced by phosphate solubilizing endophytic *Pseudomonas* isolates. Front Microbiol 6:745

Pastor N, Rosas S, Luna V, Rovera M (2014) Inoculation with *Pseudomonas putida* PCI2, a phosphate solubilizing rhizobacterium, stimulates the growth of tomato plants. Symbiosis 62:157–167

Patil S, Bheemaraddi MC, Shivannavar CT, Gaddad SM (2014) Biocontrol activity of siderophore producing *Bacillus subtilis* CTS-G24 against wilt and dry root rot causing fungi in chickpea. IOSR-JAVS 7:63–68

Pavlo A, Leonid O, Iryna Z, Natalia K, Maria PA (2011) Endophytic bacteria enhancing growth and disease resistance of potato (*Solanum tuberosum* L.) Biol Control 56:43–49

Peix A, Rivas-Boyero AA, Mateos PF, Rodriguez-Barrueco C, Martínez-Molina E, Velazquez E (2001) Growth promotion of chickpea and barley by a phosphate solubilizing strain of *Mesorhizobium mediterraneum* under growth chamber conditions. Soil Biol Biochem 33:103–110

Penrose DM, Glick BR (2003) Methods for isolating and characterizing ACC deaminase-containing plant growth-promoting rhizobacteria. Physiol Plant 118:10–15

Philippot L, Hallin S, Schloter M (2007) Ecology of denitrifying prokaryotes in agricultural soil. In: Donald LS (ed) Advances in agronomy, vol 96. Elsevier Inc., Netherlands

Pilet PE, Chanson A (1981) Effect of abscisic acid on maize root growth: a critical examination. Plant Sci Lett 21:99–106

Pilet PE, Saugy M (1987) Effect on root growth of endogenous and applied IAA and ABA: a critical reexamination. Plant Physiol 83:33–38

Pishchik VN, Chernyaeva II, Kozhemaykov AP, Vorobyov NI, Lazarev AM, Kozlov LP (1998) Effect of inoculation with nitrogen-fixing *Klebsiella* on potato yield. In: Malik KA et al (eds) Nitrogen fixation with non-legumes. Kluwer Academic Publishers, Great Britain, pp 223–235

Porcel R, Zamarreño AM, García-Mina JM, Aroca R (2014) Involvement of plant endogenous ABA in *Bacillus megaterium* PGPR activity in tomato plants. BMC Plant Biol 14:36. doi:10.1186/1471-2229-14-36

Prasad R, Kumar M, Varma A (2015) Role of PGPR in soil fertility and plant health. In: Egamberdieva D et al (eds) Plant-growth-promoting rhizobacteria (PGPR) and medicinal plants. Soil biology, vol 42. Springer International Publishing, Switzerland, pp 247–260

Qurashi AW, Sabri AN (2012) Bacterial exopolysaccharide and biofilm formation stimulate chickpea growth and soil aggregation under salt stress. Braz J Microbiol 43:1183–1191

Radzki W, Gutierrez Mañero FJ, Algar E, Lucas García JA, García-Villaraco A, Ramos Solano B (2013) Bacterial siderophores efficiently provide iron to iron-starved tomato plants in hydroponics culture. Antonie Van Leeuwenhoek 104:321–330

Rana A, Saharan B, Nain L, Prasanna R, Shivay YS (2012) Enhancing micronutrient uptake and yield of wheat through bacterial PGPR consortia. Soil Sci Plant Nutr 58:573–582

Raupach GS, Liu L, Murphy JF, Tuzun S, Kloepper JW (1996) Induced systemic resistance in cucumber and tomato against cucumber mosaic cucumovirus using plant growth promoting rhizobacteria. Plant Dis 80:891–894

Reddy PP (2013) Plant growth-promoting rhizobacteria (PGPR). In: Reddy PP (ed) Recent advances in crop protection. Springer, India, pp 131–158

Reddy PP (2014) Potential role of PGPR in agriculture. In: Reddy PP (ed) Plant growth promoting rhizobacteria for horticultural crop protection. Springer, India, pp 17–34

Renshaw JC, Robson GD, Trinci APJ, Wiebe MG, Livens FR, Collison D, Taylor RJ (2002) Fungal siderophores: structures, functions and applications. Mycol Res 106:1123–1142

Rizvi A, Khan MS, Ahmad E (2014) Inoculation impact of phosphate-solubilizing microorganisms on growth and development of vegetable crops. In: Khan MS et al (eds) Phosphate solubilizing microorganisms. Springer International Publishing, Switzerland, pp 287–297

Rondon MR, Ballering KS, Thomas MG (2014) Identification and analysis of a siderophore biosynthetic gene cluster from *Agrobacterium tumefaciens* C58. Microbiology 150:3857–3866

Saleem M, Arshad M, Hussain S, Bhatti AS (2007) Perspective of plant growth promoting rhizobacteria (PGPR) containing ACC deaminase in stress agriculture. J Ind Microbiol Biotechnol 34:635–648

Sánchez C, Tortosa G, Granados A, Delgado A, Bedmar EJ, Delgado MJ (2011) Involvement of *Bradyrhizobium japonicum* denitrification in symbiotic nitrogen fixation by soybean plants subjected to flooding. Soil Biol Biochem 43:212–217

Sarig S, Kapulnik Y, Okon Y (1986) Effect of *Azospirillum* inoculation on nitrogen fixation and growth of several winter legumes. Plant Soil 90:335–342

Sashidhar B, Podile AR (2010) Mineral phosphate solubilization by rhizosphere bacteria and scope for manipulation of the direct oxidation pathway involving glucose dehydrogenase. J Appl Microbiol 109:1–12

Sayyed RZ, Chincholkar SB (2010) Growth and siderophores production in *Alcaligenes faecalis* is regulated by metal ions. Indian J Microbiol 50:179–182

Schwarz D, Rouphael Y, Colla G, Venema JH (2010) Grafting as a tool to improve tolerance of vegetables to abiotic stresses: thermal stress, water stress and organic pollutants. Sci Hortic 127:162–171

Selvakumar G, Bhatt RM, Upreti KK, Bindu GH, Shweta K (2015) *Citricoccus zhacaiensis* B-4 (MTCC 12119) a novel osmotolerant plant growth promoting actinobacterium enhances onion (*Allium cepa* L.) seed germination under osmotic stress conditions. World J Microbiol Biotechnol 31:833–839

Selvakumar G, Bindu GH, Bhatt RM, Upreti KK, Paul AM, Asha A, Shweta K, Sharma M (2016) Osmotolerant cytokinin producing microbes enhance tomato growth in deficit irrigation conditions. Proc Natl Acad Sci, India, Sect B Biol Sci. doi:10.1007/s40011-016-0766-3

Seyedsayamdost MR, Cleto S, Carr G, Vlamakis H, João Vieira M, Kolter R, Clardy J (2012) Mixing and matching siderophores clusters: structure and biosynthesis of serratiochelins from *Serratia* sp. V4. J Am Chem Soc 134:13550–13553

Shaharoona B, Arshad M, Waqas R, Khalid A (2012) Role of ethylene and plant growth-promoting rhizobacteria in stressed crop plants. In: Venkateswarlu B et al (eds) Crop stress and its management: perspectives and strategies. Springer Science + Business Media B.V, Dordrecht, Netherlands, pp 429–446

Shahbaz M, Ashraf M (2013) Improving salinity tolerance in cereals. Crit Rev Plant Sci 32:237–249

Shalaby ME, Kamal EG, El-Diehi MA (2013) Biological and fungicidal antagonism of *Sclerotium cepivorum* for controlling onion white rot disease. Ann Microbiol 63:1579–1589

Sharma SB, Sayyed RZ, Trivedi MH, Gobi TA (2013) Phosphate solubilizing microbes: sustainable approach for managing phosphorus deficiency in agricultural soils. SpringerPlus 2:587

Shrivastava P, Kumar R (2015) Soil salinity: a serious environmental issue and plant growth promoting bacteria as one of the tools for its alleviation. Saudi J Biol Sci 22:123–131

Siddikee MA, Glick BR, Chauhan PS, Yim WJ, Sa T (2011) Enhancement of growth and salt tolerance of red pepper seedlings (*Capsicum annuum* L.) by regulating stress ethylene synthesis with halotolerant bacteria containing 1-aminocyclopropane-1-carboxylic acid deaminase activity. Plant Physiol Biochem 49:427–434

Sindhu SS, Dadarwal KR (2001) Chitinolytic and cellulolytic *Pseudomonas* sp. antagonistic to fungal pathogens enhances nodulation by *Mesorhizobium* sp. *Cicer* in chickpea. Microbiol Res 156:353–358

Singh PP, Shin YC, Park CS, Chung YR (1999) Biological control of *Fusarium* wilt of cucumber by chitinolytic bacteria. Phytopathology 89:93–99

Sivasakthi S, Usharani G, Saranraj P (2014) Biocontrol potentiality of plant growth promoting bacteria (PGPR)–*Pseudomonas fluorescens* and *Bacillus subtilis*: a review. Afr J Agric 9:1265–1277

Skiba U, Smith KA, Fowler D (1993) Nitrification and denitrification as sources of nitric oxide and nitrous oxide in a sandy loam soil. Soil Biol Biochem 25:1527–1536

Song OR, Lee SJ, Lee YS, Lee SC, Kim KK, Choi YL (2008) Solubilization of insoluble inorganic phosphate by *Burkholderia cepacia* Da23 isolated from cultivated soil. Braz J Microbiol 39:151–156

Spaepen S, Vanderleyden J, Remans R (2007a) Indole-3-acetic acid in microbial and microorganism-plant signaling. FEMS Microbiol Rev 31:425–448

Spaepen S, Versées W, Gocke D, Pohl M, Steyaert J, Vanderleyden J (2007b) Characterization of phenylpyruvate decarboxylase, involved in auxin production of *Azospirillum brasilense*. J Bacteriol 189:7626–7633

Spaepen and Vanderleyden (2010) Auxin and Plant-Microbe Interactions. Cold Spring Harb Perspect Biol. doi: 10.1101/cshperspect.a001438

Tailor AJ, Joshi BH (2012) Characterization and optimization of siderophore production from *Pseudomonas fluorescens* strain isolated from sugarcane rhizosphere. J Environ Res Dev 6(3A):688–694

Takahashi N, Phinney BO, MacMillan J (1991) Gibberellins, with 176 illustrations. Springer-Verlag New York Inc., New York

Taller BJ, Wong TY (1989) Cytokinins in *Azotobacter vinelandii* culture medium. Appl Environ Microbiol 55:266–267

Tian F, Ding Y, Zhu H, Yao L, Du B (2009) Genetic diversity of siderophore-producing bacteria of tobacco rhizosphere. Braz J Microbiol 40:276–284

Timmusk S, Nicander B, Granhall U, Tillberg E (1999) Cytokinin production by *Paenibacillus polymyxa*. Soil Biol Biochem 31:1847–1852

Tokala RK, Strap JL, Jung CM, Crawford DL, Salove MH, Deobald LA, Bailey JF, Morra MJ (2002) Novel plant-microbe rhizosphere interaction involving *Streptomyces lydicus* WYEC108 and the pea plant (*Pisum sativum*). Appl Environ Microbiol 68:2161–2171

Toklikishvili N, Dandurishvili N, Vainstein A, Tediashvili M, Giorgobiani N, Lurie S, Szegedi E, Glick BR, Chernin L (2010) Inhibitory effect of ACC deaminase-producing bacteria on crown gall formation in tomato plants infected by *Agrobacterium tumefaciens* or *A. vitis*. Plant Pathol 59:1023–1030

Tomić S, Gabdoullin RR, Kojić-Prodić B, Wade RC (1998) Classification of auxin plant hormones by interaction property similarity indices. J Comput Aid Mol Des 12:63–79

Tortora ML, Díaz-Ricci JC, Pedraza RO (2011) *Azospirillum brasilense* siderophores with antifungal activity against *Colletotrichum acutatum*. Arch Microbiol 193:275–286

Tuomi T, Rosenquist H (1995) Detection of abscisic, gibberellic and indole-3-acetic acid from plant and microbes. Plant Physiol Biochem 33:725–734

Tran LSP, Pal S (2014) Phytohormones: a window to metabolism, signaling and biotechnological applications. Springer Science + Business Media, New York

Ullah U, Ashraf M, Shahzad SM, Siddiqui AR, Piracha MA, Suleman M (2016) Growth behavior of tomato (*Solanum lycopersicum* L.) under drought stress in the presence of silicon and plant growth promoting rhizobacteria. Soil Environ 35:65–75

United Nations, Department of Economic and Social Affairs, Population Division (2004) World population prospects: world population to 2300. Working paper no. ST/ESA/SER.A/236, New York

United Nations, Department of Economic and Social Affairs, Population Division (2015) World population prospects: the 2015 revision, key findings and advance tables. Working paper no. ESA/P/WP.241, New York

Valencia-Cantero E, Hernández-Calderón E, Velázquez-Becerra C, Joel E, López-Meza A-CR, López-Bucio J (2007) Role of dissimilatory fermentative iron-reducing bacteria in Fe uptake by common bean (*Phaseolus vulgaris* L.) plants grown in alkaline soil. Plant Soil 291:263–273

Valenzuela-Soto JH, Estrada-Hernàndez MG, Ibarra-Laclette E, Délano-Frier JP (2010) Inoculation of tomato plants (*Solanum lycopersicum*) with growth-promoting *Bacillus subtilis* retards white fly *Bemisia tabaci* development. Planta 231:397–410

Verma JP, Yadav J, Tiwari KN, Kumar A (2013) Effect of indigenous *Mesorhizobium* spp. and plant growth promoting rhizobacteria on yields and nutrients uptake of chickpea (*Cicer arietinum* L.) under sustainable agriculture. Ecol Eng 51:282–228

Vessey JK (2003) Plant growth promoting rhizobacteria as biofertilizers. Plant Soil 255:571–586

Vidhyasekaran P (2015) Auxin signaling system in plant innate immunity. In: Vidhyasekaran P (ed) Plant hormone signaling systems in plant innate immunity, signaling and communication in plants, vol 2. Springer Science + Business Media, Dordrecht, pp 311–357

Vogel JP, Woeste KE, Theologis A, Kieber JJ (1998) Recessive and dominant mutations in the ethylene biosynthetic gene ACS5 of *Arabidopsis* confer cytokinin insensitivity and ethylene overproduction, respectively. Plant Biol 95:4766–4771

Volpiano CG, Estevam A, Saatkamp K, Furlan F, Vendruscolo ECG, Dos Santos MF (2014) Physiological responses of the co-cultivation of PGPR with two wheat cultivars in vitro under stress conditions. BMC Proceedings 2014 8(Suppl 4):P108

Walpola BC, Yoon MH (2013) Isolation and characterization of phosphate solubilizing bacteria and their co-inoculation efficiency on tomato plant growth and phosphorous uptake. Afr J Microbiol Res 7:266–275

Wang C, Wang C, Gao YL, Wang YP, Guo JH (2015) A consortium of three plant growth-promoting rhizobacterium strains acclimates *Lycopersicon esculentum* and confers a better tolerance to chilling stress. J Plant Growth Regul. doi:10.1007/s00344-015-9506-9

Wang CJ, Yang W, Wang C, Gu C, Niu DD, Liu HX, Wang YP, Guo JH (2012) Induction of drought tolerance in cucumber plants by a consortium of three plant growth-promoting rhizobacterium strains. PLoS ONE 7:1–10

Weisbeek P, Marugg J, van der Hofstad G, Bakker P, Schippers B (1987) Siderophore biosynthesis, uptake and effect on potato growth of rhizosphere strains. In: Verma DPS et al (eds) Molecular genetics of plant-microbe interactions. Martinus Nijhoff Publishers, Dordrecht, pp 51–53

Williams M, Stout J, Roth B, Cass S, Papa V, Rees B (2014) Environmental implications of legume cropping. Legume Futures Report 3.7. www.legumefutures.de

Wong WS, Tan SN, Ge L, Chen X, Yong JWH (2015) The importance of phytohormones and microbes in biofertilizers. In: Maheshwari DK (ed) Bacterial metabolites in sustainable agro-ecosystem, sustainable development and biodiversity, vol 12. Springer International Publishing, Switzerland, pp 105–158

Xu S, Kim BS (2016) Evaluation of *Paenibacillus polymyxa* strain SC09-21 for biocontrol of *Phytophthora* blight and growth stimulation in pepper plants. Trop Plant Pathol 41:162

Xun F, Xie B, Liu S, Guo C (2015) Effect of plant growth-promoting bacteria (PGPR) and arbuscular mycorrhizal fungi (AMF) inoculation on oats in saline-alkali soil contaminated by petroleum to enhance phytoremediation. Environ Sci Pollut Res 22:598–608

Yamaguchi M (1983) World vegetables: principles, production and nutritive values. AVI Publishing Company, Inc., Westport, CT

Yandigeri MS, Malviya N, Solanki MK, Shrivastava P, Sivakumar G (2015) Chitinolytic *Streptomyces vinaceus drappus* S5MW2 isolated from Chilika Lake, India enhances plant growth and biocontrol efficacy through chitin supplementation against *Rhizoctonia solani*. World J Microbiol Biotechnol 31:1217–1225

Zehr JP, Jenkins BD, Short SM, Steward GF (2003) Nitrogenase gene diversity and microbial community structure: a cross-system comparison. Environ Microbiol 5:539–554

Role of Nitrogen-Fixing Plant Growth-Promoting Rhizobacteria in Sustainable Production of Vegetables: Current Perspective

3

Almas Zaidi, Mohammad Saghir Khan, Saima Saif,
Asfa Rizvi, Bilal Ahmed, and Mohammad Shahid

Abstract

Vegetables due to high nutritional value comprising of carbohydrates, proteins, vitamins and several other essential elements are considered one of the important dietary constituents. In order to achieve optimum yields, agrochemicals are frequently used in vegetable cultivation. However, the excessive and inappropriate use of agrochemicals has been found deleterious for both soil fertility and vegetable production. The negative impact of agrochemicals in vegetable production practices can be avoided by the use of biofertilizers involving nitrogen-fixing plant growth-promoting rhizobacteria. The use of non-pathogenic nitrogen-fixing plant growth-promoting rhizobacteria to enhance vegetable production is, therefore, currently considered as a safe, viable and inexpensive alternative to chemical fertilization. Even though there are no direct connections between nitrogen-fixing organisms and vegetables, both symbiotic and asymbiotic/associative nitrogen-fixing bacteria have been used to facilitate the growth and yield of non-legume crops like vegetables through mechanisms other than nitrogen fixation. Indeed, there are numerous reports on the effect of plant growth-promoting rhizobacteria on vegetable production, but the information on nitrogen-fixing bacteria employed in vegetable production is scarce. Considering these gaps and success of nitrogen-fixing bacteria application in vegetable production achieved so far, efforts have been directed to highlight the impact of nitrogen fixers on the production of vegetables. Here, efforts will be made to identify most suitable nitrogen fixers which could be used to improve the health and quality of vegetables grown in different regions. The use of nitrogen fixers is also likely to reduce the use of chemicals in vegetable production.

A. Zaidi (✉) • M.S. Khan • S. Saif • A. Rizvi • B. Ahmed • M. Shahid
Faculty of Agricultural Sciences, Department of Agricultural Microbiology, Aligarh Muslim University, Aligarh 202002, Uttar Pradesh, India
e-mail: zaidi29@rediffmail.com

© Springer International Publishing AG 2017
A. Zaidi, M.S. Khan (eds.), *Microbial Strategies for Vegetable Production*,
DOI 10.1007/978-3-319-54401-4_3

49

3.1 Introduction

Nitrogen-fixing plant growth-promoting rhizobacteria including both symbiotic and asymbiotic/associative bacteria have been used in agricultural practices to promote growth and yield of many crops (Ahmad et al. 2013) including vegetables (Antoun et al. 1998; Lamo 2009; Vikhe 2014; Ziaf et al. 2016). Of these, bacteria that form root nodules on leguminous plants and transform atmospheric nitrogen (N) into usable form of N are collectively known as rhizobia: a general term used to denote all rhizobial genera together (Lindstrom and Martinez-Romero 2005). Beijerinck (1888) first of all isolated a bacterium from root nodules, which he identified as *Bacillus radicicola*. In the late nineteenth century, Frank (1889) named this bacterium *Rhizobium leguminosarum* and identified other species belonging to the same group. The term 'rhizobia' was originally used to name bacteria belonging to the genus *Rhizobium*, but nowadays, rhizobia also include other genera, for example, *Bradyrhizobium, Sinorhizobium, Azorhizobium, Mesorhizobium*, etc. (Sahgal and Johri 2003; Graham 2008). The designation rhizobia currently includes more than 70 species distributed over 13 genera including some *Betaproteobacteria* such as *Burkholderia* and *Cupriavidus* (Chen et al. 2007; Barrett and Parker 2006). Other nitrogen-fixing bacteria are free-living nitrogen-fixing bacteria such as *Azotobacter* and *Azospirillum*. They also have the ability to fix nitrogen and to release certain phytohormones, i.e. GA3, IAA and cytokinins (Vikhe 2014) which could stimulate plant growth and increase the availability of nutrients for plant roots. Traditionally, nitrogen fixers have largely been used to supply nitrogen to plants. However, more recently, some nitrogen fixers including both symbiotic (e.g. rhizobia) and asymbiotic (e.g. *Azotobacter*) have also attracted the attention of vegetable growers due to their positive effects on nonlegumes (Antoun et al. 1998; Bhadoria et al. 2005; Lamo 2009; Ramakrishnan and Selvakumar 2012). Vegetable growers on the contrary have long been using agrochemicals (Guertal 2009) in order to obtain maximum yields. The extensive use of fertilizers in vegetable production is, however, at present under debate due to environmental distress and problems to consumer health. Consequently, there has recently been a growing level of interest in environmentally friendly sustainable vegetable practices. In this regard, the integrated use of biofertilizers and chemical fertilizers is considered as the best choice not only to reduce the intensive consumption of chemical fertilizers but also to sustain soil with minimum undesirable impacts and to maximize fertilizer use efficiency in soil (Singh et al. 1999; Palm et al. 2001). Accordingly, soil microorganisms especially PGPR become important in horticultural practices because they are inexpensive and do not cause soil pollution. Among nitrogen-fixing PGPR, rhizobia are reported to possess many desirable plant growth-promoting traits (Ghosh et al. 2015) apart from their normal nitrogen fixation ability. When applied properly, they have been found to exert diverse positive effects on many important nonlegume crops (García-Fraile et al. 2012) including vegetables (Islam et al. 2013; Silva et al. 2014). Mechanistically, nitrogen-fixing PGPR can improve the growth and development of vegetables by producing compounds such as the phytohormone indole acetic acid (Sahasrabudhe 2011) or the enzyme ACC deaminase (Bhattacharjee et al. 2012) involved in the metabolism of 1-aminocyclopropane-1-carboxylic acid (ACC), a precursor of ethylene. They can also mobilize certain major nutrients to the plants such as phosphorous via solubilization of soil insoluble phosphates (Singh et al. 2014a). Nitrogen-fixing PGPR expressing one or multiple plant growth-promoting activities can directly or indirectly promote vegetable growth. Also, some nitrogen-fixing PGPR secrete antimicrobial compounds like siderophores (Singh et al. 2014a), a low-molecular iron-chelating molecules, which restrict the growth of phytopathogens in soils with low content of this ion promoting indirectly the plant growth (Bhattacharjee et al. 2008; Lugtenberg and Kamilova 2009). Considering the importance of nitrogen-fixing PGPR in vegetable production, efforts are made here to collect information on the impact of nitrogen-fixing PGPR on different vegetables grown in different ecological niches.

3.2   Rationale for Using Nitrogen Fixers in Vegetable Production

Vegetables are one of the most important food commodities that significantly affect human health. Due to constantly increasing health awareness among masses, there is greater demand of quality vegetables on regular basis. In order to fulfil the growing demands of vegetarians, vegetable growers have increased the use of synthetic fertilizers to achieve optimum vegetable yields (Abayomi and Adebayo 2014; Guo et al. 2011). The intensive use of chemical fertilizers, however, is reported to cause soil/underground water pollution, destructs microbial composition and their functions, reduces soil fertility and human health (via food chain) problems, makes plant more susceptible to the attack of diseases (Abdelaziz et al. 2007) and leads to ecological risks and poor quality and lesser vegetable yields (Olowoake and Adeoye 2010). Furthermore, higher rates of fertilizer application in vegetable cultivation result in reduced ascorbic acid (vitamin C) content, accumulation of higher level of nitrates especially in leafy vegetables, altered flavour, delayed maturity and increased weight loss. Considering the deleterious effects of fertilizers, and challenge to produce fresh and healthy vegetables, there is urgent need to find suitable alternatives that could help to implement need-based nutrient management (NBNM) practices in order to achieve optimum quality vegetables without any dangerous impact of such chemicals on vegetables. In this context, the use of microbial preparations often called biofertilizers (Dixit et al. 2007) has been found safe for supplying the nutrients to crops besides limiting the problems associated with the use of conventional chemical fertilizers. Biofertilizer is essentially a natural product carrying living microorganisms recovered from various sources including rhizospheres or cultivated soils. Indeed, biofertilizers prepared from nitrogen-fixing PGPR don't have any ill effect on soil fertility and environment instead they improve the soil quality. A small dose of biofertilizer is sufficient to produce desirable results because each gram of carrier of biofertilizers contains at least 10 million viable cells of a specific strain (Anandaraj and Delapierre 2010). Taking into consideration the success of PGPR achieved so far with other crops (Ahmad et al. 2013; Zaidi et al. 2015), different workers have applied nitrogen-fixing PGPR including rhizobia (García-Fraile et al. 2012), *Azotobacter* (Bhadoria et al. 2005) and *Azospirillum* (Ramakrishnan and Selvakumar 2012) along with (Bhadoria et al. 2005) or without (Sharafzadeh 2012) fertilizers for enhancing the production of different vegetables. Apart from their main role in nitrogen fixation, they also stimulate plant growth by other mechanisms such as providing hormones, better nutrient uptake and increased tolerance towards drought and moisture stress. Other major problem in vegetable production is the occurrence of diseases caused by many phytopathogens such as *Pythium aphanidermatum* causing damping-off disease of cucumber (Elazzazy et al. 2012), *Ralstonia solanacearum* causing wilt of brinjal (Chakravarty and Kalita 2012), *Fusarium oxysporum* f.sp. *lycopersici* causing tomato wilt (Loganathan et al. 2014), etc. Traditionally, such diseases are controlled by agrochemicals (pesticides), using sanitary/cultural practices and developing resistant varieties (Sharma and Saikia 2013; Sahar et al. 2013). These disease control measures have, however, neither been promising nor successful. Therefore, the secretion of physiologically active biomolecules such as siderophores (Panhwar et al. 2014), antibiotics (Keel et al. 1992), cyanogenic compounds (Ruangsanka 2014) and lytic enzymes (Nabti et al. 2014) by some nitrogen-fixing PGPR such as rhizobia (Datta and Chakrabartty 2014), *Azotobacter* (Shimaa et al. 2015) or *Azospirillum* (Tortora et al. 2011) has been considered a viable, inexpensive and most effective option for controlling such lethal diseases. More importantly, the use of nitrogen fixers has been found safe for human health after several decades of crop inoculation ensuring that they are optimal bacteria for biofertilization.

3.3 Nitrogen Fixers-Vegetable Interactions: How Nitrogen Fixers Enter Vegetables

Nitrogen fixers in general have widely been used as biofertilizer to supply nitrogen to legumes or other associated crops. Among nitrogen fixers, the members of family *Rhizobiaceae* have also been found to form non-specific associative interactions with roots of other plants without forming nodules (Reyes and Schmidt 1979). Associative symbiosis refers to a wide variety of nitrogen-fixing species that colonize the root surface of nonleguminous plants without formation of differentiated structures (Elmerich and Newton 2007). In other words, these nitrogen-fixing soil bacteria possess the ability to promote the growth of nonlegumes by acting as PGPR (Noel et al. 1996). Indeed, rhizobia can attach to the surface of monocots in the same manner as they attach to dicot hosts (Shimshick and Hebert 1979; Terouchi and Syono 1990). Also, rhizobia grow readily in the presence of germinating seeds and developing root systems in a similar manner with legumes and nonlegumes (Pena-Cabriales and Alexander 1983). It is also interesting to note that the endophytic interaction of rhizobia and nonlegumes occurs without the involvement of genetic signals as observed between rhizobia and legumes during nodulation process (Reddy et al. 1997). Generally, the nitrogen fixers, for instance, rhizobia, enter inside nonlegume plant tissues mainly through cracks in epidermal cells of the roots and in fissure sites where lateral roots have emerged (Dazzo and Yanni 2006; Prayitno et al. 1999). Summarily, the rhizobial endophytic establishment being a dynamic process begins with root colonization which is followed by crack entry into the root interior through separated epidermal cells. Thereafter, endophytes consistently travel up to the stem base, leaf sheath and leaves where they grow rapidly to high population densities (Chi et al. 2005). After they enter inside the plant tissues and attain high population densities, they may influence plant growth by different PGPR mechanisms. Both rhizobia and *Azotobacter* species, apart from supplying N to their respective host plants, secrete some compounds like auxins, cytokinins and antibiotics which directly or indirectly promote the growth of nonlegume plants. For example, Sarhan (2008) indicated a positive effect of *Azotobacter* on growth and yield of potato plants.

3.4   Mechanism of Vegetable Growth Promotion by Nitrogen-Fixing Plant Growth-Promoting Rhizobacteria

Nitrogen fixers like many conventional free-living PGPR can affect plant growth via direct or indirect mechanisms. The direct mechanisms by which nitrogen fixers promote the growth of nonlegumes including vegetable include the solubilization of insoluble P by rhizobia (Singh et al. 2014a; Abd-Alla 1994; Halder and Chakrabarty 1991) and species of *Azotobacter* (Nosrati et al. 2014). Symbiotic rhizobia are advantageous than free-living PGPR in P solubilization as these bacteria are well protected inside the nodule tissues and face little/no competition from indigenous soil microbiota. Another important growth regulator that directly promotes the growth of vegetables is indole acetic acid secreted both by rhizobia (Kumar and Ram 2012; Sahasrabudhe 2011) and *Azotobacter* (Kumar et al. 2014). Indole acetic acid has been reported to play a central role in plant growth and development and acts as a signal molecule which is involved in plant signal processing, motility or attachment of bacteria in root which help in legume-*Rhizobium* symbiosis (Spaepen et al. 2009). On the contrary, the indirect mechanisms of plant growth promotion by rhizobia/*Azotobacter* involve the secretion of compounds that lessen or prevent the deleterious effects of one or more phytopathogenic organisms (Gandhi Pragash et al. 2009). Productions of siderophores (Greek for iron carrier), a low-molecular (500–1000 daltons) iron-chelating substance by *Azotobacter* (Muthuselvan and Balagurunathan 2013) or rhizobia (Ahmad et al. 2013; Datta and Chakrabartty 2014), may be considered a direct factor, since siderophores solubilize and sequester iron from soil and provide it to plant cells. But it can also be considered an indirect factor, since it is associated with suppression of plant pathogens by depriving them of iron uptake. Moreover, siderophore-producing ability helps in the sustenance of rhizobia in iron-deficient soils (Lesueur et al. 1995). The growth-promoting substances involved in vegetable production synthesized by various rhizobia/*Azotobacter* are summarized in Table 3.1.

Table 3.1 Examples of plant growth-promoting substances released by some commonly employed nitrogen-fixing plant growth-promoting rhizobacteria

Rhizobia	Source	Plant growth regulators	Reference
Rhizobium undicola, *Rhizobium* spp.	Nodules of aquatic legume	ACC deaminase, indole acetic acid	Ghosh et al. (2015), Bhagat et al. (2014)
Mesorhizobium, *R. leguminosarum*, *Bradyrhizobium*, *Sinorhizobium meliloti*	*Neptunia oleracea*, *Pisum sativum*, *Trifolium alexandrinum* L., *Cicer arietinum* L., *Trigonella foenum-graecum* L., *Medicago sativa* L., *Indigofera* spp. birdsfoot trefoil (*Lotus corniculatus*)	Exopolysaccharides, N_2 fixation, P solubilization, siderophores, ammonia, hydrogen cyanide, antifungals, volatile antifungal compounds, protease	Machado et al. (2013), Bhattacharjee et al. (2012), Sahasrabudhe (2011), Ma et al. (2004)
Azotobacter	Rhizosphere soil	P solubilization, siderophores, ammonia, hydrogen cyanide, IAA	Prasad et al. (2014)
Sinorhizobium sp. strain MRR101-KC428651, *Rhizobium* sp. strain 103-JX576499, *Sinorhizobium kostiense* strain MRR104-KC428653	Root nodules of *Vigna trilobata* plants	P solubilization, antifungal activity	Kumar et al. (2014)
Azotobacter	Rhizosphere soil	IAA	Kumar et al. (2014)
Azotobacter	Rhizosphere soil	Siderophores	Muthuselvan and Balagurunathan (2013)
Rhizobium psm6	Agricultural soil	P solubilization	Karpagam and Nagalakshmi (2014)
Mesorhizobium	Tunisian soils	P solubilization	Imen et al. (2015)
Rhizobium BICC 651	Root nodule of chickpea	Siderophores	Datta and Chakrabartty (2014)
Mesorhizobium spp.	Native isolates	HCN, siderphores, protease, cellulose, volatile antifungal compounds	Bhagat et al. 2014
Azospirillum brasilense	–	Siderophores, IAA antifungal activity	Tortora et al. (2011), Zakharova et al. (1999)

3.5   Nitrogen-Fixing Plant Growth-Promoting Rhizobacteria Improve Vegetable Production: A General Perspective

Conventional growers in order to achieve high yield and quality vegetables apply higher rates of chemical fertilizers, which are expensive and destructive to environment (Orhan et al. 2006). Considering the threat of the excessive use of fertilizers to soil fertility and vegetable production, vegetable growers have shown interest in applying environmentally friendly and sustainable nitrogen-fixing PGPR (Dixit et al. 2007; Shukla et al. 2012; Ziaf et al. 2016). Generally, the application of nitrogen-fixing PGPR in vegetable production has been found as an attractive alternative to replace chemical fertilizer, pesticides and other supplements. Nitrogen fixers including both symbiotic rhizobia and asymbiotic/associative nitrogen fixers, for example, *Azotobacter* or *Azospirillum*, have traditionally been used as biofertilizer to supply N to legumes and cereals/other crops. Among non-symbiotic N-fixing bacteria, *Azotobacter* and *Azospirillum* have widely been used for enhancing the production of vegetables (Doifode and Nandkar 2014; Solanki et al. 2010). The beneficial effects of *Azotobacter* and *Azospirillum* are attributed mainly to an improvement in root development, an increase in the rate of water and mineral uptake by roots, displacement of fungi and plant pathogenic bacteria and, to a lesser extent, biological nitrogen fixation (Okon and Itzisohn 1995). Besides N_2 fixation, *Azotobacter* synthesizes and secretes considerable amounts of biologically active substances like B vitamins, nicotinic acid, pantothenic acid, biotin, heteroxins, gibberellins, etc. which enhance root growth of plants (Rao 1986). Another important characteristic of *Azotobacter* association with crop improvement is secretion of ammonia in the rhizosphere in the presence of root exudates, which helps in modification of nutrient uptake by the plants (Narula and Gupta 1986). The ability of *Azospirillum* to produce plant growth regulatory substances along (Tahir et al. 2013) with N_2 fixation stimulates plant growth and thereby productivity. Considering these, nitrogen-fixing PGPR for non-legumes especially vegetable production (Table 3.2) have attracted greater attention

Table 3.2 Some examples of vegetable inoculation with nitrogen-fixing plant growth-promoting rhizobacteria

Host vegetables	Botanical name	Inoculant nitrogen fixers	Reference
Potato	*Solanum tuberosum*	*Rhizobium* sp. TN42, *Azotobacter chroococcum*	Naqqash et al. (2016), Meshram (1984), Hussain et al. (1993)
Radish	*Raphanus sativus*	*Azotobacter* + PSB	Ziaf et al. (2016)
Tomato	*Solanum lycopersicum*	*Bradyrhizobium japonicum*; *Azotobacter*	Parveen et al. (2008), El-Sirafy et al. (2010),
Okra	*Abelmoschus esculentus*	*Rhizobium meliloti*	Tariq et al. (2007)
Eggplant	*Solanum melongena*	*Azotobacter* and *Bacillus polymyxa*	Doifode and Nandkar (2014), Bhadoria et al. (2005),
Cabbage	*Brassica oleracea*	*Azotobacter*, *Azospirillum* and VAM	Sharma et al. (2013)

in recent times. Nitrogen-fixing PGPR have been found to colonize and survive in the rhizosphere of the nonlegumes plant to act as PGPR in the rhizosphere of non-host legumes and nonlegumes (Wiehe and Höflich 1995). Nitrogen-fixing plant growth-promoting rhizobacteria when used alone or in combination with other free-living PGPR have also caused a dramatic increase in vegetable production (Noel et al. 1996; Antoun et al. 1998). Mechanistically, as inoculant, nitrogen-fixing PGPR facilitate the vegetable growth by mechanisms other than nitrogen fixation (Trabelsi et al. 2012). When used as mixture, the composite nitrogen fixers provide multiple benefits to crops in addition to their normal physiological activity of N fixation (Iqbal et al. 2012). And hence, the synergistic effects of nitrogen fixer and other free-living PGPR/AM fungi have been found more effective than single inoculation and massively increase vegetable production largely due to enhanced synthesis of phytohormones and nutrient absorption and mobilization (Reimann et al. 2008; Yu et al. 2012). As an example, the composite application of rhizobia (*Bradyrhizobium japonicum*), *Pseudomonas aeruginosa* and mineral fertilizers (urea and potash) has been reported to suppress the deleterious impact of root-rotting fungi and root-knot nematode leading consequently to enhanced tomato production (Parveen et al. 2008). Conclusively, due to their variable growth-promoting activities, nitrogen fixers can be used either alone or in combination with other free-living PGPR/AM fungi for enhancing the production of vegetable in different vegetable production systems.

3.6   Effects of Nitrogen-Fixing Plant Growth-Promoting Rhizobacteria on Important Vegetable Crops

Vegetables are one of the most important food commodities which have occupied a central place in human dietary systems. Production of fresh and quality vegetables is, therefore, required in order to fulfil the demands of vegetarian around the world. Therefore, considering the importance of nitrogen-fixing plant growth-promoting rhizobacteria in vegetable growth, an attempt is made in the following section to highlight the impact of nitrogen-fixing PGPR on some vegetables grown in different production systems.

3.6.1   Potato (*Solanum tuberosum*)

Potato is a starchy and tuberous crop of the Solanaceae family. Potato, ranking fourth in production among vegetables, is a high-yielding, nutrient-exhaustive and short-duration crop. Potato requires higher quantities of nitrogen and phosphorus fertilizers for optimum production (Igual et al. 2001). Therefore, to reduce fertilizer application, nitrogen-fixing PGPR have been employed as a biofertilizer or as bacterial inoculum in potato production (Sidorenko et al. 1996; Kumar et al. 2001; Shafeek et al. 2004). For example, in order to investigate the effects of natural and chemical fertilizers on yield and quality of potato, Mohammadi et al. (2013) conducted a study at the Agricultural Research Farm of Razi University, Kermanshah,

Iran. The experiment included three factors: (1) nitragin biofertilizer (a combination of *Azotobacter* species and *Azospirillum* species), (2) HB-101 (a completely organic natural extract) and (3) chemical urea fertilizer (500 kg/ha). Generally, all factors showed significant effects on tuber yield, tuber weight, number of tuber per plant, biological yield, harvest index and tuber nitrate content of potato. However, the highest tuber yield and the number of tuber per plant were obtained when tubers were inoculated jointly with nitragin, urea and HB-101. On the contrary, the lowest tuber nitrate content was obtained when HB-101 was sprayed two times and the tubers were inoculated with nitragin biofertilizer. From this study, it was concluded that the composite application of natural and biological fertilizers along with urea can be useful to enhance potato yield and quality. In a similar study, Verma et al. (2011) conducted an experiment on potato variety Kufri Jawahar to assess the effect of organic components on growth, yield and economic return in potato. The results revealed that combination of crop residues + *Azotobacter* + phosphobacteria + bio-dynamic approach was the best among all the treatments for most of the growth and yield parameters and gave highest net return and B:C (benefit/cost) ratio. Thus, it can be concluded that the biofertilizers (*Azotobacter*, phosphobacteria, microbial culture and biodynamic approach) are an advantageous source for sustainable organic agriculture, especially for heavy feeder crops like potato. Zahir et al. (1997) also conducted a pot experiment to evaluate the effects of an auxin precursor L-tryptophan (L-TRP) and *Azotobacter* inoculation on yield and chemical composition of potato grown with varying rates of fertilizers. Inoculated (with *Azotobacter*) and uninoculated potato tubers were sown in fertilized (with NPK 250:150:150 kg/ha, respectively) pots, and 1-week-old seedlings were treated with different concentrations of L-TRP (10^{-4}–10^{-7} g/kg soil). Results revealed that L-TRP application alone had no significant effect on tuber and straw yield and PK uptake; however, N uptake and NPK concentrations in the potato tubers were significantly increased at some of the L-TRP levels. *Azotobacter* inoculation significantly increased tuber yield by 28.5%, N uptake and NPK concentrations relative to control. Also, *Azotobacter* inoculation in the presence of L-TRP was found more effective and considerably increased the tuber and straw yield by 62.9 and 47.8%, respectively, and NPK uptake compared to sole application of *Azotobacter*. Hussain et al. (1993) conducted a field experiment to assess the ability of *Azotobacter* inoculation for enhancing yield and other growth parameters on a sandy loam soil treated with NPK (250:125:125 kg/ha, respectively). Shoot, root, single tuber weight, tuber yield plant and R/S ratio increased significantly following inoculation with all *Azotobacter* strains, and maximum tuber yield (18.13% higher than control) was observed with *Azotobacter* strain. The increase in potato growth was possibly due to the production of plant growth regulators since there was no possibility of N_2 fixation in the presence of such a high dose of N. Similar increase in growth and yield and other components of potato due to inoculation with biofertilizer (*Azotobacter chroococcum* with *Azospirillum brasilense*) is reported (Osman 2007). Mirshekari and Alipour (2013) evaluated the bio-priming effect of three different types of biofertilizers: *Azotobacter*, super nitro plus and super nitro on three potato cultivars—Agria, Satina and Kuzima—grown under field conditions. The number of tubers per plant in potato inoculated with *Azotobacter* and super nitro was 8.2, while non-inoculated seeds produced seven tubers per plant. However, seed inoculation with biofertilizers reduced the tubers size considerably over control. Among all treatments, seeds inoculated with *Azotobacter* had higher tuber yield (18,840 kg/ha), while the lowest was recorded for control (15,380 kg/ha). The stepwise regression analysis further verified that the tubers with diameter of greater than 40 mm and mean of tuber weight per plant had a marked increasing effect on the seed yield of potato. The present findings suggested that the tested biofertilizers could be used by farmers before sowing for enhancing potato production. In a follow-up study, Naqqash et al. (2016) inoculated potato with five bacteria belonging to genera *Rhizobium*, *Azospirillum*, *Agrobacterium*, *Pseudomonas* and *Enterobacter* under axenic conditions and observed differential growth responses of potato. Of these, associative nitrogen fixer *Azospirillum* sp. TN10 showed the highest increase in fresh and dry weight of potato over control plants. Also, the N contents of shoot and roots were found maximum following *Azospirillum* sp. TN10 application. Additionally, bacterial strains did colonize and maintained their population densities in the potato rhizosphere for up to 60 days, with *Azospirillum* sp. and *Rhizobium* sp. showing the highest survival. Since all strains showed variable impact, it was suggested that *Azospirillum* and *Rhizobium* could be used to develop biofertilizer for the production of potato.

Apart from directly affecting the growth and yield of potato, nitrogen-fixing PGPR have also been used to facilitate the growth of potato indirectly by secreting siderophores (Muthuselvan and Balagurunathan 2013), HCN (Prasad et al. 2014) or antifungal metabolites (Bhosale et al. 2013). As an example, Meshram (1984) reported that isolates of *Azotobacter chroococcum* were found to be promising for the control of infestation of potato plants with *Rhizoctonia solani*. Inoculation with an isolate of *Verticillium biguttatum* in combination with isolates of *A. chroococcum* effectively protected sprouts, stems and stolons against infestation with *R. solani*. The effect of inoculation, however, varied with soil temperature. No sclerotia were formed on potatoes harvested from soil in pots inoculated with isolates of *A. chroococcum* plus *V. biguttatum* under glasshouse conditions, and the yield increased significantly over the control.

3.6.2   Tomato (*Lycopersicum esculentum* Mill.)

Tomato, the second-most important vegetable crops (Dorais et al. 2008), is cultivated throughout the world occupying an area of 3.5×106 ha with the production of 1×10^6 tons (FAO 2010). In India, it occupies an area of 0.54 million ha with a production of 7.60 million ton with an average yield of 14.074 tons per ha (Anonymous 2006). Tomato is a tasty and nutritious vegetable containing vitamins A and C and lycopene content. Due to these nutritive properties, the efforts to produce safe and quality tomatoes both in developing and developed countries have increased (Mahajan and Singh 2006; Flores et al. 2010). In order to reduce the cost and to avoid toxic impact of synthetic fertilizers on tomato production, Ramakrishnan and Selvakumar (2012) applied different biofertilizers to assess their effect on growth and yield of tomato plants. For this, 20-day-old seedlings were transplanted into field until the fruit ripening period. After transplanting, tomato seedlings were bacterized with *Azotobacter*, *Azospirillum* and mixture of both *Azotobacter* and *Azospirillum*. Microbial inoculations, in general, significantly enhanced the whole plant dry weight, plant height, number of leaves per plant, number of fruits per plant, yield per plant, average fruit weight per plant, chlorophyll and protein content. Among all treatments, the composite application of *Azotobacter* and *Azospirillum* showed maximum yield relative to single inoculations and control. The overall results suggest that biofertilizer inoculation improves plant mineral concentration through nitrogen fixation and thereby alters fruit production in tomato plants.

In a similar study, Islam et al. (2013) used 13 nitrogen-fixing bacterial strains belonging to 11 different genera which were positive for 1-aminocyclopropane-1-carboxylate deaminase (ACCD), IAA, salicylic acid and ammonia production. The strains RFNB3 of *Pseudomonas* sp. and RFNB14 of *Serratia* sp. most effectively solubilized both tricalcium phosphate and zinc oxide. In addition, all strains except *Pseudomonas* sp. RFNB3 oxidized sulphur, and six strains were positive for siderophore synthesis, and each strain expressed at least four PGP properties in addition to N_2 fixation. Of these, nine strains were selected based on their multiple PGP potential and evaluated for their effects on early growth of tomato and red pepper under gnotobiotic conditions. Bacterial inoculation considerably influenced root and shoot length, seedling vigour and dry biomass of the two crop plants. Three strains demonstrating substantial performance were further selected for greenhouse trials with red pepper. Of the selected strains, *Pseudomonas* sp. RFNB3 resulted in significantly higher plant height (26%) and dry biomass (28%) compared to control. The highest rate of N_2 fixation as determined by acetylene reduction assay (ARA) occurred in *Novosphingobium* sp. RFNB21-inoculated red pepper root (49.6 nM of ethylene/h/g of dry root) and rhizosphere soil (41.3 nM of ethylene/h/g of dry soil). Moreover, the inoculation with nitrogen-fixing bacteria significantly increased chlorophyll content and the uptake of different macro- and micronutrient contents leading to enhanced red pepper shoots compared to uninoculated controls. The findings of this study suggest that certain nitrogen-fixing strains possessing multiple PGP traits could be used as biofertilizers for enhancing the production of vegetables. Likewise, Bhadoria et al. (2005) conducted a field trial to assess the effect of three *Azotobacter* inoculation (without inoculation, soil inoculation and seedling inoculation) and five levels of N (0, 25, 50, 75, and 100 kg/ha) on tomato and red pepper. A basal dose of P (80 kg/ha) and K (80 kg/ha) along with 50% N was applied at the time of field preparation. Remaining dose of N was top-dressed after 30 days of transplanting. *Azotobacter* culture was used as soil inoculant (5 kg/ha) and seedling inoculant (2 kg/ha), and fresh and dry weight of fruit, ascorbic acid content, total soluble solids (TSS) and cracking percentage of fruits were recorded. Maximum fresh and dry weight, ascorbic acid, TSS (%) and minimum percentage of fruit cracking were observed under the seedling treatment with *Azotobacter* culture over soil inoculation and without inoculation. Favourable environments like proper aeration around roots and considerably

phosphate-solubilizing bacteria. The application of *Azotobacter* biofertilizer caused a significant increase in plant height, branch number, fruit number per plant and yield/ha compared to the uninoculated control. The inoculation effect was maximum in the treatment containing 100% recommended dose of NPK (NPK 100:50:50 kg/ha) + *A. chroococcum* biofertilizer. Subsequently, the *A. chroococcum* inoculation resulted in consistent increase in yield attributes with gradual increase in the level of N. The yield obtained with 75% RDN+ *A. chroococcum* was almost equal to control. From this study, it was obvious that 25 kg N/ha could be saved if supplemented with *A. chroococcum* inoculation. Similarly, the application of *Azospirillum* and *Azotobacter* along with recommended dose of fertilizer resulted in maximum plant height, number of branches per plant, number of fruits per plant, fruit yield per plant and per ha and TSS in brinjal plants. Whereas days to initiation of flowering, fruit weight and crude protein did not change significantly (Solanki et al. 2010).

In a recent study, Latha et al. (2014) observed that the sole/composite application of microbial and chemical fertilizers had a great effect on the measured stages of eggplant growth. However, the total biomass differed significantly among treatments. Among all treatments, the maximum biomass was observed for treatment containing urea, super phosphate, muriate of potash, *Azospirillum*, phosphobacteria and potassium mobilizer (each 5 g/pot), and the fresh weight was 89.67 g/plant and dry weight 6.15 g/plant at harvest. Maximum chlorophyll content (1.7490 mg/g), protein content (18.2 mg/g), phenol content (19.6 mg/g) and carbohydrates (92 mg/g) in inoculated eggplant were recorded at flowering stage. In a follow-up study, Doifode and Nandkar (2014) evaluated the effect of biofertilizer like *Azotobacter* and *Bacillus polymyxa* (PSB) used alone and in different combinations with recommended dose of chemical fertilizer (NPK) on brinjal crop during kharif season to explore the possibility of reducing doses of chemical fertilizers and for better soil health. The growth characters such as height of plant (11.03–37.54%), stem diameter (6.38–23.79%), length of root (5.56–36.93%), number of functional leaves (5.67–51.51%), weight of fresh shoot (7.90–35.91%) and weight of dry shoot (7.14–46.94%) were significantly improved following microbial inoculations over control. Similarly, number of fruits per plant (11.3–52.81%) and yield of fruits (11.89–54.61%) were more in inoculated crop, and the attack of shoot-root borer, fruit borer and little leaf infestation was less (26.71–50.14%) as compared to uninoculated condition.

3.6.4 Cabbage (*Brassica oleracea*)

Cabbage is yet another important vegetable that requires proper nutrients for optimum production. And hence, nutrient management involving the use of chemical fertilizers coupled with inexpensive biofertilizers and environmentally safe organic manures in balanced proportion may be effective in augmenting the cabbage production (Hussein and Joo 2011). Considering this strategy, Sharma (2002) in a field trial assessed the impact of nitrogen-fixing PGPR (*Azospirillum* and *Azotobacter*) and different levels of N (0.30, 45 and 60 kg N/ha) on growth and yield of cabbage. *Azospirillum* application significantly increased the number and weight of non-wrapper leaves/plant, head length and width, gross and net weight of head/plant and yield/ha. Similarly, N at 60 kg/ha produced maximum number and weight of non-wrapper leaves/plant, head length and width, gross and net weight of head/plant and yield/ha. In addition, *Azospirillum* in the presence of 60 kg N/ha resulted in maximum yield/ha with benefit:cost ratio of 2.9. Similarly, Sarkar et al. (2010) assessed the influence of varying dose of N (0, 60, 80 and 100 kg/ha) and biofertilizer (*Azotobacter*) on growth and yield of cabbage grown at Horticulture Research Station, Mondouri of Bidhan Chandra Krishi Viswavidyalaya, Mohanpur, Nadia, West Bengal, using a plot size of 4.2 × 3.6 m. Application of both N and biofertilizer in general displayed a significant impact on growth and yield attributes of cabbage. In terms of plant improvement, 100 kg N/ha was found to be superior which was followed by 80 kg N/ha. *Azotobacter*-inoculated cabbage plants performed better than non-inoculated plants, and statistical differences were noted in this respect except the number of outer leaves. Plants inoculated with *Azotobacter* had head yield of 31.77 t/ha which was 19.66% higher than non-inoculated plants. The increase in plant growth has been attributed to the fact that N increases the chlorophyll content of the leaves which in turn ensure production of more carbohydrates and, hence, accelerated the growth and head yield of cabbage (Sharma 2002 and Lopandic and Zaric 1997). Other factors by which *Azotobacter* might have promoted the growth and development of cabbage could be the synthesis of auxin, vitamins, growth substances, antifungal and antibiotics by *Azotobacter*. The better results obtained due to *Azotobacter* inoculation are also supported by the findings of Jeevajohti et al. (1993) in cabbage where they reported that growth-promoting substances secreted by microbial inoculants might have led to better root development, transport of water, uptake and deposition of nutrients. The composite application of N and biofertilizer however resulted in significant increase in head weight and head yield of cabbage. The combined application of 100 kg N/ha and biofertilizer recorded highest head yield of 37.80 t/ha which was significantly higher than the other combination treatments. Verma et al. (1997) also recorded highest vegetable and seed yield of cabbage due to application of 60 kg N/ha along with *Azotobacter* inoculation. These studies together suggest that *Azotobacter* in the presence of 100 kg N/ha could be the best option to achieve highest head yield of cabbage. In other report, Sharma et al. (2013) observed the effects of single and composite culture of *Azotobacter*, *Azospirillum* and VAM on cabbage crop. The results showed that 4 kg/ha dose of each biofertilizer resulted in maximum plant height, number of leaves per plant, diameter of stem, length and width of longest leaf and plant spread compared to other doses. Among biofertilizers, *Azospirillum* was found superior and significantly enhanced the growth and fresh weight of green leaves per plant to the extent of 25.85 and 15.24% over *Azotobacter* and VAM, respectively. Also, *Azospirillum* significantly enhanced the total production of trimmed head of cabbage to the extent of 7.06% compared to those observed with *Azotobacter* application. Among various doses of biofertilizers, 4 kg/ha dose of each biofertilizer demonstrated greatest favourable effect on field-grown cabbage production than 2 kg/ha or even 6 kg/ha dose of *Azotobacter* and *Azospirillum*. In yet other microbial approach, Ishfaq et al. (2009) applied vermicompost (0, 5 and 10 t/ha) and *Azotobacter* (0, 5 and 10 kg/ha) against cabbage cv. 'Pride of India'. Application of vermicompost at 10 t/ha resulted in the tallest plant, maximum plant spread, largest size of head and highest yield of heads per plant and per hectare. The number of leaves/plant and number of wrapper leaves/head were, however, maximum with 5 t Vc/ha. Among various levels of biofertilizer inoculation, 10 kg/ha of *Azotobacter* application gave maximum plant height and diameter of head, maximum number of leaves/plant and number of wrapper leaves/head, while the length of head and head yield/plant were maximum with 5 kg *Azotobacter*/ha. Results by Hussein and Joo (2011) showed that seedling inoculation with bacterial (*A. chroococcum*) and fungal effective microorganisms (EM) significantly enhanced Chinese cabbage growth. Shoot dry and fresh weight and leaf length and width were significantly increased by both bacterial and fungal inoculation. However, the NPK chemical fertilizer decreased microflora inhabiting the soil, while the effective microorganisms either fungi or bacteria increased the microbial density significantly. This study implies that both fungal and bacterial EM are effective for the improvement of the Chinese cabbage growth and enhance the microorganisms in soil.

3.6.5 Broccoli (*Brassica oleracea*)

Broccoli is an important winter season vegetable crop which is cultivated widely in many European and American countries. It is an edible green vegetable belonging to cabbage family Brassicaceae whose large flowering head is eaten as a vegetable. Broccoli has many nutritional and medicinal values due to its high content of vitamins (A, B1, B2, B5, B6, C and E), minerals (Ca, Mg, Zn and Fe) and a number of antioxidants (Talalay and Fahey 2001; Rangkadilok et al. 2002; Rozek and Wojciechowska 2005; Wojciechowska et al. 2005). Broccoli is a rich source of sulphoraphane, a potent anticarcinogenic compound. It is a low-sodium, fat-free and low-calorie food (Decoteau 2000). Due to its variable use and great nutritional value, broccoli has attracted greater attention in recent times. For enhancing the growth, yield and head quality of broccoli, higher rates of plant nutrients are applied (Brahma and Phookan 2006). In order to reduce the usage of fertilizers in broccoli production, Abou El-Magd et al. (2014) conducted two field experiments in newly reclaimed land during two winter seasons in Egypt to study the effect of bio-nitrogen (*Azospirillum brasilense* and *A. chroococcum*) and different levels of mineral N [60, 90 and 120 kg N per feddan (one feddan = 0.42 ha)] on vegetative growth, yield and head quality of broccoli (cv. Hybrid Decathlon). Plants treated with nitrogen-fixing PGPR *A. brasilense* and *A. chroococcum* (bio-nitrogen) had higher vegetative growth, i.e. plant length, number of leaves, fresh weight of leaves, stems and total plant. The dry matter accumulation in leaves and heads, main head yield and physical head quality (weight and diameter) as well as N, P and K contents of leaves and heads were greater in nitrogen-fixing PGPR-inoculated broccoli plants compared to those found in untreated control plants. Of the two inoculants, *A. chroococcum* was found superior and resulted in dramatic increase in vegetative growth, main head yield and physical head quality (weight and diameter), as well as N, P and K content of leaves and heads of broccoli compared to those recorded for *A. brasilense* or non-inoculated control plants. Varying levels of N, however, differed statistically in their effects on the measured parameters of broccoli plants. Among N levels, 120 kg N/feddan showed the highest vegetative growth which was followed by 90 kg N/feddan. The lowest vegetative growth, main head yield, physical head quality and N, P and K of broccoli leaves and heads were, however, obtained by 60 kg N/feddan application. The present findings showed that the composite application of nitrogen-fixing PGPR and mineral N caused statistically a significant positive impact on vegetative growth, yield and nutrient uptake of broccoli. However, among all single or multiple inoculation treatments, the combined application of 120 kg N/feddan with bio-nitrogen *A. brasilense* resulted in the highest vegetative growth, yield and chemical contents of broccoli. Yildirim et al. (2011), on the contrary, investigated the effects of root inoculations with *B. cereus* (N_2 fixing), *Brevibacillus reuszeri* (P solubilizing) and *Rhizobium rubi* (both N_2 fixing and P solubilizing) on growth, nutrient uptake and yield of broccoli, grown in field soils, treated with manure and some fertilizers. Bacterial inoculations with manure significantly increased the yield, plant weight, head diameter, chlorophyll content and N, K, Ca, S, P, Mg, Fe, Mn, Zn and Cu contents of broccoli over control. Among different treatments, manure with sole culture of *B. cereus*, *R. rubi* and *B. reuszeri* increased the yield by 17, 20.2 and 24.3%, respectively, and chlorophyll content by 14.7, 14 and 13.7%, respectively, over control. It was suggested from this study that seedling inoculation with P solubilizing (*B. reuszeri*) and both N_2 fixing and P solubilizing (*R. rubi*) could be employed as an alternative to partially reduce the use of costly fertilizers in broccoli production.

Biofertilizers prepared from *Azospirillum*, PSB, *Azotobacter* and VAM applied alone and in combinations with/without inorganic fertilizer had variable impact on yield and quality of broccoli (Singh et al. 2014a). The composite application of *Azospirillum* + *Azotobacter* (50% each) significantly increased the curd size (15.17 cm diameter) and curd yield (1.17 kg and 0.93 kg curd with and without guard leaves, respectively) of broccoli, and this combination was found superior compared to other microbial or fertilizer applications. The results further revealed and showed that 100% application each of *Azospirillum*, PSB and *Azotobacter* also had better performance than the recommended dose of fertilizers. However, all other treatment combinations except *Azospirillum* + *Azotobacter* (50% each) performed poor than the recommended dose of fertilizer. Among the biofertilizer, the coculture of *Azospirillum* and *Azotobacter* (50% each) increased the protein and lipid profile along with phosphate and sulphate content of broccoli curd. Conclusively, the composite application of *Azospirillum* + *Azotobacter* applied each at 50% level was found better for enhancing the curd yield of broccoli and its active biomolecules.

3.6.6   Okra (*Abelmoschus esculentus* L.)

Okra is an annual flowering vegetable grown for its edible pods which can be used as fresh, canned, frozen or dried food worldwide. The approximate nutrient content of the edible okra pods is as follows: water, 88%; protein, 2.1% m; fat, 0.2%; carbohydrate, 8.0%; fibre, 1.7%; and ash, 0.2% (Tindall 1983). Besides these, okra also contains minerals and vitamins. For production and maintenance, okra requires nutrients such as N, P, Ka, Ca, Na and S (Ahmed et al. 2015; Hooda et al. 1980). Deficiency of any of these nutrients resulted in poor growth and leads to a lower yield (Shukla and Nalk 1993). Therefore, an integrated approach involving bio-inoculants/bioagents and fertilizers has been practised over the years for okra production (Singh et al. 2010). The biological potential of different microbial antagonists like, *Bacillus thuringiensis*, nitrogen-fixing PGPR *Rhizobium meliloti*, *Aspergillus niger* and *Trichoderma harzianum* in the suppression of root-rotting fungi like *Macrophomina phaseolina*, *Rhizoctonia solani* and *Fusarium* spp. inflicting losses to okra and sunflower plants, was evaluated by Dawar et al. (2008). All biocontrol agents enhanced the germination, growth, length of plant organs (shoot and root) and dry matter accumulation in shoot and root of both okra and sunflower compared to control. The length and weight of shoot and root were significantly increased in sunflower and okra when seeds were coated with *R. meliloti* and *B. thuringiensis*. Also, *Rhizobium* used alone as seed dressing also significantly improved plant growth and reduced disease intensity of plants. *Rhizobium meliloti* significantly inhibited the infection of *R. solani* on okra plant when *R. meliloti* was multiplied on leaves powder of *Rhizophora mucronata* plant (Tariq et al. 2007). Rhizobia which are good rhizosphere organism for leguminous or nonleguminous plants presumably prevent the contact of pathogenic fungi on roots by covering the hyphal tips of the fungus and parasitizing it. Maximum plant height was observed where seeds of okra and sunflower were coated with *T. harzianum* using 2% of glucose followed by gum arabic, mollases and sugar solution. Gum arabic was found more effective in reducing infection by root-rotting fungi, viz. *M. phaseolina*, *R. solani* and *Fusarium* spp. Of the different microbial antagonists used, *T. harzianum* was found more effective followed by *B. thuringiensis*, *R. meliloti* and *A. niger* in the control of root-rotting fungi. Similarly, Ehteshamul-Haque and Ghaffar (2008) reported that *Rhizobium meliloti* inhibited growth of *M. phaseolina*, *R. solani* and *Fusarium solani* while *B. japonicum* inhibited *M. phaseolina* and *R. solani* producing zones of inhibition. In field, *R. meliloti*, *R. leguminosarum* and *B. japonicum* used either as seed dressing or as soil drench reduced infection of *M. phaseolina*, *R. solani* and *Fusarium* spp., in both leguminous (soybean, mung bean) and nonleguminous (sunflower and okra) plants. Likewise, the antagonistic effects of *Bacillus subtilis*, *B. thuringiensis*, *B. cereus* and *R. meliloti* against the control of root-infecting fungi on mash bean and okra were reported by Tariq et al. (2007). Germination of seeds, shoot and root length and shoot and root weight of okra and mung bean were significantly improved following *B. subtilis*, *B. thuringiensis*, *B. cereus* and *R. meliloti* application. Infection of *R. solani* was significantly inhibited on okra when *R. meliloti* was used at 1% w/w, whereas all biocontrol bacteria, viz. *B. subtilis*, *B. thuringiensis*, *B. cereus* and *R. meliloti*, completely suppressed the infection of *R. solani* and *M. phaseolina* on mung bean.

3.6.7 Onion (*Allium cepa*)

Onion, a widely cultivated commercial bulbous vegetable and spice of the genus *Allium*, is grown worldwide. Among onion-producing countries, India ranks second and occupies 756,200 ha area with a production of 12.15 MT and productivity of 16.1 tons/ha (Anonymous 2010). Onion has stimulant, diuretic and expectorant properties and is considered useful in flatulence and dysentery. The shallow-rooted onion plants require large amounts of N for better growth, development and quality of bulb and consequently optimum production (Gamiely et al. 1991; Drost et al. 2002 and Woldetsadik et al. 2003). On the contrary, the inadequate or low N supply increases the incidence of onion bolting and limits bulb yield (Diaz-Perez et al. 2003). The application of super-optimal N has been reported to overstimulate growth and results in (1) extensive foliage growth, (2) delayed crop maturity and (3) poor bulb quality with increased storage losses (Brown et al. 1988; Brewster 1994 and Woldetsadik et al. 2003). Therefore, since both the lower and higher rates of N adversely affect the quality and quantity of onion, the careful application of N fertilizer becomes extremely important in order to improve the yielding ability and bulb quality of onion plants. Under these circumstances, synthetic nitrogenous fertilizer must be supplemented with biofertilizers especially those prepared from PGPR so that the cost of production could be reduced and quality of onion be maintained.

Like other vegetables, the production of onion is also greatly influenced by biofertilizers, organic manures and inorganic fertilizers (Banjare et al. 2015; Yeptho et al. 2012; Yadav et al. 2004). For example, the impact of single and composite culture of *B. circulans*, *Azospirillum lipoferum*, *A. chrococcoum* (nitrogen-fixing PGPR), *B. polymyxa*, *Rhizobium* sp. and AM fungi on growth and quality of onion bulbs was found favourable (El-Batanomy 2009). Vegetative growth and total bacterial populations in onion rhizosphere were increased due to PGPR inoculations. Additionally, the mixture of all cultures showed highest increase in dry matter and bulb diameter. The composite microbial cultures resulted in maximum nitrogenase activity (41.98 μmole C_2H_4/h/g RDW) and mycorrhizal infection (95%) in onion roots. The mixture of *B. circulans*, *A. lipoferum*, *A. chrococcoum*, *B. polymyxa*, *Rhizobium* sp. and AM fungi showed maximum NPK (4:1.97:2.91%) in dry onion shoots relative to fertilized control. Also, the total carbohydrate was highest (29.23 mg/g) in onion plants inoculated with six cultures together which was followed by co-inoculation of *Rhizobium* sp. and AM fungi (28.77 mg/g) and *B. circulans* used alone (24.9 mg/g). Similarly, Ghanti and Sharangi (2009) studied the effect of combinations of six biofertilizers [(1) *Azotobacter* + PSB, (2) *Azotobacter* + AM fungi, (3) *Azotobacter* + *Azospirillum*, (4) *Azospirillum* + PSB, (5) *Azospirillum* + AM fungi and (6) PSB + AM fungi)] and two levels of chemical fertilizers (NPK 100% and 50%) on onion cv. Sukhsagar under field experiment, carried out during the winter season. The co-inoculation of *Azotobacter* + VAM showed the maximum height (43.46 cm) of plants, while number of leaves, number of inflorescence/plot and bulb diameter were maximum due to inoculation with *Azotobacter* + *Azospirillum*. The composite application of *Azotobacter* and *Azospirillum* in the presence of 100% NPK produced maximum length of bulbs (6.03 cm), and the maximum number of scale per bulb (9.81) was recorded with 50% NPK. The plants grown with 100% NPK had maximum bulb weight of 67.45 g, maximum and TSS (12.29%), but the plants fertilized with 50% NPK had the highest reducing sugar (1.420%) and starch (6.27%). It was concluded from this study that the combination of *Azotobacter* and *Azospirillum* could be developed as an effective microbial pairing for enhancing the growth, yield and quality of onion. Furthermore, even though the 100% NPK fertilizer (recommended dose) produced the best result relative to combinations of biofertilizers, the application of biofertilizer should be preferred in order to achieve sustainable and safe production of onion. Balemi (2006) conducted a field experiment using four levels of N (0, 25, 50 and 75% recommended doses) and three strains (CBD-15, AS-4 and M-4) of *Azotobacter* with two uninoculated controls, one with the full dose of N and the other without NPK during summer season against onion cultivar Pusa Madhvi to identify a suitable *Azotobacter* strain and N level for better yield and quality of onion. Application of 75% recommended N along with *Azotobacter* CBD-15 or M-4 significantly increased the marketable yield and the N content in both leaves and bulbs, over control (full dose of N), whereas only 75% recommended N + *Azotobacter* CBD-15 significantly increased the total yield. However, total soluble solids and neck thickness were significantly reduced by 50% recommended N applied with CBD-15 or M-4 compared with the uninoculated control (full N dose). *Azotobacter* strains in the presence of 50 or 75% recommended N significantly reduced the sprouting loss during storage, while nitrogen-fixing PGPR in the presence of 50 or 25% recommended N doses significantly reduced rotting and total losses. Inoculation with a mixture of N-fixing bacteria (*Azospirillum*, *Azotobacter* and *Klebsiella*), biofertilizer (Halex 2) alone or combined with four levels of N (00, 30, 60 and 90 kg N/fed.) had a variable impact on growth, yield components and bulb quality of onion (Yaso et al. 2007). A significant increase in plant height and number of leaves, average bulb weight and marketable and total bulb yield were observed following consistent increase in N levels. Inoculation of onion transplants with Halex 2 significantly improved onion bulb yield and its components (average bulb weight and marketable yield), in both seasons, and accelerated the maturity of onion bulbs in the first season but did not significantly influence vegetative growth and bulb quality characters (plant height, number of leaves and percentages of single and double bulbs, bolters, TSS and sprouted bulbs). Among all treatments, combination of 60 kg N/fed and biofertilizer (Halex 2) was found as the best combination which gave the maximum marketable yield and total bulb yield. The use of Halex 2 could replace one-third of the used chemical N fertilizer and, consequently, improve the economics of onion production. In other study, significant increase in growth and yield of onion plants due to the synthesis of IAA, siderophores and P-solubilizing activity of *B. subtilis* and *A. chroococcum* is reported (Colo et al. 2014). The longest seedling was observed due to inoculation with *A. chroococcum*, while all inoculated plants had maximum height recorded 60 days after sowing. The onion yield was highest when plants were bacterized with *B. subtilis* and *A. chroococcum*.

3.6.8 Radish and Daikon (*Raphanus sativus*)

Radish, a native of Europe and Asia (Gill 1993), is a popularly grown root vegetable which belongs to the family Cruciferae. In many countries like India, it is grown almost everywhere throughout the year. The fusiform roots of radish are eaten raw as salad or as cooked vegetable. Its leaves are rich in minerals and vitamins A and C and are also cooked as leafy vegetable. Like many other nonlegumes, growth and development of radish are also influenced by some nitrogen-fixing PGPR. For instance, *B. japonicum* strain Soy 213 among 266 PGPR strains tested by Antoun et al. (1998) showed the highest stimulatory effect on radish plant. A maximum of 60% increase in stimulatory effect was obtained with *B. japonicum*, while about 25% of all strains of rhizobia and bradyrhizobia, in general, increased radish growth by 20% or more. Similarly, strain Tal 629 of *B. japonicum* significantly increased the dry matter yield of radish by 15% over control in a second plant inoculation assay. It was concluded from these experiments that rhizobia like many other PGPR could also be used as traditional PGPR for enhancing the production of vegetables. In a follow-up study, Basavaraju et al. (2002) reported the effect of asymbiotic nitrogen-fixing PGPR *Azotobacter* strains C_1 and C_2 on germination and seedling development of radish grown under controlled conditions. Of the two strains, strain C_2 of *A. chroococcum* maximally enhanced the germination percentage by 9.33%, radical length by 90.47% and plumule length by 54.37% over uninoculated control. Furthermore, inoculation of radish seeds with *Azotobacter* showed increase in plant height, number of leaves, leaf area, root girth, root length, fresh and dry weights of root and leaf and root N contents over uninoculated control. However, *Azotobacter* in the presence of 75% recommended dose of N per ha was found to be more advantageous and helped to reduce dependence on nitrogenous fertilizers while maintaining good yields. Shukla et al. (2012) carried out an experiment with radish cv. Chinese pink using synthetic fertilizers (N, P and K) and biofertilizers (*Azospirillum*, phosphorus-solubilizing bacteria and AM fungi). Seed yield (10.2 q per ha), 1000 seed weight and seedling vigour index-II were recorded maximum with the combined application of *Azospirillum* + recommended rates of NPK. Ziaf et al. (2016) in a recent study evaluated the effect of nitrogen-fixing PGPR like *Azotobacter* spp., PSB, germinator (Ger, a synthetic germination and early growth enhancer) and PSB + Ger in combination with full (recommended dose of fertilizer), half dose of N and half dose of P on yield of radish cv. 'Mino Early'. The results revealed that *Azotobacter* spp. improved plant- and yield-related attributes, while germinator negatively affected them. The combined application of PSB and recommended dose of fertilizer resulted in maximum number of leaves per plant, root fresh weight and marketable yield. On the contrary, the application of *Azotobacter* spp. in combination with half dose of N and half dose of P showed the highest leaf fresh weight, above ground plant biomass, biological yield, agronomic efficiency and yield response. Moreover, root diameter increased when PSB or *Azotobacter* spp. was applied with recommended dose of fertilizer, while plants treated with *Azotobacter* spp. along with a half dose of P had the longest roots. Correlation analysis revealed that marketable yield of radish was dependent on root fresh weight.

3.6.9   Lettuce (*Lactuca sativa* L.)

Lettuce is considered as one of the most important vegetable crops grown in many countries. It is reported that 100 g of lettuce contain 95% water, 1 g of protein, 3 g carbohydrate, Ca (22 mg), P (25 mg) and vitamin A (Work and Carew 1955). Among many factors, fertilizer application is the most important factor that affects greatly the quantity and quality of lettuce. However, the excessive use of fertilizers negatively affects its production, and hence, the combination of chemical and biological fertilizers is recommended for this crop so that the quality of lettuce is maintained while preserving the soil fertility (Forlin et al. 2008; Sarhan 2008). For example, Sarhan (2012) carried out an experiment during winter season to investigate the effects of nitrogen-fixing bacterium (*Azotobacter*) with different levels of N (100, 200, 300 kg/ha) and without *Azotobacter* (N alone) on growth, yield quantity and quality of lettuce. The results revealed a significant increase in measured characteristics such as plant height, leaves number, length of stem, fresh and dry weight of head, head diameter and head yield following application of *Azotobacter* with low levels of N. Chabot et al. (1996a) on the other examined the single and composite effect of symbiotic nodule forming *R. leguminosarum* bv. phaseoli strains P31 and R1, *Serratia* sp. strain 22b, *Pseudomonas* sp. strain 24 and *Rhizopus* sp. strain 68 on lettuce and forage maize, grown in field conditions having high to low amounts of available P. The composite inoculation of strains R1 of *R. leguminosarum* and 22b of *Serratia* sp. significantly increased the dry matter yield of lettuce shoots where lettuce inoculated with *R. leguminosarum* R1 had a 6% higher P concentration than the uninoculated control. Similarly, at other experimental site (poorly fertile soil), the dry matter of lettuce shoots was significantly increased by inoculation of *R. leguminosarum* strain P31 and *Pseudomonas* sp. 24 along with 35 kg/ha P superphosphate or with *Rhizopus* sp. strain 68 plus 70 kg/ha P superphosphate. The present findings clearly demonstrated that rhizobia expressing P solubilization activity can also function as PGPR with nonlegumes especially lettuce and maize. In a follow-up experiment, Chabot et al. (1996b) assessed the effects of two strains of *R. leguminosarum* bv. phaseoli and three other PGPR on maize and lettuce root colonization. Maize and lettuce seeds were treated with derivatives of all strains marked with lux genes for bioluminescence and resistance to kanamycin and rifampin prior to planting in non-sterile Promix and natural soil. The introduced bacterial strains were quantified on roots by dilution plating on antibiotic media together with observation of bioluminescence. Rhizobia were found as superior colonizers compared with other tested bacteria; rhizobial populations were 4.1 CFU/g (fresh weight) on maize roots 4 weeks after seeding, while 3.7 CFU/g (fresh weight) was found on lettuce roots 5 weeks after seeding. The average populations of the recovered PGPR strains were 3.5 and 3 CFU/g (fresh weight) on maize and lettuce roots, respectively. Bioluminescence also revealed in situ root colonization in rhizoboxes and showed the ability of rhizobial strains to colonize and survive on maize and lettuce roots. In a study, Galleguillos et al. (2000) observed that the rhizobial strains increased very efficiently the lettuce biomass and also induced modifications on root morphology, particularly in mycorrhizal plants suggesting that these strains behaved as PGPR. However, rhizobial strains differed in mycorrhizal plants with regard to (1) the biomass production, (2) the length of axis and lateral roots and (3) the number of lateral roots formed; effects which were, in turn, affected by the AM fungus are involved. Microbial treatments were more effective in terms of growth and morphology of roots at 20 days of plant growth, but after 40 days, the microbial inoculation profoundly increased plant biomass. The interaction between the AM fungi (*Glomus mosseae*) and rhizobial strain had the maximum growth-promoting effect (476% over control) despite the fact that *G. intraradices* showed a quicker and higher colonization ability than *G. mosseae*. Flores-Félix et al. (2013) assessed the impact of *R. leguminosarum* strain PEPV16 on crops like lettuce and carrot and observed a significant increase in macro- and micronutrients of both lettuce and carrots. Also, the rhizobial inoculation enhanced the N and P uptake by lettuce and carrot plants. The P uptake in lettuce shoots was increased by 15, while 40% increase in P concentration was recorded in carrot roots. Increase in Fe content of both crops was attributed to the production of siderophores by *R. leguminosarum* strain PEPV16.

3.6.10 Spinach (*Spinacia oleracea*)

Spinach, an annual member of Chenopodiaceae family, is a valuable leafy vegetable. It is a rich source of chlorophyll, which gives spinach a dark-green colour, good quality and consumer acceptance. Also, spinach is a low-calorie vegetable but contains unusually high minerals like iron, vitamin A and vitamin C contents, which add nutritive value to it. For enhancing growth, yield, seed production and quality of spinach, nitrogenous and phosphorus fertilizers are frequently applied. However, like other vegetables, the quantity and quality of spinach also suffer from uncontrolled application of such fertilizers. And hence, like many crops, the use of biofertilizers has also been suggested as a cheap and viable option for optimizing the production of spinach. For example, the application of nitrogen-fixing PGPR such as *Azotobacter chroccocum* and phosphorein when used singly or in combination with different rates of N and P fertilizers showed a variable effect on growth, yield, sex ratio and seeds (yield and quality) of spinach plants cv. Dokki (El-Assiouty and Abo-Sedera 2005). Seed inoculation with 300 g phosphorein inoculum/fed. in the presence of 40 kg N/fed. (100% of the recommended N dose) + 15 or 7.5 kg P/fed. (66.7 or 33% of the recommended dose of P_2O_5) and seeds inoculated with 300 g *Azotobacter* inoculum in the presence of the full dose of P_2O_5 (22.5 kg P_2O_5/fed.) + 50% of the full dose of N (20 kg/fed) demonstrated the optimum favourable impact on growth, yield, sex ratio and higher seed yield with the best quality relative to control (40 kg N + 22.5 kg P_2O_5 fed.). The populations of inoculated microbes were higher in spinach rhizosphere when seeds were inoculated with *Azotobacter* and phosphorein compared with uninoculated control. Among all treatments, application of 40 kg N + 15 kg P_2O_5 + 300 g phosphorein increased plant fresh yield by 27.2 and 42.3% and 16.3 and 10.4% seed yield over control in the first and second seasons, respectively.

Conclusion

Nitrogen fixers are well known for their beneficial effect resulting from the symbiotic and asymbiotic nitrogen fixation with legumes and other crops including vegetables. In this work, we have tried to showcase the beneficial activity of two contrasting nitrogen fixers on the overall performance of different vegetables grown distinctively in different agroclimatic regions of the world. The advantages of using nitrogen fixers as PGPR are the easy availability of the technology for inocula production and seed inoculation and the better understanding of the functional diversity and genetics of these bacteria. In addition, they have been used in agronomic practices since very long without any adverse impact, they can, therefore, be considered as environmentally benign PGPR for nonlegumes. The work presented here is likely to help vegetable growers to optimize the vegetable production through the use of inexpensive and environmentally safe nitrogen-fixing PGPR while reducing the dependence on chemical input in vegetable production system across the globe.

References

Abayomi OA, Adebayo OJ (2014) Effect of fertilizer types on the growth and yield of *Amaranthus caudatus* in Ilorin, Southern Guinea, Savanna Zone of Nigeria. Adv Agric 2014: 5. Article ID: 947062

Abd-Alla MH (1994) Use of organic phosphorus by *Rhizobium leguminosarum* biovar. viceae phosphatases. Biol Fertil Soils 18:216–218

Abdelaziz M, Pokluda R, Abdelwahab M (2007) Influence of compost, microorganisms and NPK fertilizer upon growth, chemical composition and essential oil production of *Rosmarinus officinalis*. Not Bot Hortic Agrobot Cluj 35:1842–4309

Abou El-Magd MM, Zaki MF, Abo Sedera SA (2014) Effect of bio-nitrogen as a partial alternative to mineral-nitrogen fertilizer on growth, yield and head quality of broccoli (*Brassica oleracea* L. var. Italica). World Appl Sci J 31:681–691

Ahmad E, Khan MS, Zaidi A (2013) ACC deaminase producing *Pseudomonas putida* strain PSE3 and *Rhizobium leguminosarum* strain RP2 in synergism improves growth, nodulation and yield of pea grown in alluvial soils. Symbiosis 61:93–104

Ahmed A, Mohsen M, Abdel-Fattah MK (2015) Effect of different levels of nitrogen and phosphorus fertilizer in combination with botanical compost on growth and yield of okra (*Abelmoschus esculentus* L.) under sandy soil conditions in Egypt. Asian J Agric Res 9:249–258

Anandaraj B, Delapierre LRA (2010) Studies on influence of bioinoculants (*Pseudomonas fluorescens, Rhizobium* sp., *Bacillus megaterium*) in green gram. J Biosci Tech 1:95–99

Anonymous (2006) Invasive species and poverty: exploring the links. GISP Global Invasive Species Programme, Cape Town, South Africa

Anonymous (2010) Annual report. National Horticulture Board, Ministry of Agriculture, Government of India

Antoun H, Beauchamp CJ, Goussard N, Chabot R, Lalande R (1998) Potential of *Rhizobium* and *Bradyrhizobium* species as plant growth promoting rhizobacteria on non-legumes: effect on radishes (*Raphanus sativus* L.) Plant Soil 204:57–67

Balakrishnan R (1988) Studies on the effect of *Azospirillum*, nitrogen and NAA on growth and yield of chilli. South Indian Hortic 36:218

Balemi T (2006) Effect of integrated use of *Azotobacter* and nitrogen fertilizer on yield and quality of onion (*Allium cepa* L.) Acta Agron Hung 54:499–505

Banjare C, Shukla N, Sharma PK et al (2015) Effect of organic substances on yield and quality of onion, *Allium cepa* L. Int J Farm Sci 5(1):30–35

Barrett CF, Parker MA (2006) Coexistence of *Burkholderia*, *Cupriavidus*, and *Rhizobium* sp. nodule bacteria on two *Mimosa* spp. in Costa Rica. Appl Environ Microbiol 72:1198–1206

Basavaraju O, Rao ARM, Shankarappa TH (2002) Effect of *Azotobacter* inoculation and nitrogen levels on growth and yield of radish (*Raphanus sativus* L.) In: Rajak RC (ed) Biotechnology of microbes and sustainable utilization. Scientific Publishers, Jabalpur, India, pp 155–160

Beijerinck MW (1888) Culture des Bacillus radicicola aus den Knollchen. Bot Ztg 46:740–750

Bhadoria SKS, Dwivedi YC, Kushwah SS (2005) Effect of *Azotobacter* inoculation with nitrogen levels on quality characters of tomato. Veg Sci 32:94–95

Bhagat D, Sharma P, Sirari A et al (2014) Screening of *Mesorhizobium* spp. for control of *Fusarium* wilt in chickpea in vitro conditions. Int J Curr Microbiol Appl Sci 3:923–930

Bhakare GC, Deokar CD, Navale AM et al (2008) Comparative performance of *Azotobacter* biofertilizers on growth and yield of brinjal. Asian J Hortic 3:377–379

Bhattacharjee RB, Singh A, Mukhopadhyay SN (2008) Use of nitrogen-fixing bacteria as biofertilizer for non-legumes: prospects and challenges. Appl Microbiol Biotechnol 80:199–209

Bhattacharjee RB, Philippe J, Clémence C, Bernard D, Aqbal S, Satya NM (2012) Indole acetic acid and ACC deaminase-producing *Rhizobium leguminosarum* bv. trifolii SN10 promote rice growth, and in the process undergo colonization and chemotaxis. Biol Fertil Soils 48:173–182

Bhosale HJ, Kadam TA, Bobade AR (2013) Identification and production of *Azotobacter vinelandii* and its antifungal activity against *Fusarium oxysporum*. J Environ Biol 34:177–182

Bobadi S, Van Damme P (2003) Effect of nitrogen application on flowering and yield of eggplant (*Solanum melongena* L.) Commun Agric Appl Biol Sci 68:5–13

Brahma S, Phookan DB (2006) Effect of nitrogen, phosphorus and potassium on yield and economics of broccoli [*Brassica oleracea* (L.) var. italica] cv. Pusa. Res Crops 7:261–262

Brewster JL (1994) Onions and other vegetable Alliums. CAB International, Wallingford, UK, p 236

Brown BD, Hornbacher AJ, Naylor DV (1988) Sulfur coated urea as a slow-release nitrogen source for onions. J Am Soc Hortic Sci 113:864–869

Chabot R, Antoun H, Cescas MP (1996a) Growth promotion of maize and lettuce by phosphate-solubilizing *Rhizobium leguminosarum* biovar phaseoli. Plant Soil 184:311–321

Chabot R, Antoun H, Kloepper JW et al (1996b) Root colonization of maize and lettuce by bioluminescent *Rhizobium leguminosarum* biovar phaseoli. Appl Environ Microbiol 62:2767–2772

Chakravarty G, Kalita MC (2012) Biocontrol potential of *Pseudomonas fluorescens* against bacterial wilt of brinjal and its possible plant growth promoting effects. Ann Biol Res 3:5083–5094

Chattoo MA, Gandroo MY, Zargar MY (1997) Effect of *Azospirillum* and *Azotobacter* on growth, yield and quality of knol khol (*Brassica oleracea* L. Var. Gongylodes). Veg Sci 24:16–19

Chen WM, de Faria SM, James EK et al (2007) *Burkholderia nodosa* sp. nov., isolated from root nodules of the woody Brazilian legumes *Mimosa bimucronata* and *Mimosa scabrella*. Int J Syst Evol Microbiol 57:1055–1059

Chi F, Shen SH, Cheng HP et al (2005) Ascending migration of endophytic rhizobia from roots to leaves, inside rice plants and assessment of benefits to rice growth physiology. Appl Environ Microbiol 71:7271–7278

Choudhary B (1976) Vegetables, 4th edn. National Book Trust, New Delhi, pp 50–58

Colo J, Hajnal-Jafari TI, Durić S et al (2014) Plant growth promotion rhizobacteria in onion production. Pol J Microbiol 63:83–88

Datta B, Chakrabartty PK (2014) Siderophore biosynthesis genes of *Rhizobium* sp. isolated from *Cicer arietinum* L. 3 Biotech 4:391–401

Dawar S, Hayat S, Anis M et al (2008) Effect of seed coating material in the efficacy of microbial antagonists for the control of root rot fungi on okra and sunflower. Pak J Bot 40(3):1269–1278

Dazzo FB, Yanni YG (2006) The natural rhizobium-cereal crop association as an example of plant-bacterial interaction. In: Uphoff N, Ball AS, Fernandes E, Herren H, Husson O, Laing M, Palm C, Pretty J, Sanchez P, Sanginga N, Thies J (eds) Biological approaches to sustainable soil systems. CRC Press, Boca Raton, pp 109–127

Decoteau DR (2000) Vegetable crops. Upper Rever Company, New Jersey, USA

Diaz-Perez JC, Purvis AC, Paulk JT (2003) Bolting, yield, and bulb decay of sweet onion as affected by nitrogen fertilization. J Am Soc Hortic Sci 128:144–149

Dixit SK, Sanjay K, Meena ML et al (2007) Bio-fertilizers may be a substitute of chemical fertilizers. Asian J Agric Sci 2(2):298–300

Doifode VD, Nandkar PB (2014) Influence of biofertilizers on the growth, yield and quality of brinjal crop. Int J Life Sci A2:17–20

Dorais M, Ehret DL, Papadopoulos AP (2008) Tomato (*Solanum lycopersicum*) health components: from the seed to the consumer. Phytochem Rev 7:231–250

Drost D, Koenig R, Tindall T (2002) Nitrogen use efficiency and onion yield increased with a polymer-coated nitrogen source. Hortic Sci 37:338–342

Ehteshamul-Haque S, Ghaffar A (2008) Use of rhizobia in the control of root rot diseases of sunflower, okra, soybean and mungbean. J Phytopathol 138:157–163

El-Assiouty FMM, Abo-Sedera SA (2005) Effect of bio and chemical fertilizers on seed production and quality of spinach (*Spinacia oleracea* L.) Int J Agric Biol 7:947–949

Elazzazy AM, Almaghrabi OA, Moussa TAA et al (2012) Evaluation of some plant growth promoting rhizobacteria (pgpr) to control *Pythium aphanidermatum* in cucumber plants. Life Sci J 9:3147–3153

El-Batanomy N (2009) Synergistic effect of plant-growth promoting rhizobacteria and arbuscular mycorrhiza fungi on onion (*Allium cepa*) growth and its bulbs quality after storage. New Egypt J Microbiol 23:163–182

Elmerich C, Newton WE (2007) Associative and endophytic nitrogen-fixing bacteria and cyano-bacterial associations. Springer, The Netherlands, p 323

El-Sirafy MZ, Sarhan SH, Abd-El Hafez AA, Baddour AGA (2010) Effect of the interaction between *Azotobacter* inoculation; organic and mineral fertilization on tomato (*Lycopersicon esculentum* Mill.) J Soil Sci Agric Eng 1:93–104

Fageria MS, Arya PS, Jagmohan K et al (1992) Effects of nitrogen levels on growth, yield and quality of tomato. Veg Sci 19:25–29

Flores FB, Sanchez-Bel P, Estan MT, Martinez-Rodriguez MM, Moyano E, Morales B, Campos JF, Garcia-Abellán JO, Egea MI, Fernández-Garcia N, Romojaro F, Bolarín M (2010) The effectiveness of grafting to improve tomato fruit quality. Sci Hortic 125:211–217

Flores-Félix JD, Menéndez E, Rivera LP et al (2013) Use of *Rhizobium leguminosarum* as a potential biofertilizer for *Lactuca sativa* and *Daucus carota* crops. J Plant Nutr Soil Sci 176:876–882

Food and Agricultural Organization (FAO) (2010) Crop water information: tomato. www.fao.org/nr/water/cropinfo_tomato.html

Forlin MB, Pastorelli R, Sarvilli S (2008) Root potentially related properties in plant associated bacteria. J Gen Breed Italy 49:343–352

Frank B (1889) Uber die Pilzsymbiose der leguminosen. Berichte der Deutschen Beru Deut Bot Ges 7:332–346

Galleguillos C, Aguirre C, Barea JM, Azcon R (2000) Growth promoting effect of two *Sinorhizobium meliloti* strains (a wild type and its genetically modified derivative) on a non-legume plant species in specific interaction with two arbuscular mycorrhizal fungi. Plant Sci 159:57–63

Gamiely S, Randle WM, Mills HA et al (1991) Onion plant growth, bulb quality, and water uptake following ammonium and nitrate nutrition. Hortic Sci 26:1061–1063

Gandhi Pragash M, Narayanan KB, Naik PR et al (2009) Characterization of *Chryseobacterium aquaticum* strain PUPC1 producing a novel antifungal protease from rice rhizosphere soil. J Microbiol Biotechnol 19:99–107

García-Fraile P, Carro L, Robledo M et al (2012) *Rhizobium* promotes non-legumes growth and quality in several production steps: towards a biofertilization of edible raw vegetables healthy for humans. PLoS One 7:38122

Ghanti S, Sharangi AB (2009) Effect of bio-fertilizers on growth, yield and quality of onion cv. Sukhsagar. J Crop Weed 5:120–123

Ghosh PK, De TK, Maiti TK (2015) Production and metabolism of indole acetic acid in root nodules and symbiont (*Rhizobium undicola*) isolated from root nodule of aquatic medicinal legume *Neptunia oleracea* Lour. J Bot 2015:11. Article ID: 575067

Gill HS (1993) Improvement of root crops. In: Chadha KL, Kalloo G (eds) Advances in horticulture, vol 5. Malhotra Book Depot, New Delhi, pp 201–206

Graham PH (2008) Ecology of the root-nodule bacteria of legumes. In: Dilworth MJ, James EK, Spent JI, Newton WE (eds) Nitrogen-fixing leguminous symbioses. Springer, Dordrecht, The Netherlands, pp 23–58

Guertal EA (2009) Slow-release nitrogen fertilizers in vegetable production: a review. Hortic Technol 19:16–19

Guo Z, He C, Ma Y et al (2011) Effect of different fertilization on spring cabbage (*Brassica oleracea* L. var. capitata) production and fertilizer use efficiencies. Agric Sci 2:208–212

Halder AK, Chakrabarty PK (1991) Solubilization of inorganic phosphates by *Rhizobium*. Folia Microbiol 38:325–330

Hooda RS, Pandita ML, Sidhu AS (1980) Studies on the effect of nitrogen and phosphorus on growth and green pod yield of okra (*Abelmoschus esculentus* (L.) Moench). Haryana J Hortic Sci 9:180–183

Hussain AM, Arshad SM, Javed M (1993) Potential of *Azotobacter* for promoting potato growth and yield under optimum fertilizer application. Pak J Agric Sci 30:217–220

Hussein KA, Joo JH (2011) Effects of several effective microorganisms (EM) on the growth of Chinese cabbage (*Brassica rapa*). Korean J Soil Sci Fert 44:565–574

Ibiene AA, Agogbua JU, Okonko IO, Nwachi GN (2012) Plant growth promoting rhizobacteria (PGPR) as biofertilizer: effect on growth of *Lycopersicon esculentus*. J Am Sci 8:318–324

Igual JM, Valverde A, Cervantes E, Velázquez E (2001) Phosphate-solubilizing bacteria as inoculants for agriculture: use of updated molecular techniques in their study. Agronomie 21:561–568

Imen H, Neila A, Adnane B et al (2015) Inoculation with phosphate solubilizing *Mesorhizobium* strains improves the performance of chickpea (*Cicer arietinum* L.) under phosphorus deficiency. J Plant Nutr 38:1656–1671

Iqbal MA, Khalid M, Shahzad SM, Ahmad M, Soleman N, Akhtar N (2012) Integrated use of *Rhizobium leguminosarum*, plant growth promoting rhizobacteria and enriched compost for improving growth, nodulation and yield of lentil (*Lens culinaris* Medik.) Chilean J Agric Res 72:104–110

Ishfaq AP, Vijai K, Faheem MM (2009) Effect of bio-organic fertilizers on the performance of cabbage under western U.P. conditions. Ann Hortic 2:204–206

Islam MR, Sultana T, Joe MM et al (2013) Nitrogen-fixing bacteria with multiple plant growth-promoting activities enhance growth of tomato and red pepper. J Basic Microbiol 53:1004–1015

Jeevajohti L, Mani AK, Pappiah CM et al (1993) Influence of N, P, K and *Azospirillum* on the yield of cabbage. South Indian Hortic 41:270–272

Kantharajah AS, Golegaonkar PG (2004) Somatic embryogenesis in eggplant review. J Sci Hortic 99:107–117

Karpagam T, Nagalakshmi PK (2014) Isolation and characterization of phosphate solubilizing microbes from agricultural soil. Int J Curr Microbiol Appl Sci 3:601–614

Keel C, Schnider U, Maurhofer M et al (1992) Suppression of root diseases by *Pseudomonas fluorescens* CHAO: importance of bacterial secondary metabolite, 2,4-diacetylphoroglucinol. Mol Plant-Microbe Interact 5:4–13

Kumar PR, Ram MR (2012) Production of indole acetic acid by *Rhizobium* isolates from *Vigna trilobata* (L) Verdc. Afr J Microbiol Res 6:5536–5541

Kumar V, Jaiswal RC, Singh AP (2001) Effect of biofertilizer on growth and yield of potato. J Indian Potato Assoc 28:60–61

Kumar A, Kumar K, Kumar P et al (2014) Production of indole acetic acid by *Azotobacter* strains associated with mungbean. Plant Archives 14:41–42

Lamo K (2009) Effect of organic and bio-fertilizers on seed production of radish (*Raphanus sativus* L) cv Chinese Pink. M.Sc. thesis, Dr YS Parmar University of Horticulture and Forestry, Nauni, Solan, India

Latha P, Jeyaraman S, Prabakaran R (2014) Effect of microbial and chemical fertilizer on egg plant (*Solanum melongena* Linn.) C.Var CO-2. Int J Pure Appl Biosci 2:119–124

Lesueur D, Carr Del Rio M, Dien HG (1995) Modification of the growth and the competitiveness of *Bradyrhizobium* strain obtained through affecting its siderophore-producing ability. In: Abadia J (ed) Iron nutrition in soil and plants. Kluwer Academic Publishers, Pays Bas, pp 59–66

Lindstrom K, Martinez-Romero ME (2005) International committee on systematics of prokaryotes; subcommittee on the taxonomy of *Agrobacterium* and *Rhizobium*: minutes of the meeting, 26 July 2004, Toulouse, France. Int J Syst Evol Microbiol 55:1383

Loganathan M, Garg R, Venkataravanappa V et al (2014) Plant growth promoting rhizobacteria (PGPR) induces resistance against Fusarium wilt and improves lycopene content and texture in tomato. Afr J Microbiol Res 8:1105–1111

Lopandic D, Zaric D (1997) Effect of nitrogen rates and application dates on cabbage yield. Acta Hortic 462:595–598

Lugtenberg B, Kamilova F (2009) Plant-growth-promoting rhizobacteria. Annu Rev Microbiol 63:541–556

Ma W, Charles TC, Glick BR (2004) Expression of an exogenous 1-aminocyclopropane-1-carboxylate deaminase gene in *Sinorhizobium meliloti* increases its ability to nodulate alfalfa. Appl Environ Microbiol 70:5891–5897

Machado RG, Sá ELS, Bruxel M, Giongo A, Santos NS, Nunes AS (2013) Indoleacetic acid producing rhizobia promote growth of Tanzania grass (*Panicum maximum*) and Pensacola grass (*Paspalum saurae*). Int J Agric Biol 15:827–834

Mahajan G, Singh KG (2006) Response of greenhouse tomato to irrigation and fertigation. Agric Water Manage 84:202–206

Marschner H (1995) Mineral nutrition of higher plants. Academic Press, London

Martinez R, Dibut B, Ganzalez R (1993) Stimulation of tomato development and yield by inoculation of red ferralitic soils with *Azotobacter chroocucum*. Memorias - 11th - Congress - Latinoamerican - de – da - suelo 5:1396–1398

Meshram SU (1984) Suppressive effect of *Azotobacter chroococcum* on *Rhizoctonia solani* infestation of potatoes. Neth J Plant Pathol 90:127–132

Mirshekari B, Alipour MH (2013) Potato (*Solanum tuberosum*) seed bio-priming influences tuber yield in new released cultivars. Int J Biosci 3:26–31

Mohammadi GR, Ajirloo AR, Ghobadi ME et al (2013) Effects of non-chemical and chemical fertilizers on potato (*Solanum tuberosum* L.) yield and quality. J Med Plant Res 7:36–42

Muthuselvan I, Balagurunathan R (2013) Siderophore production from *Azotobacter* sp. and its application as biocontrol agent. Int J Curr Res Rev 5(11):23–35

Nabti E, Bensidhoum L, Tabli N et al (2014) Growth stimulation of barley and biocontrol effect on plant pathogenic fungi by a *Cellulosimicrobium* sp. strain isolated from salt-affected rhizosphere soil in northwestern Algeria. Eur J Soil Biol 61:20–26

Nanthakumar S, Veeragavathatham D (2000) Effect of integrated nutrient management on growth parameters and yield of brinjal (*Solanum melongena* L.) cv. PLR-1. South Indian Hortic 48:31–35

Naqqash T, Hameed S, Imran A et al (2016) Differential response of potato toward inoculation with taxonomically diverse plant growth promoting rhizobacteria. Front Plant Sci 7:144

Narula N, Gupta KG (1986) Ammonia excretion by *Azotobacter chroococcum* in liquid culture and soil in the presence of manganese and clay minerals. Plant Soil 93:205–209

Noel TC, Sheng C, Yost CK et al (1996) *Rhizobium leguminosarum* as plant growth-promoting rhizobacterium: direct growth promotion of canola and lettuce. Can J Microbiol 42:279–283

Nosrati R, Owlia P, Saderi H et al (2014) Phosphate solubilization characteristics of efficient nitrogen fixing soil *Azotobacter* strains. Iran J Microbiol 6:285–295

Okon Y, Itzisohn R (1995) The development of *Azospirillum* as a commercial inoculant for improving crop yields. Biotechnol Adv 13:414–424

Olowoake AA, Adeoye GO (2010) Comparative efficacy of NPK fertilizer and composted organic residues on growth, nutrient absorption and dry matter accumulation in maize. Int J Org Agric Res Dev 2:43–53

Orhan E, Esitken A, Ercisli S et al (2006) Effects of plant growth promoting rhizobacteria (PGPR) on yield, growth and nutrient contents in organically growing raspberry. Sci Hortic 111:38–43

Osman AS (2007) Effect of partial substitution of mineral-N by bio-fertilization on growth, yield and yield components of potato. In: The third conference of sustainable agriculture development, Faculty of Agriculture, Fayoum University, 12–14 Nov, pp 381–396

Palm CA, Gachengo CN, Delve RJ et al (2001) Organic inputs for soil fertility management in tropical agro ecosystems: application of an organic resource database. Agric Ecosyst Environ 83:27–42

Panhwar QA, Naher UA, Jusop S, Othman R, Latif MA, Ismail MR (2014) Biochemical and molecular characterization of potential phosphate solubilizing bacteria in acid sulphate soils and their beneficial effects on rice growth. PLoS One 9(10):PMC4186749

Parveen G, Ehteshamul-Haque S, Sultana V et al (2008) Suppression of root pathogens of tomato by rhizobia, *Pseudomonas aeruginosa*, and mineral fertilizers. Int J Veg Sci 14:205–215

Pena-Cabriales JJ, Alexander M (1983) Growth of *Rhizobium* in unamended soil. Soil Sci Soc Am J 47:81–84

Prasad JS, Reddy RS, Reddy PN et al (2014) Isolation, screening and characterization of *Azotobacter* from rhizospheric soils for different plant growth promotion (pgp) & antagonistic activities and compatibility with agrochemicals: an *in vitro* study. Ecol Environ Conserv 20:959–966

Prayitno J, Stefaniak J, Mciver J et al (1999) Interactions of rice seedlings with bacteria isolated from rice roots. Aust J Plant Physiol 26:521–535

Ramakrishnan K, Selvakumar G (2012) Effect of biofertilizers on enhancement of growth and yield on tomato (*Lycopersicum esculentum* Mill.) Int J Res Bot 2:20–23

Rangkadilok N, Nicolas ME, Bennett RN et al (2002) Determination of sinigrin and glucoraphanin in *Brassica* species using a simple extraction method combined with ion-pair HPLC analysis. Sci Hortic 96:27–41

Rao DLN (1986) Nitrogen fixation in free living and associative symbiotic bacteria. In: Subba Rao NS (ed) Soil microorganisms and plant growth. Oxford and IBH Pub. Co., New Delhi

Reddy PM, Ladha JK, So RB et al (1997) Rhizobial communication with rice roots: induction of phenotypic changes, mode of invasion and extent of colonization. Plant Soil 194:81–98

Reimann S, Hauschild R, Hildebrandt U, Sikora RA (2008) Interrelationships between *Rhizobium etli* G12 and *Glomus intraradices* and multitrophic effects in the biological control of the root-knot nematode *Meloidogyne incognita* on tomato. J Plant Dis Protect 115:108–113

Reyes VG, Schimidt EL (1979) Population densities of *Rhizobium japonicum* strain 123 estimated directly in soil and rhizosphere. Appl Environ Microbiol 37:854–858

Rozek S, Wojciechowska R (2005) Effect of urea foliar application and different levels of nitrogen on broccoli head yield and its quality in autumn growing cycle. Sodininkyste-ir Darzininkyste 24:291–301

Ruangsanka S (2014) Identification of phosphate-solubilizing fungi from the asparagus rhizosphere as antagonists of the root and crown rot pathogen *Fusarium oxysporum*. ScienceAsia 40:16–20

Sahar P, Sahi ST, Jabbar A, Rehman A, Riaz K, Hannan A (2013) Chemical and biological management of *Fusarium oxysporum* f.sp. melongenae. Pak J Phytopathol 25:155–159

Sahasrabudhe MM (2011) Screening of rhizobia for indole acetic acid production. Ann Biol Res 2:460–468

Sahgal M, Johri BN (2003) The changing face of rhizobial systematics. Curr Sci 84:43–48

Sarhan TZ (2008) Effect of biological fertilizers, animal residues and urea on growth and yield of potato plant cv. desiree *Solanum tuberosum* L. Ph.D. thesis, Horticulture Science and Landscape Design (Vegetable), University of Mosul, College of Agriculture and Forestry, Iraq

Sarhan TZ (2012) Effect of biofertilizer and different levels of nitrogen (urea) on growth, yield and quality of lettuce (*Lactuca sativa* L.) Ramadi cv. J Agric Sci Technol B2:137–141

Sarkar A, Mandal AR, Prasad PH et al (2010) Influence of nitrogen and biofertilizer on growth and yield of cabbage. J Crop Weed 6:72–73

Shafeek MR, Fatin S, El-Al A et al (2004) The productivity of broad been plant as affected by chemical and/or natural phosphorus with different biofertilizer. J Agric Sci 29:2727–2740

Sharafzadeh S (2012) Effects of PGPR on growth and nutrients uptake of tomato. Int J Adv Eng Technol 2:27–31

Sharma SK (2002) Effect of *Azospirillum*, *Azotobacter* and nitrogen on growth and yield of cabbage (*Brassica oleracea* var capitata). Indian J Agric Sci 72:555–557

Sharma P, Saikia MK (2013) Management of late blight of potato through chemicals. IOSR J Agric Vet Sci 2:23–36

Sharma D, Singh RK, Parmar AS (2013) Effect of doses of biofertilizers on the growth and production of cabbage (*Brassica oleracea* L. Var. Capitata). TECHNOFAME-J Multidiscipl Adv Res 2:30–33

Shimaa D, Marwa AS, Doaa K et al (2015) Production of hydroxamate siderophores by *Azotobacter chroococcum* bacterium. Menofia J Agric Res 40:409

Shimshick EJ, Hebert RR (1979) Binding characteristics of N2-fixing bacteria to cereal roots. Appl Environ Microbiol 38:447–453

Shukla V, Nalk LB (1993) Agro-technique for malvaceae vegetables. In: Shukla V, Nalk LB (eds) Ifovance in horticulture, vol 5. Malhotra Publishing House, New Delhi, India, pp 399–425

Shukla YR, Mehta S, Sharma R (2012) Effect of integrated nutrient management on seed yield and quality of radish (*Raphanus sativus* L) cv Chinese Pink. Int J Farm Sci 2:47–53

Sidorenko O, Storozhenko V, Kokharen Kova O (1996) The use of bacterial preparation in potato cultivation. Mezhdunarodngi, Sel: Skokhozyaistvenny Zhurnal 6:36–38

Sihachkr D, Chaput MH, Serraf L et al (1993) Regeneration of plants from protoplasts of eggplant (*Solanum melongena* L.) In: Bajaj YPS (ed) Biotechnology in agriculture and forestry, plant protoplasts and genetic engineering. Springer, Berlin, pp 108–122

Silva LR, Azevedo J, Pereira MJ et al (2014) Inoculation of the nonlegume *Capsicum annum* (L.) with *Rhizobium* strains: effect on bioactive compounds, antioxidant activity, and fruit ripeness. J Agric Food Chem 62:557–564

Singh AB, Singh SS (1992) Effect of various levels of nitrogen and spacing on growth, yield and quality of tomato. Veg Sci 19:1–6

Singh NP, Sachan RS, Pandey PC (1999) Effect of a decade long fertilizer and manure application on soil fertility and productivity of rice-wheat system in Molisols. J Indian Soc Soil Sci 47:72–80

Singh JK, Bahadur A, Singh NK, Singh TB (2010) Effect of using varying level of NPK and biofertilizers on vegetative growth and yield of okra (*Abelmoschus esculentus* (L.) Moench). Veg Sci 37:100–101

Singh S, Gupta G, Khare E et al (2014a) Phosphate solubilizing rhizobia promote the growth of chickpea under buffering conditions. Int J Pure Appl Biosci 2:97–106

Singh BK, Singh S, Singh BK (2014b) Some important plant pathogenic disease of brinjal (*Solanum melongena* L.) and their management. Plant Pathol J 13:208–213

Solanki MP, Patel BN, Tandel YN (2010) Growth, yield and quality of brinjal as affected by use of bio-fertilizers. Asian J Hortic 5:403–406

Spaepen S, Das F, Luyten E et al (2009) Indole-3-acetic acid-regulated genes in *Rhizobium etli* CNPAF512: research letter. FEMS Microbiol Lett 291:195–200

Tahir M, Mirza MS, Zaheer A, Dimitrov MR, Smidt H, Hameed S (2013) Isolation and identification of phosphate solubilizer *Azospirillum*, *Bacillus* and *Enterobacter* strains by 16SrRNA sequence analysis and their effect on growth of wheat (*Triticum aestivum* L.) Aust J Crop Sci 7:1284–1292

Talalay P, Fahey JW (2001) Phytochemicals from cruciferous plants protect against cancer by modulating carcinogen metabolism. Am Soc Nutr Sci 23:3027–3033

Tariq M, Dawar S, Mehdi FS (2007) Antagonistic activity of bacterial inoculum multiplied on *Rhizophora mucronata* Lamk., in the control of root infecting fungi on mash bean and okra. Pak J Bot 39:2159–2165

Terouchi N, Syono K (1990) *Rhizobium* attachment and curing in asparagus, rice and oat plants. Plant Cell Physiol 31:119–127

Tindall HD (1983) Vegetables in the tropics, 1st edn. Macmillan Press, London, UK, p 533

Tortora ML, Díaz-Ricci JC, Pedraza RO (2011) *Azospirillum brasilense* siderophores with antifungal activity against *Colletotrichum acutatum*. Arch Microbiol 193:275–286

Trabelsi D, Ammar HB, Mengoni A, Mhamdi R (2012) Appraisal of the crop-rotation effect of rhizobial inoculation on potato cropping systems in relation to soil bacterial communities. Soil Biol Biochem 54:1–6

Verma TS, Thakur PC, Ajeet S (1997) Effect of biofertilizers on vegetable and seed yield of cabbage. Veg Sci 24:1–3

Verma SK, Asati BS, Tamrakar SK et al (2011) Effect of organic components on growth, yield and economic returns in potato. Potato J 38:51–55

Vikhe PS (2014) *Azotobacter* species as a natural plant hormone synthesizer. Res J Recent Sci 3:59–63

Wiehe W, Höflich G (1995) Survival of plant growth promoting rhizosphere bacteria in the rhizosphere of different crops and migration to non-inoculated plants under field conditions in north-East Germany. Microbiol Res 150:201–206

Wojciechowska R, Rozek S, Rydz A (2005) Broccoli yield and its quality in spring growing cycle as dependent on nitrogen fertilization. Folia Hortic 17:141–152

Woldetsadik K, Gertsson U, Ascard J (2003) Response of shallots to mulching and nitrogen fertilization. Hortic Sci 38:217–221

Work P, Carew J (1955) Vegetable production and marketing, 2nd edn. John Wiley and Sons Inc., New York, p 537

Yadav BD, Khandelwal RB, Sharma YK (2004) Use of bio-fertilizer (*Azospirillum*) in onion. Haryana J Hortic Sci 33:281–283

Yaso IA, Abdel-Razzak HS, Wahb-Allah MA (2007) Influence of biofertilizer and mineral nitrogen on onion growth, yield and quality under calcareous soil conditions. J Agric Environ Sci 6:245–264

Yeptho KA, Singh AK, Kanaujia SP et al (2012) Quality production of Kharif onion (*Allium cepa*) in response to biofertilizers inoculated organic manures. Indian J Agric Sci 82:236–240

Yildirim E, Karlidag H, Turan M et al (2011) Growth, nutrient uptake and yield promotion of broccoli by plant growth promoting rhizobacteria with manure. Hortic Sci 46:932–936

Yu X, Liu X, Zhu T, Liu G, Mao C (2012) Co-inoculation with phosphate-solubilizing and nitrogen-fixing bacteria on solubilization of rock phosphate and their effect on growth promotion and nutrient uptake by walnut. Eur J Soil Biol 50:112–117

Zahir ZA, Arshad M, Azam M et al (1997) Effect of an auxin precursor tryptophan and *Azotobacter* inoculation on yield and chemical composition of potato under fertilized conditions. J Plant Nutr 20:745–752

Zaidi A, Ahmad E, Khan MS, Saif S, Rizvi A (2015) Role of plant growth promoting rhizobacteria in sustainable production of vegetables: current perspective. Sci Hortic 193:231–239

Zakharova E, Shcherbakov A, Brudnik V, Skripko N, Bulkhin N, Ignatov V (1999) Biosynthesis of indole-3-acetic acid in *Azospirillum brasilense*. Insights from quantum chemistry. Eur J Biochem 259:572–576

Zenia M, Halina B (2008) Content of microelements in eggplant fruits depending on nitrogen fertilization and plant training method. J Elementol 13:269–274

Ziaf K, Latif U, Amjad M et al (2016) Combined use of microbial and synthetic amendments can improve radish (*Raphanus sativus*) yield. J Environ Agric Sci 6:10–15

Role of Plant Growth-Promoting Rhizobacteria (PGPR) in the Improvement of Vegetable Crop Production Under Stress Conditions

4

Srividya Shivakumar and Sasirekha Bhaktavatchalu

Abstract

Biotic and abiotic stresses are major constrains to agricultural production. Among abiotic stress, drought and salinity are the major environmental factors limiting growth and productivity of many crops including vegetables, particularly in arid and semiarid areas of the world. Abiotic stress causes more than 50% average yield loss worldwide. Globally, demand for vegetables is increasing, and this has boosted the vegetable production in recent times. The substantial increase in production of key vegetables such as tomato, onion, cucumber, eggplant, cauliflower, pepper, lettuce, carrot, and spinach has been recorded. However, vegetables are generally considered more vulnerable than staple crops to stressful environmental conditions including extremes of temperature, drought, salinity, water logging, mineral nutrient excess and deficiency, and changes in soil pH which are likely to be exacerbated by the prevalent climatic change in many parts of the world. Plant growth under stress conditions on the contrary may be enhanced by the application of microbial inoculation including plant growth-promoting rhizobacteria (PGPR). These microbes promote plant growth by regulating nutritional and hormonal balance, producing plant growth regulators, solubilizing nutrients, and inducing resistance against plant pathogens. In addition to their interactions with plants, these microbes exhibit synergistic as well as antagonistic interactions with other soil microbiota. These interactions are vital to maintain soil fertility and concurrently the growth and development

S. Shivakumar (✉)
Department of Microbiology, Centre for Post Graduate Studies, Jain University,
Bangalore, Karnataka 560011, India
e-mail: sk2410@yahoo.co.uk

S. Bhaktavatchalu
Department of Microbiology, Acharya Bangalore B School,
Off Magadi Road, Bangalore, Karnataka 560091, India

© Springer International Publishing AG 2017
A. Zaidi, M.S. Khan (eds.), *Microbial Strategies for Vegetable Production*,
DOI 10.1007/978-3-319-54401-4_4

of vegetables under stress conditions. The present literature comprehensively discusses recent developments on the effectiveness of PGPR in enhancing vegetable growth under stressful environments.

4.1   Introduction

Population of the world is predicted to increase beyond 8 billion by 2030 which is likely to pose major challenges for agricultural sector to secure food availability (Smol 2012). In the developing countries, abiotic stresses such as soil salinization, soil sodification, drought, soil pH, and environmental temperature are major limiting factors in crop production. Of these, soil salinization and drought are the two major factors endangering the potential use of soils and leading to soil degradation and soil desertification (Ladeiro 2012). The Global Assessment of Soil Degradation (GLASOD) estimated that about 13% (or 850 million ha) of the land in Asia and the Pacific is degraded due to soil salinization, soil sodification, and drought (Ladeiro 2012). Abiotic stress is the primary cause of worldwide crop loss, leading to more than 50% crop yield reduction (Shahbaz and Ashraf 2013). Plants as sessile organisms are constantly exposed to changes in environmental conditions; when these changes are rapid and extreme, plants generally perceive them as "stress" (Carillo et al. 2011). Drought and salinity are the two most devastating environmental stress, which is increasing day by day and reducing the agricultural productivity in large areas of the world (Hasanuzzaman et al. 2013). According to the United States Department of Agriculture (USDA), onions are highly sensitive to saline soils, while cucumbers, eggplants, peppers, and tomatoes are sensitive to salinity (Nandakumar et al. 2012). The majority of horticultural and cereal crops cultivated are susceptible to excessive concentrations of dissolved ions (30 mM or 3.0 dS/m) in the rhizosphere (Ondrasek et al. 2010). Horneck et al. (2007) have reported 50% yield reduction in potato, corn, onion, and bean when soil EC is increased to 5 dS/m. Sibomana et al. (2013) reported 69% tomato yield reduction due to water stress. Also, many studies have shown that salinity reduces microbial activity and microbial biomass and changes microbial community structure (Andronov et al. 2012). On the other hand, soil water content controls microbial activity and is a major factor that determines the rates of mineralization (Paul et al. 2003).

To minimize crop loss, scientists have attempted to develop salt-tolerant crop through breeding (Araus et al. 2008; Witcombe et al. 2008). However, gaps in understanding the complex physiological, biochemical, developmental, and genetic basis of environmental stress tolerance, and the subsequent difficulty in combining favorable alleles to create improved high yielding genotypes, are the major constraint to improve crop yield under abiotic stress (Dwivedi et al. 2010). Apart from the development of some salt-tolerant plant species, a wide range of salt-tolerant plant growth-promoting rhizobacteria such as *Rhizobium*, *Azospirillum*, *Pseudomonas*, *Flavobacterium*, *Arthrobacter*, and *Bacillus* have also shown beneficial interactions with plants in stressed environments (Egamberdieva 2011).

Here, an attempt is made to overview the effects of drought and salinity on crop plants especially vegetables and to identify/develop management strategies to overcome such effects on vegetables grown distinctly in different agroecological regions.

4.2 Stress Factors

4.2.1 Soil Salinization

Soil in which the electrical conductivity (EC) of the saturation extract (ECe) in the root zone exceeds 4 dS/m (approximately 40 mM NaCl) at 25 °C and has exchangeable sodium of 15% is referred to as salinity in soil (Jamil et al. 2011). Agricultural losses caused by salinity are difficult to assess, but these have been estimated to be substantial and expected to increase with time. It has been estimated that 20% of the total cultivated and 33% of the irrigated agricultural lands are affected by high salinity (Shrivastava and Kumar 2015). Annually salinized areas are increasing at a rate of 10% due to various reasons such as high surface evaporation, low precipitation, saline water irrigation, and poor cultural practices, and it has been estimated that >50% of the cultivatable land would be salinized by 2050 (Jamil et al. 2011). Soil salinity can also be a consequence of natural causes, such as (1) weathering of parent rocks and minerals in the soil, which releases various ions (e.g., Na, Ca, K and Mg, sulfates, and carbonates) to the soil solution (Moreira-Nordemann 1984); (2) seawater intrusion into coastal areas leading to increased salinity levels in the soil and channel water, which may be the major factor causing reduction in crop production (Kotera et al. 2008; Mahajan and Tuteja 2005); (3) rainwater containing 50 mg/l NaCl (Munns and Tester 2008) which can result in the precipitation of 250 kg NaCl per ha for every 500 mm of annual rainfall; and (4) wind-borne materials from lake or land surfaces. Nevertheless, the more significant proportion of saline soils is attributed to intensive agricultural cultivation (FAO 2008). The removal of natural perennial vegetation and its replacement with annual agricultural crops was perhaps the first factor in man-induced salinity (Manchanda and Garg 2008). Use of salt-rich irrigation water is undoubtedly one of the foremost factors responsible for soil salinity. In addition, improper irrigation management, which might be responsible for rise in the water table, known as secondary salinization, is an important contributor to soil salinity. Based on the salinity development, salt-affected soils can be classified into (1) primary salinity, which occurs naturally where the soil parent material is rich in soluble salts or geochemical processes resulting in salt-affected soil, and (2) secondary salinity, salinization of land and water resources due to human activities. Human activities like poor irrigation management, insufficient drainage, improper cropping patterns and rotations, and chemical contamination can also induce salinization.

In India, approximately 7 million hectares of land is covered by saline soil (Patel et al. 2011), most of which occurs in Indo-Gangetic Plain that covers the states of Punjab, Haryana, Uttar Pradesh, Bihar, and some parts of Rajasthan. Arid tracts of Gujarat and Rajasthan and semiarid tracts of Gujarat, Madhya Pradesh, Maharashtra,

Table 4.1 Degree of salinity in soil and plant response

Degree of salinity	Electrical conductivity (dS/m)	Level of effect	Use	Plant response	Salt-tolerant crops
Nonsaline	0–2	Salinity effects are negligible	Cropping	Very little effect on plant	Carrot Okra Radish
Low salinity	2–4	Salinity effects are minimal	Cropping	Yields of sensitive crops may be restricted	Celery Common beans Pea
Moderate salinity	4–8	Yield of the plant is restricted	Crop-pasture rotation	Some effect on salt-sensitive crops	Cabbage Tomato Potato Onion Peas Squash Cucumber Cauliflower Eggplant
High salinity	8–16	Only salt-tolerant plants yield satisfactory	Grazing or revegetation	Considerable effect on salt-sensitive crops	Artichoke Beetroot
Very high salinity	<16	Few salt-tolerant plants yield satisfactory	Very few plants will tolerate and grow	Some effect on salt-tolerant crops	Asparagus

Adapted from Abou-Baker and El-Dardiry (2015)

Karnataka, and Andhra Pradesh are also largely affected by saline lands (Shrivastava and Kumar 2015). When salt accumulates in soil, excessive sodium from salt destroys soil structure and hydraulic properties of soil, increases soil pH, and reduces infiltration of water and aeration in soil leading to soil compaction, soil erosion, and water runoff (Ondrasek et al. 2010). Further, sodium is the most pronounced destructor of secondary clay minerals by dispersion. Dispersed clay particles undergo leaching through the soil and may accumulate and block pores, especially in fine-textured soil horizons (Burrow et al. 2002). The soil becomes unsuitable for proper root growth and plant development (Table 4.1).

Saline soil inhibits plant growth, firstly by reducing the ability of the plant to take up water which in turn reduces the growth rate. This is referred to as "osmotic or water-deficit effect of salinity." Secondly, if excessive amounts of salt enter the plant in the "transpiration stream," there will be injury to cells in the transpiring leaves. This is called the "salt-specific or ion-excess effect of salinity" (Greenway and Munns 1980). Salinity is often caused by rising water tables, and it can be accompanied by water logging. Water logging itself inhibits plant growth and also reduces the ability of the roots to exclude salt, thus increasing the uptake rate of salt and its accumulation in shoots.

4.2.1.1 Effect of Salinity on Plant Growth

Most of the widely used crops in human or animal nutrition such as cereals (rice, maize), forages (clover), or horticultural crops (potatoes, tomatoes) require extensive irrigation practices, but are also susceptible to excessive concentration of salts either dissolved in irrigation water or already present naturally in soil (rhizosphere) (Ondrasek et al. 2010). When present in excess, salts cause osmotic and ionic stress such as toxicity of Na^+ in plants. These stresses result in complete or partial stomata closure, C assimilation reduction, reduced leaf area and chlorophyll content, accelerated defoliation (Shannon and Grieve 1999), nutritional imbalance (reduced intake N, Ca, K, P, Fe, Zn), alteration of metabolic processes, membrane disorganization, reduction of cell division and expansion, and genotoxicity (Carillo et al. 2011). Salt stress like other abiotic stress also leads to oxidative stress due to increased production of reactive oxygen species (ROS), such as singlet oxygen, superoxide anion, hydrogen peroxide, and hydroxyl radical. These ROS are highly reactive and can alter normal cellular metabolism with oxidative damage to carbohydrates, proteins, and nucleic acids and cause peroxidation of membrane lipids (Azevedo Neto et al. 2008). Soil salinity significantly reduces plant phosphorus (P) uptake because phosphate ions precipitate with Ca ions (Bano and Fatima 2009). Together, these effects reduce plant growth, development, and survival. Salinity adversely affects reproductive development by inhabiting microsporogenesis and stamen filament elongation, enhancing programmed cell death in some tissue types, ovule abortion, and senescence of fertilized embryos (Shrivastava and Kumar 2015). In the rhizosphere, excess sodium and more importantly chloride competitively interacts with other nutrient ions (K^+, NO_3^-, and $H_2PO_4^-$) for binding sites and transport proteins (Tester and Davenport 2003). Uptake and accumulation of Cl^- inhibit nitrate reductase activity, thereby disrupting photosynthetic function (Xu et al. 2000). Once the capacity of cells to store salts is exhausted, salts build up in the intercellular space leading to cell dehydration and death. Salinity has an adverse effect on cell cycle and differentiation. Salinity arrests the cell cycle by reducing the expression and activity of cyclins and cyclin-dependent kinases that result in fewer cells in the meristem, thus limiting growth (Javid et al. 2011).

In a study on vegetables, Bojovic et al. (2010) reported that seed germination of cabbage (*Brassica oleracea*), tomato (*Solanum lycopersicum*), and bell pepper (*Capsicum annuum*) was inhibited at higher concentrations (400–800 Mm) of NaCl. Similarly, Ramazani et al. (2009) and Asaadi (2009) reported decrease in seed germination of fenugreek (*Trigonella foenum-graecum*) due to salinity-induced disturbance of metabolic process leading to increase in phenolic compounds. On the contrary, salinity significant reduction in the leaf area, total root dry weight, photosynthesis, and stomatal conductance of sugar beet was reported by Dadkhah (2011), while Taffouo et al. (2010) confirmed the inhibitory effect of salinity on photosynthesis and photosynthetic pigments in cowpea. A similar study reported an inverse relationship between salt concentration and chlorophylls "a" and "b," and total chlorophyll content is reported for bean plant (Qados 2011). Kapoor and Srivastava (2010) demonstrated decrease in protein content in black gram plants treated with different salt concentrations. Takagi et al. (2009) reported decreased whole plant biomass with reduced leaf photosynthesis and transport of carbon assimilates as an effect of salinity (100 mM NaCl) in *S. lycopersicum*. Egamberdieva (2011) observed that increasing salt content reduced the shoot length (50%) and root length (7%) of bean seedling grown in a gnotobiotic sand system. Adolf et al. (2013) showed relatively low stomatal conductance (67%) in salt-treated *Titicaca* plant. Several studies have reported reduced nitrogen absorption and accumulation in plants under saline conditions (Silvera et al. 2001). In eggplant, accumulation of Cl^- in leaves was accompanied with decreased concentration of NO_3^- (Savvas and Lenz 2000).

4.2.1.2 Impact of Salinity on Microorganisms

Even though soil microorganisms constitute less than 0.5% (w/w) of the soil mass, they play a key role in maintaining soil fertility (Tate 2000). Microbial biomass is an important labile fraction of the soil organic matter which participates in oxidation, nitrification, ammonification, nitrogen fixation, and other processes which lead to decomposition of soil organic matter and hence to the transformation of nutrients. They can also store C and other nutrients in their biomass which are mineralized after cell death by surviving microbes (Anderson and Domsch 1980). Stress factors are detrimental for beneficial soil microorganisms and have been reported to adversely affect the activity of surviving cells (Chowdhury et al. 2011). Soluble salts in the soil increase the osmotic potential, drawing water out of microbial cells. Low osmotic potential also makes it more difficult for roots and microbes to remove water from the soil (Oren 1999). Soil microbes, however, can adapt to low osmotic potential by accumulating osmolytes. High bioenergetic taxation to maintain osmotic equilibrium between the cytoplasm and the surrounding medium, excluding sodium ions from inside the cell, leads to reduction in growth and activity of the surviving microbes (Chowdhury et al. 2011; Ibekwe et al. 2010). The presence of loose, flexible surface appendage surrounding the bacteria under low electrolyte concentration condition acts as a protective barrier, thereby attenuating the impact of changes in extracellular ionic strength and lowering the osmotic pressure constraint (Francius et al. 2011). With an increase in the salinity level above 5%, the total count of bacteria and *Actinobacteria* were drastically reduced in a study conducted by Wichern et al. (2006). Azam and Ifzal (2006) reported nitrogen immobilization (remineralization and nitrification) process retardation in the presence of NaCl. Soil salinity also inhibits the enzyme activities of benzoyl argininamide, alkaline phosphatase, β-glucosidase, amylase, invertase, catalase, phosphatase, urease, and also microbial respiration (Ghollaratta and Raiesi 2007). A study by Nelson and Mele (2007) concluded a significant decrease in diversity and species richness in rhizosphere microbial community structure indirectly through root exudates quantity and/or quality rather than directly through microbial toxicity as an effect of salinity. Soil salinity has also been reported to disturb the symbiotic interaction between legumes and rhizobia. Singleton and Bohlool (1984) and Rabie et al. (2005) reported decrease in nodulation and nitrogen fixation with reduced nitrogenase activity in legumes such as soybean, common bean, and faba bean.

4.3   Drought

A shortfall in precipitation coupled with high evapotranspiration demand leads to agricultural drought (Mishra and Cherkauer 2010). Drought severity, however, depends on many factors, namely, (1) occurrence and distribution of rainfall, (2) evaporative demands, and (3) moisture storing capacity of soils (Wery et al. 1994). Three main mechanisms which reduce crop yield by soil water deficit are (1) reduced canopy absorption of photosynthetically active radiation, (2) decreased radiation use efficiency, and (3) reduced harvest index (Earl and Davis 2003).

4.3.1   Influence of Drought Stress on Morphological Characteristics of Plants

The effects of drought range from morphological to molecular levels and are evident at all phenological stages of plant growth. Under water stress conditions, which are related to water depletion and/or high atmospheric vapor pressure deficit, photosynthesis decreases through several mechanisms including stomata closure, reduced mesophyll conductance to CO_2 and feedback regulation by end-product accumulation (Nikinmaa et al. 2013). In response to a water deficit stress, ion and water transport systems across membranes function to control turgor pressure changes in guard cells and stimulate stomatal closure (Osakabe et al. 2014). The physiological, biochemical, and molecular responses of plants to drought stress are presented in Table 4.2.

Drought stress impairs mitosis, cell elongation, and expansion resulting in reduced plant height, leaf area, and crop growth (Hussain et al. 2008). Moisture stress during early reproductive growth phase usually reduces yield by reducing the

Table 4.2 Physiological, biochemical, and molecular response of plants to drought stress

Drought stress		
Physiological responses	Biochemical responses	Molecular responses
Recognition of root signal	Transient decrease in photochemical efficiency	Stress-responsive gene expression
Loss of turgor and osmotic adjustment	Decreased efficiency of Rubisco	Increased expression in ABA biosynthetic genes
Decrease in stomatal conductance to CO_2	Accumulation of stress metabolites like MDHA, glutathione, proline, glycine betaine polyamines, and α-tocopherol	Expression of ABA responsive genes
Reduced internal CO_2 concentration	Increase in antioxidants	Synthesis of specific proteins like late embryogenesis abundant
Decline in net photosynthesis	Reduced ROS accumulation	Desiccation stress protein, dehydrins, etc.
Reduced growth rates		

Adapted from Reddy et al. (2004)

Table 4.3 Drought stress and its impact on vegetable crops

Vegetable crops	Critical period of watering	Water stress impact
Brinjal	Flowering and fruit development	Reduced seed viability and yield
Cauliflower, cabbage, and broccoli	Head formation and enlargement	Browning and buttoning in cauliflower
Onion	Bulb formation and enlargement	Splitting and doubling of bulb
Carrot, radish, and turnip	Root enlargement	Distorted, rough, and poor growth of roots
Tomato	Early flowering, fruit set, and enlargement	Flower shedding, lack of fertilization, reduced fruit size, fruit splitting, puffiness
Asparagus	Spear production	Reduce spear quality and increased fiber content
Leafy vegetables	Growth and development of the plant	Toughness of leaves, poor foliage growth
Vegetable pea	Flowering and pod filling	Reduction in root nodulation and plant growth, poor pod filling, poor seed viability
Sweet potato	Root enlargement	Reduced root enlargement with poor yield, growth crack
Sweet corn	Silking, tasseling, and ear development	Crop may tassel and shed pollen before silks on ears are ready for pollination; lack of pollination may result in missing rows of kernels, reduced yields, and poor seed viability or even eliminate ear production

Adapted from Bahadur et al. (2011) and Kumar et al. (2012)

number of fruits/seeds in vegetables, while during flowering and fruit-setting stage, drought stress reduces fruit quality, number of fruits, size of fruits, and finally yield loss (Chatterjee and Solankey 2015). Water stress, mostly at critical period of growth, may drastically reduce productivity and quality of vegetables (Table 4.3).

In a study conducted by Okcu et al. (2005), drought stress impaired the germination and early seedling growth of five pea cultivars tested, while stem length of potato and okra was significantly affected under water stress (Sankar et al. 2008). In a similar study, water stress decreased the growth, total plant dry weight, leaf water potential, leaf relative water content, and leaf pigment of *C. annuum*, whereas contents of malondialdehyde, proline, superoxide dismutase, and peroxidize activity were increased (Qiu-shi et al. 2009).

4.4 Alleviation of Drought and Salinity Stress by Plant Growth-Promoting Rhizobacteria

Abiotic factors and stress have always played a major role in reducing agricultural crop yield. However, to circumvent these effects, scientists have developed several strategies to produce stress-tolerant crops which involve plant breeding and plant genetic engineering, but little success has been achieved so far due to genetic and physiological complexity of the stress trait. On the other hand, better agricultural land management, use of fertilizers, safe and efficient pesticides and herbicides, farm mechanization, and transgenic crop usage (Glick 2014) are the solutions to increase the agricultural productivity, but they give only short-term benefits. For an effective and long-term solution to provide food for the world, sustainable and eco-friendly biological solutions have to be implemented. Promising measures include use of microbial inoculants which can ameliorate stress, promote growth, control diseases, and contribute to the development of sustainable agriculture (Berg et al. 2013). Plant growth-promoting rhizobacteria (PGPR) colonize the rhizosphere/endorhizosphere of plants and promote growth of the plants through various direct and indirect mechanisms (Ramadoss et al. 2013).

The use of PGPR as an alternative to alleviate abiotic plant stress is gaining importance (Dodd and Perez-Alfocea 2012) (Table 4.4). The ability of PGPR to induce stress tolerance is often attributed by various processes that involve physiological and biochemical changes. It includes modifications in phytohormonal content (Kaushal and Wani 2016), antioxidant defense (Jyothsna and Murthy 2016), osmolyte production (Diby and Harshad 2014), ACC deaminase activity (Yang et al. 2009), and biofilm formation (Vanderlinde et al. 2010; Yang et al. 2009). Rhizobacteria often induce modifications in phytohormone signaling, which mediates effects on meristem activity (Hayat et al. 2010). Saravanakumar et al. (2011) and Sandhya et al. (2010) have suggested the possible role of PGPR to alleviate the oxidative damage elicited by abiotic stress through the manipulation of antioxidant enzymes in different crops. The beneficial effects of PGPR to reduce adverse effects of salinity have been demonstrated in tomatoes (Kidoglu et al. 2008), bell peppers, cucumbers (Kidoglu et al. 2008), radish (Yildirim et al. 2008), barley (Cakmakci et al. 2007), tobacco, mustard (Asghar et al. 2002), and eggplant (Bochow et al. 2001). Several studies indicate that plants require microbial association for stress tolerance (Egamberdieva and Jabborova 2013). Rabie et al. (2005) reported increased N and P nutrition, increased nodulation, and nitrogenase activity in AM fungi and N-fixer *Azospirillum brasilense*-treated cowpea plant at different NaCl salinity levels. Application of *Pseudomonas chlororaphis* (TSAU13) to tomato and cucumber promoted growth and fruit yield in saline soil and also reduced the incidence of disease caused by *Fusarium solani* (Egamberdieva 2012). Basha and Vivekanandan (2000) isolated a salt-tolerant rhizobial strain from tannery sludge which successfully nodulated cowpea in saline soils (250 mM NaCl).

High salinity suppresses the phosphorus (P) uptake by plant roots and reduces the available P by sorption processes (Vivekanandan et al. 2015). PGPR strains having efficient P solubilizing ability even under high saline (60 g/l NaCl) conditions have been reported (Upadhyay et al. 2011). Gibberellins secreting *Pseudomonas putida* H-2-3 improved plant growth in soybean under drought conditions (Sang-Mo et al. 2014). The other mechanism by which PGPR facilitates the growth of plants is the secretion of 1-aminocyclopropane-1-carboxylate (ACC) deaminase. This enzyme decreases plant ethylene level and diminishes negative effects caused by stress condition (Glick 2014). The ACC deaminase activity of *Achromobacter piechaudii* was shown to confer drought tolerance in

Table 4.4 Role of PGPR in alleviating salinity and drought stress

PGPR strain	Plants	Stress factor	PGP activity	Reference
Enterobacter sp.	Okra	Salinity	ACC deaminase	Habib et al. (2016)
Phyllobacterium	Strawberries	Salinity and drought	Phosphate solubilization, siderophore production	Flores-Felix et al. (2015)
Burkholderia cepacia and *Promicromonospora* sp.	Cucumber	Salinity and drought	Increased gibberellic acid, salicylic acid	Sang-Mo et al. (2014)
Streptomyces sp. strain PGPA39	Tomato	Salinity	Increased ACC deaminase activity, IAA production, and phosphate solubilization	Palaniyandi et al. (2014)
Chryseobacterium	Tomato	Salinity and drought	Siderophore production	Radzki et al. (2013)
B. licheniformis K11	*Capsicum annuum*	Drought	Ethylene concentration reduction	Lim and Kim (2013)
Brevibacterium iodinum, *Bacillus licheniformis*, *Zhihengliuela alba*	Red pepper	Salinity	ACC deaminase	Siddikee et al. (2011)
Bacillus	Alfalfa	Salinity and drought	Antibiotic production	Sokolova et al. (2011)
P. putida UW4 and *Gigaspora rosea* BEG9	Cucumber	Salinity	ACC deaminase	Gamalero et al. (2010)
Pseudomonas sp.	Eggplant	Salinity	Antioxidant enzymes	Fu et al. (2010)
P. mendocina	*Lactuca sativa*	Salinity	Water content was greater in leaves of plants, higher concentrations of foliar K, and lower concentrations of foliar Na	Kohler et al. (2009)
B. subtilis	Tomato	Drought	Cytokinin signaling	Arkhipova et al. 2007)

tomato and pepper, resulting in significant increases in fresh and dry weights with a decrease in ethylene level (Mayak et al. 2004). Another ACC deaminase-positive PGPR strain, *A. piechaudii* ARV8, conferred IST (induced systemic resistance) to drought stress in pepper and tomato plants (Mayak et al. 2004). Furthermore, tomato plants inoculated with ACC deaminase producing

Pseudomonas fluorescens YsS6 and *P. migulae* 8R6 had higher fresh and dry biomass and higher chlorophyll content than the ACC deaminase-negative bacteria-treated tomato plants, grown with 165 mM and 185 mM of salt (Ali et al. 2014). In a similar investigation, cucumber plants inoculated with *P. putida* UW4 and *Gigaspora rosea* BEG9 and grown at 72 mM salt concentration showed significantly higher root and shoot fresh biomass than uninoculated plants (Gamalero et al. 2010). Co-inoculation of lettuce with PGPR *Pseudomonas mendocina* and AM fungi (*Glomus intraradices* or *G. mosseae*) augmented the antioxidant catalase activity under severe drought conditions suggesting that they can be used as inoculants to alleviate the oxidative damage elicited by drought (Kohler et al. 2008). Also, *A. piechaudii*, which produced ACC, increased the growth of tomato seedlings by 66% in the presence of high salt contents (Choudhary et al. 2011). Tank and Saraf (2010) reported phosphate solubilization, phytohormones, and siderophore production in tomato plant under 2% NaCl stress by PGPR.

Proline is often synthesized by plants in response to various abiotic and biotic stresses, mediating osmotic adjustment, free radical scavenging, and subcellular structure stabilization (Hare and Cress 1997). Increased proline synthesis has been shown in abiotically stressed plants in the presence of beneficial bacteria. Modifications of plant morphogenetic parameters and increased efficiency of photosynthesis was induced by AM fungi in salt conditions (Gamalero et al. 2009). Kohler et al. (2009) investigated the influence of inoculation with a PGPR, *P. mendocina*, alone or in combination with an AM fungus, *G. intraradices* or *G. mosseae*, on growth and nutrient uptake and other physiological activities of *Lactuca sativa* affected by salt stress.

Chookietwattana and Maneewan (2012) selected 84 halotolerant bacterial strains and assessed their phosphate-solubilizing activity. Of these, *Bacillus megaterium* A12 was selected as the efficient halotolerant PSB because it demonstrated the highest phosphate solubilization activity under saline conditions. The *B. megaterium* A12 significantly increased the germination percentage and germination index of tomato seeds grown with NaCl concentrations between 30 and 90 mM and increased the seedling dry weight at NaCl up to 120 mM. Their results suggest that the halotolerant PSB may be used to alleviate the effects of salts and provide great potential for use as biofertilizers in the arid and salt-affected areas. Wang et al. (2012), when tested the effect of PGPR strain BSS on cucumber plant against drought tolerance, reported induction of systemic resistance in cucumber plant. Other workers have also reported that the stomatal conductance of plant leaf was higher in PGPR *Pseudomonas aeruginosa*-inoculated mung bean plants than non-PGPR inoculated ones under drought conditions (Ahmad et al. 2013, Sarma and Saikia 2014, and Naveed et al. 2014). Nautiyal et al. (2008) demonstrated that the *Bacillus lentimorbus* strain increased the antioxidant capacity of the edible parts of spinach, carrots, and lettuce, as well as increase in growth. Yildirim et al. (2008) studied the ameliorative effect of *Staphylococcus kloosii* strain EY37 and *Kocuria erythromyxa* strain EY43 on radish growing in saline soil. They observed that bacterial inoculants significantly increased shoot/root dry weight, leaf number per plant, relative water content of the leaf, and chlorophyll content of radish fruit. Corn, beans, and clover inoculated with AM fungi had increased proline content which resulted in salinity resistance (Grover et al. 2011).

Conclusion

Salinity and drought stress are the serious environmental issues which drastically reduce the productivity of vegetables. The use of plant growth-promoting micro-organism in vegetable crop production has received little attention. Enhancement in the use of PGPR is one of the newly emerging options to meet the agricultural challenges imposed in the stresses in soil environment. Few reports which are available today have shown that PGPR could improve plant productivity even in stressed environment by counteracting the negative effects of saline and water stresses on plant growth. PGPR promote the growth of plants through variety of mechanisms like triggering osmotic response, providing growth hormones and nutrients, acting as biocontrol agents, and modifying root to shoot signaling in plants. Developing salt-tolerant crops is still in the pipeline, and therefore, the only viable alternative seems to be the use of PGPR for enhancing vegetable production under stressed environment. The complex and dynamic interactions between microorganisms and plant roots under conditions of abiotic stress affect not only the plants but also the physical, chemical, and structural properties of soil. Selection of microorganisms from stressed ecosystems and their possible application under stressed conditions to mitigate the impact of abiotic stresses are likely to improve the production of vegetables in soils stressed with different abiotic factors.

References

Abou-Baker NHA, El-Dardiry EA (2015) Integrated management of salt affected soils in agriculture: incorporation of soil salinity control methods. Academic Press, New York

Adolf VI, Jacobsen SE, Shabala S (2013) Salt tolerance mechanisms in quinoa (*Chenopodium quinoa* Willd). Environ Exp Bot 92:43–54

Ahmad M, Zahir ZA, Khalid M (2013) Efficacy of *Rhizobium* and *Pseudomonas* strains to improve physiology, ionic balance and quality of mung bean under salt-affected conditions on farmer's fields. Plant Physiol Biochem 63:170–176

Ali S, Charles TC, Glick BR (2014) Amelioration of high salinity stress damage by plant growth-promoting bacterial endophytes that contain ACC deaminase. Plant Physiol Biochem 80:160–167

Anderson JPE, Domsch KH (1980) Quantities of plant nutrients in the microbial biomass of selected soils. Soil Sci 130:211–216

Andronov EE, Petrova SN, Pinaev AG, Pershina EV, Rakhimgalieva SZ, Akhmedenov KM, Gorobets AV, Sergaliev NK (2012) Analysis of the structure of microbial community in soils with different degrees of salinization using T-RFLP and real time PCR techniques. Eurasian Soil Sci 45:147–156

Araus JL, Slafer GA, Royo C, Serret MD (2008) Breeding for yield potential and stress adaptation in cereals. Crit Rev Plant Sci 27:377–412

Arkhipova TN, Prinsen E, Veselov SU, Martinenko EV, Melentiev AI, Kudoyarova GR (2007) Cytokinin producing bacteria enhance plant growth in drying soil. Plant Soil 292:305–315

Asaadi AM (2009) Investigation of salinity stress on seed germination of *Trigonella foenum-graecum*. Res J Biol Sci 4:1152–1155

Asghar HN, Zahir ZA, Arshad M, Khaliq K (2002) Relationship between in vitro production of auxins by rhizobacteria and their growth-promoting activities in *Brassica juncea* L. Biol Fertil Soils 35:231–237

Azam F, Ifzal M (2006) Microbial populations immobilizing NH_4^+-N and NO_3^--N differ in their sensitivity to sodium chloride salinity in soil. Soil Biol Biochem 38:2491–2494

Azevedo Neto AD, Gomes-Filho E, Prisco JT (2008) Salinity and oxidative stress. In: Khan NA, Sarvajeet S (eds) Abiotic stress and plant responses. IK International, New Delhi, pp 58–82

Bahadur A, Chatterjee A, Kumar R, Singh M, Naik PS (2011) Physiological and biochemical basis of drought tolerance in vegetables. Veg Sci 38:1–16

Bano A, Fatima M (2009) Salt tolerance in *Zea mays* (L) following inoculation with *Rhizobium* and *Pseudomonas*. Biol Fertil Soils 45:405–413

Basha MG, Vivekanandan M (2000) Potency of Rhizobial strains from different environments to increase economic productivity in some legumes. Philipp J Sci 129:131–134

Berg G, Zachow C, Müller H, Phillips J, Tilcher R (2013) Next-generation bio-products sowing the seeds of success for sustainable agriculture. Agronomy 3:648–656

Bochow H, El-Sayed SF, Junge H, Stauropoulou A, Schmieeknecht G (2001) Use of *Bacillus subtilis* as biocontrol agent. IV. Salt-stress tolerance induction by *Bacillus subtilis* FZB24 seed application in tropical vegetable field crops, and its mode action. J Plant Dis Prot 108:21–30

Bojovic B, Delic G, Topuzovic M, Stankovic M (2010) Effects of NaCl on seed germination in some species from families *Brassicaceae* and *Solanaceae*. Kragujevac J Sci 32:83–87

Burrow DP, Surapaneni A, Rogers ME, Olsson KA (2002) Groundwater use in forage production: the effect of saline-sodic irrigation and subsequent leaching on soil sodicity. Aust J Exp Agric 42:237–247

Cakmakci R, Donmez MF, Erdogan U (2007) The effect of plant growth promoting rhizobacteria on barley seedling growth, nutrient uptake, some soil properties, and bacterial counts. Turk J Agric For 31:189–199

Carillo P, Annunziata MG, Pontecorvo G, Fuggi A, Woodrow P (2011) Salinity stress and salt tolerance. In: Arun S (ed) Abiotic stress in plants—mechanisms and adaptations. InTech, Croatia, pp 22–38

Chatterjee A, Solankey SS (2015) Climate dynamics in horticultural science. Apple Academic Press Inc, Oakville, ON

Chookietwattana K, Maneewan K (2012) Screening of efficient halotolerant phosphate solubilizing bacterium and its effect on promoting plant growth under saline conditions. World Appl Sci J 16:1110–1117

Choudhary DK, Sharma KP, Gaur RK (2011) Biotechnological perspectives of microbes in agro-ecosystems. Biotechnol Lett 33:1905–1910

Chowdhury N, Marschner P, Burns RG (2011) Soil microbial activity and community composition: impact of changes in matric and osmotic potential. Soil Biol Biochem 43:1229–1236

Dadkhah A (2011) Effect of salinity on growth and leaf photosynthesis of two sugar beet (*Beta vulgaris* L.) cultivars. J Agric Sci Technol 13:1001–1012

Diby P, Harshad L (2014) Plant-growth-promoting rhizobacteria to improve crop growth in saline soils: a review. Agron Sustain Dev 34:737–752

Dodd IC, Perez-Alfocea F (2012) Microbial amelioration of crop salinity stress. J Exp Bot 63:3415–3428

Dwivedi S, Upadhyaya H, Subudhi P, Gehring C, Bajic V, Ortiz R (2010) Plant breeding reviews. John Wiley & Sons, Inc., Hoboken, NJ

Earl HJ, Davis RF (2003) Effect of drought stress on leaf and whole canopy radiation use efficiency and yield of maize. Agron J 95:688–696

Egamberdieva D (2011) Survival of *Pseudomonas extremorientalis* TSAU20 and *P. chlororaphis* TSAU13 in the rhizosphere of common bean (*Phaseolus vulgaris*) under saline conditions. Plant Soil Environ 57:122–127

Egamberdieva D (2012) *Pseudomonas chlororaphis*: a salt-tolerant bacterial inoculant for plant growth stimulation under saline soil conditions. Acta Physiol Plant 34:751–756

Egamberdieva D, Jabborova D (2013) Biocontrol of cotton damping-off caused by *Rhizoctonia solani* in salinated soil with rhizosphere bacteria. Asian Australas J Plant Sci Biotechnol 7(2):31–38

FAO (2008) Land and plant nutrition management service. www.fao.org/ag/agl/agll/spush

Flores-Felix JD, Silva LR, Rivera LP (2015) Plants probiotics as a tool to produce highly functional fruits: the case of *Phyllobacterium* and vitamin C in strawberries. PLoS One 10:e0122281

Francius G, Polyakov P, Merlin J, Abe Y, Ghigo JM, Merlin C, Beloin C, Duval JF (2011) Bacterial surface appendages strongly impact nanomechanical and electrokinetic properties of *Escherichia coli* cells subjected to osmotic stress. PLoS One 6:e20066

Fu Q, Liu C, Ding N, Guo B (2010) Ameliorative effects of inoculation with the plant growth-promoting rhizobacterium *Pseudomonas* sp. DW1 on growth of eggplant (*Solanum melongena* L.) seedlings under salt stress. Agric Water Manage 97:1994–2000

Gamalero E, Berta G, Massa N, Glick BR, Lingua G (2010) Interactions between *Pseudomonas putida* UW4 and *Gigaspora rosea* BEG9 and their consequences for the growth of cucumber under salt-stress conditions. J Appl Microbiol 108:236–245

Gamalero E, Lingua G, Berta G, Glick BR (2009) Beneficial role of plant growth promoting bacteria and arbuscular mycorrhizal fungi on plant responses to heavy metal stress. Can J Microbiol 55:501–514

Ghollaratta M, Raiesi F (2007) The adverse effects of soil salinization on the growth of *Trifolium alexandrinum* L. and associated microbial and biochemical properties in a soil from Iran. Soil Biol Biochem 39:1699–1702

Glick BR (2014) Bacteria with ACC deaminase can promote plant growth and help to feed the world. Microbiol Res 169:30–39

Greenway H, Munns R (1980) Mechanisms of salt tolerance in non-halophytes. Annu Rev Plant Physiol 31:149–190

Grover M, Ali SZ, Sandhya V, Rasul A, Venkateswarlu B (2011) Role of microorganisms in adaptation of agriculture crops to abiotic stress. World J Microbiol Biotechnol 27:1231–1240

Habib SH, Kausar H, Saud HM (2016) Plant growth-promoting rhizobacteria enhance salinity stress tolerance in okra through ROS-scavenging enzymes. Biomed Res Int. doi:10.1155/2016/6284547

Hare PD, Cress WA (1997) Metabolic implications of stress-induced proline accumulation in plants. Plant Growth Regul 21:79–102

Hasanuzzaman M, Nahar K, Alam MM, Roychowdhury R, Fujita M (2013) Physiological, biochemical, and molecular mechanisms of heat stress tolerance in plants. Int J Mol Sci 14(5):9643–9684

Hayat R, Ali S, Amara U, Khalid R, Ahmed I (2010) Soil beneficial bacteria and their role in plant growth promotion: a review. Ann Microbiol 60:579–598

Horneck DA, Ellsworth JW, Hopkins BG, Sullivan DM, Stevens RG (2007) Managing salt affected soils for crop production, PNW 601-E. http://extension.oregonstate.edu/catalog/pdf/pnw/pnw601-e.pdf

Hussain M, Malik MA, Farooq M, Ashraf MY, Cheema MA (2008) Improving drought tolerance by exogenous application of glycinebetaine and salicylic acid in sunflower. J Agron Crop Sci 194:193–199

Ibekwe AM, Poss JA, Grattan SA, Grieve CM, Suarez D (2010) Bacterial diversity in cucumber (*Cucumis sativus*) rhizosphere in response to salinity, soil pH, and boron. Soil Biol Biochem 42:567–575

Jamil A, Riaz S, Ashraf M, Foolad MR (2011) Gene expression profiling of plants under salt stress. Crit Rev Plant Sci 30:435–458

Javid MG, Sorooshzadeh A, Moradi F, Sanavy Seyed AMM, Allahdadi I (2011) The role of phytohormones in alleviating salt stress in crop plants. AJCS 5:726–734

Jyothsna P, Murthy SDS (2016) A review on effect of senescence in plants and role of phytohormones in delaying senescence. Int J Plant Animal Environ Sci 6:152–161

Kapoor K, Srivastava A (2010) Assessment of salinity tolerance of *Vigna mungo* var. Pu-19 using *ex vitro* and *in vitro* methods. Asian J Biotechnol 2:73–85

Kaushal M, Wani SP (2016) Plant-growth-promoting rhizobacteria: drought stress alleviators to ameliorate crop production in drylands. Ann Microbiol 66:35–42

Kidoglu F, Gul A, Ozaktan H, Tuzel Y (2008) Effect of rhizobacteria on plant growth of different vegetables. Acta Hortic 801:1471–1477

Kohler J, Hernandez JA, Caravaca F, Roldan A (2008) Plant-growth-promoting rhizobacteria and arbuscular mycorrhizal fungi modify alleviation biochemical mechanisms in water-stressed plants. Funct Plant Biol 35:141–151

Kohler J, Hernandez JA, Caravaca F, Roldan A (2009) Induction of antioxidant enzymes is involved in the greater effectiveness of a PGPR versus AM fungi with respect to increasing the tolerance of lettuce to severe salt stress. Environ Exp Bot 65:245–252

Kotera A, Sakamoto T, Nguyen DK, Yokozawa M (2008) Regional consequences of seawater intrusion on rice productivity and land use in coastal areas of the Mekong river delta. Jpn Agric Res Q 42:267–274

Kumar R, Solankey SS, Singh M (2012) Breeding for drought tolerance in vegetables. Veg Sci 39:1–15

Ladeiro B (2012) Saline agriculture in the 21st century: using salt contaminated resources to cope food requirements. J Bot 2012:1–7

Lim JH, Kim SD (2013) Induction of drought stress resistance by multi-functional PGPR *Bacillus licheniformis* K11 in pepper. Plant Pathol J 29(2):201–208

Mahajan S, Tuteja N (2005) Cold salinity and drought stresses: an overview. Arch Biochem Biophys 444:139–158

Manchanda G, Garg N (2008) Salinity and its effect on the functional biology of legumes. Acta Physiol Plant 30:595–618

Mayak S, Tirosh T, Glick BR (2004) Plant growth-promoting bacteria that confer resistance to water stress in tomatoes and peppers. Plant Sci 166:525–530

Mishra V, Cherkauer KA (2010) Retrospective droughts in the crop growing season: implications to corn and soybean yield in the Midwestern United States. Agric For Meteorol 150:1030–1045

Moreira-Nordemann LM (1984) Salinity and weathering rate of rocks in a semi-arid region. J Hydrol 71:131–147

Munns R, Tester M (2008) Mechanism of salinity tolerance. Annu Rev Plant Biol 59:651–681

Nandakumar K, Kiran K, Prashant F (2012) Vegetable crops (chilli pepper and onion): approaches to improve crop productivity and abiotic stress tolerance. In: Narendra T, Sarvajeet SG, Antonio FT, Renu T (eds) Improving crop resistance to abiotic stress, vol 1. Wiley Blackwell, Hoboken, NJ, pp 953–978

Nautiyal CS, Govindarajan R, Lavania M, Pushpangadan P (2008) Novel mechanisms of modulating natural antioxidants in functional foods: involvement of plant growth promoting rhizobacteria NRRL B-30488. J Agric Food Chem 56:4474–4481

Naveed M, Hussain MB, Zahir ZA, Mitter B, Sessitsch A (2014) Drought stress amelioration in wheat through inoculation with *Burkholderia phytofirmans* strain PsJN. Plant Growth Regul 73:121–131

Nelson DR, Mele PM (2007) Subtle changes in the rhizosphere microbial community structure in response to increased boron and sodium chloride concentrations. Soil Biol Biochem 39:340–351

Nikinmaa E, Holtta T, Hari P, Kolari P, Makela A, Sevanto S, Vesala T (2013) Assimilate transport in phloem sets conditions for leaf gas exchange. Plant Cell Environ 36:655–669

Okcu G, Kaya MD, Atak M (2005) Effects of salt and drought stresses on germination and seedling growth of pea (*Pisum sativum* L.) Turk J Agric For 29:237–242

Ondrasek G, Rengel Z, Romic D, Savic R (2010) Environmental salinisation processes in agroecosystem of Neretva river estuary. Novenytermeles 59:223–226

Oren A (1999) Bioenergetic aspects of halophilism. Microbiol Mol Biol Rev 63:334–340

Osakabe Y, Osakabe K, Shinozaki K, Lam-Son PT (2014) Response of plants to water stress. Front Plant Sci 5:86–93

Palaniyandi SA, Damodharan K, Yang SH, Suh JW (2014) *Streptomyces* sp. strain PGPA39 alleviates salt stress and promotes growth of 'Micro Tom' tomato plants. J Appl Microbiol 117:766–773

Patel BB, Patel Bharat B, Dave RS (2011) Studies on infiltration of saline–alkali soils of several parts of Mehsana and Patan districts of north Gujarat. J Appl Technol Environ Sanitation 1:87–92

Paul KI, Polglase PJ, O'Connell AM, Carlyle JC, Smethurst PJ, Khanna PK (2003) Defining the relation between soil water content and net nitrogen mineralization. Eur J Soil Sci 54:39–47

Qados AMSA (2011) Effect of salt stress on plant growth and metabolism of bean plant *Vicia faba* (L.) J Saudi Soc Agric Sci 10:7–15

Qiu-shi FU, Cui J, Zhao B, Guo Y-d (2009) Effects of water stress on photosynthesis and associated physiological characters of *Capsicum annuum* L. Sci Agric Sin 42:1859–1866

Rabie GH, Aboul-Nasr MB, Al-Humiany A (2005) Increase salinity tolerance of cowpea plants by dual inoculation of Am fungus *Glomus clarum* and Nitrogen-fixer *Azospirillum brasilense*. Mycobiology 33:51–61

Radzki W, Gutierrez Manero FJ, Algar E (2013) Bacterial siderophores efficiently provide iron to iron-starved tomato plants in hydroponics culture. Antonie Van Leeuwenhoek 104:321–330

Ramadoss D, Lakkineni VK, Bose P, Ali S, Annapurna K (2013) Mitigation of salt stress in wheat seedlings by halotolerant bacteria isolated from saline habitats. SpringerPlus 2:1–7

Ramazani M, Taghavei M, Masoudi M, Riahi A, Bahbahani N (2009) The evaluation of drought and salinity effects on germination and seedling growth caper (*Capparis spinosa* L.) Iran J Rangeland 2:411–420

Reddy AR, Chaitanya KV, Vivekanandan M (2004) Drought-induced responses of photosynthesis and antioxidant metabolism in higher plants. J Plant Physiol 161:1189–1202

Sandhya V, Ali SKZ, Grover M, Reddy G, Venkateswarlu B (2010) Effect of plant growth promoting *Pseudomonas* spp. on compatible solutes, antioxidant status and plant growth of maize under drought stress. Plant Growth Regul 62:21–30

Sang-Mo K, Radhakrishnan R, Latif Khan A, Kim M-J, Jae-Man P, Bo-Ra K, Dong-Hyun S, In-Jung L (2014) Gibberellin secreting rhizobacterium, *Pseudomonas putida* H-2-3 modulates the hormonal and stress physiology of soybean to improve the plant growth under saline and drought conditions. Plant Physiol Biochem 84:115–124

Sankar B, Jaleel CA, Manivannan P, Kishorekumar A, Somasundaram R, Panneerselvam R (2008) Relative efficacy of water use in five varieties of *Abelmoschus esculentus* (L.) Moench. under water limited conditions. Colloids Surf B Biointerfaces 62:125–129

Saravanakumar D, Kavino M, Raguchander T, Subbian P, Samiyappan R (2011) Plant growth promoting bacteria enhance water stress resistance in green gram plants. Acta Physiol Plant 33:203–209

Sarma RK, Saikia RR (2014) Alleviation of drought stress in mung bean by strain *Pseudomonas aeruginosa* GGRK21. Plant Soil 377:111–126

Savvas D, Lenz F (2000) Effects of NaCl or nutrient-induced salinity on growth, yield, and composition of eggplants grown in rock wool. Sci Hortic 84:37–47

Shahbaz M, Ashraf M (2013) Improving salinity tolerance in cereals. Crit Rev Plant Sci 32:237–249

Shannon MC, Grieve CM (1999) Tolerance of vegetable crops to salinity. Sci Hortic 78:5–38

Shrivastava P, Kumar R (2015) Soil salinity: a serious environmental issue and plant growth promoting bacteria as one of the tools for its alleviation. Saudi J Biol Sci 22:123–131

Sibomana IC, Aguyoh JN, Opiyo AM (2013) Water stress affects growth and yield of container grown tomato (*Lycopersicon esculentum* Mill) plants. Global J Biosci Biotechnol 2:461–466

Siddikee MA, Glick BR, Chauhan PS, Yim WJ, Sa T (2011) Enhancement of growth and salt tolerance of red pepper seedlings (*Capsicum annuum* L.) by regulating stress ethylene synthesis with halotolerant bacteria containing 1-aminocyclopropane-1-carboxylic acid deaminase activity. Plant Physiol Biochem 49:427–434

Silvera JAG, Melo ARB, Viegas RA, Oliveira JT (2001) Salinity-induced effects on nitrogen assimilation related to growth in cowpea plants. Environ Exp Bot 46:171–179

Singleton PW, Bohlool BB (1984) Effect of salinity on nodule formation by soybean. Plant Physiol 74:72–76

Smol JP (2012) Climate change: a planet in flux. Nature 483:S12–S15

Sokolova MG, Akimova GP, Vaishlia OB (2011) Effect of phytohormones synthesized by rhizo-sphere bacteria on plants. Prikl Biokhim Mikrobiol 47:302–307

Taffouo VD, Wamba OF, Yombi E, Nono GV, Akoe A (2010) Growth, yield, water status and ionic distribution response of three bambara groundnut (*Vigna subterranean* (L.) verdc.) landraces grown under saline conditions. Int J Bot 6:53–58

Takagi M, El-Shemy HA, Sasaki S, Toyama S, Kanai S, Saneoka H, Fujita K (2009) Elevated CO_2 concentration alleviates salinity stress in tomato plant. Acta Agr Scand B—S P 59:87–96

Tank N, Saraf M (2010) Salinity resistant plant growth promoting rhizobacteria ameliorates sodium chloride stress on tomato plants. J Plant Interact 5:51–58

Tate RL III (2000) Soil microbiology. John Wiley & Sons, New York

Tester M, Davenport R (2003) Na^+ tolerance and Na^+ transport in higher plants. Ann Bot 91:503–527

Upadhyay SK, Singh JS, Saxena AK, Singh DP (2011) Impact of PGPR inoculation on growth and antioxidant status of wheat under saline conditions. Plant Biol 14:605–611

Vanderlinde EM, Harrison JJ, Muszynski A, Carlson RW, Turner RJ, Yost CK (2010) Identification of a novel ABC transporter required for desiccation tolerance, and biofilm formation in *Rhizobium leguminosarum* bv. *viciae* 3841. FEMS Microbiol Ecol 271:327–340

Vivekanandan M, Karthik R, Leela A (2015) Improvement of crop productivity in saline soils through application of saline-tolerant rhizosphere bacteria—current perspective. Int J Adv Res 3:1273–1283

Wang CJ, Yang W, Wang C, Gu C, Niu DD, Liu HX, Wang YP, Guo JH (2012) Induction of drought tolerance in cucumber plants by a consortium of three plant growth promoting rhizo-bacterium strains. PLoS One 7:e52565

Wery J, Silim SN, Knights EJ, Malhotra RS, Cousin R (1994) Screening techniques and sources and tolerance to extremes of moisture and air temperature in cool season food legumes. Euphytica 73:73–83

Wichern J, Wichern F, Joergensen RG (2006) Impact of salinity on soil microbial communities and the decomposition of maize in acidic soils. Geoderma 137:100–108

Witcombe JR, Hollington PA, Howarth CJ, Reader S, Steele KA (2008) Breeding for abiotic stresses for sustainable agriculture. Philos Trans R Soc B Biol Sci 363:703–716

Xu ZH, Saffigna PG, Farquhar GD, Simpson JA, Haines RJ, Walker S, Osborne DO, Guinto D (2000) Carbon isotope discrimination and oxygen isotope composition in clones of the F (1) hybrid between slash pine and Caribbean pine in relation to tree growth, water-use efficiency and foliar nutrient concentration. Tree Physiol 20:1209–1217

Yang J, Kloepper JW, Ryu CM (2009) Rhizosphere bacteria help plants tolerate abiotic stress. Trends Plant Sci 14:1–4

Yildirim E, Donmez MF, Turan M (2008) Use of bioinoculants in ameliorative effect on radish (*Raphanus sativus* L.) plants under salinity stress. J Plant Nutr 31:2059–2074

Growth Improvement and Management of Vegetable Diseases by Plant Growth-Promoting Rhizobacteria

5

Asfa Rizvi, Almas Zaidi, Mohd. Saghir Khan, Saima Saif, Bilal Ahmed, and Mohammad Shahid

Abstract

Vegetables are an important part of human dietary systems. They contain several important nutrients including vitamins, antioxidants, etc. and affect immensely the human health. Vegetables are cultivated and consumed globally on a large scale and serve as the food of choice for millions of people across the globe. During cultivation, most of the vegetable crops are, however, often attacked by various insect pests and pathogenic microorganisms, thereby causing severe diseases, leading to huge yield losses. The agricultural practitioners depend heavily on chemical fertilizers to supply nutrients to vegetables while they apply pesticides to manage insect pests and to concurrently enhance vegetable production. The injudicious application of agrochemicals including pesticides into vegetable production practices adversely affects the soil fertility and consequently the plant health, thus making it unfit for human consumption. In order to protect the crops and to minimize yield losses due to phytopathogens, an alternate and inexpensive approach involving the use of plant growth-promoting rhizobacteria (PGPR) has been introduced into the vegetable production system. The application of PGPR formulations into the vegetable production strategies has been found to protect them from various diseases leading to improved yield and quality of the vegetables. The present chapter focuses on the disease incidence among some of the popularly grown vegetables and the role of PGPR in suppression of common vegetable diseases.

A. Rizvi (✉) • A. Zaidi • M.S. Khan • S. Saif • B. Ahmed • M. Shahid
Faculty of Agricultural Sciences, Department of Agricultural Microbiology, Aligarh Muslim University, Aligarh 202001, Uttar Pradesh, India
e-mail: asfarizvi09@gmail.com

© Springer International Publishing AG 2017
A. Zaidi, M.S. Khan (eds.), *Microbial Strategies for Vegetable Production*,
DOI 10.1007/978-3-319-54401-4_5

5.1   Introduction

The population of the world is expanding consistently. It has been projected to increase up to nearly 8.2 billion by the year 2025 and is expected to reach around 9.3 billion in 2050 (DESA 2000). With limited resources available, it has become extremely difficult to feed such a hugely expanding human population. Among various food items, supplying vegetables to human population is also a major challenge. So, in order to overcome the vegetable demands, efforts are directed toward enhancing the production of vegetables worldwide. Vegetables being rich in various nutrients are consumed by millions of people globally. The field-grown vegetable crops are, however, highly prone to attack by several fungal and bacterial phytopathogens leading to huge economic losses to the growers. To overcome the nuisance caused by the phytopathogens, the vegetable growers adopt many strategies such as the use of disease-resistant varieties (Witek et al. 2016), crop rotation (Ikeda et al. 2015) and other disease control measures, but all these methods have not been successful and effective. Apart from such methods, vegetable growers also apply various agrochemicals to avoid yield losses due to phytopathogens (Srivastava and Sharma 2014). Such chemicals, however, cause serious environmental pollution and consequently result in a deleterious impact onto the vegetables (Gafar et al. 2013). Therefore, to minimize/reduce the use of chemicals in vegetable production practices and to improve the yield and quality of vegetables, growers are advised to use plant growth-promoting rhizobacteria: an inexpensive and sustainable approach for vegetable production (Zaidi et al. 2015). Although literature on disease management of vegetables using PGPR is very limited, some bioformulations comprising various PGPR, having biocontrol potential, have been tried against some vegetable diseases in order to minimize the severity of the diseases (Loganathan et al. 2014) while simultaneously maximizing the yield of vegetable crops. In this chapter, an attempt has been made to highlight the diseases affecting the commonly grown vegetables and their management by plant growth-promoting rhizobacteria.

5.2   Rationale for Using PGPR in the Management of Vegetables Diseases

Bacterial and fungal pathogens, in general, are a major threat to the sustainability, quality and yield of vegetables. Therefore, to minimize the yield losses caused by phytopathogens and hence to optimize vegetable production, the vegetable growers adopt various practices such as proper sanitation of the planting fields, crop rotation, use of disease-resistant cultivars and indiscriminate spraying of pesticides, etc. without considering their toxic impact on plants and via food chain on human health. Despite adopting so many methods including the excessive use of chemicals in vegetable production, considerable success has not been achieved in combating plant diseases. Therefore, to enhance the production of healthy vegetables and to reduce the yield losses due to pathogen attack, focus in recent times has been shifted toward the use of inexpensive, eco-friendly and viable alternative like PGPR in the management of vegetable diseases. By following this, the growth, yield and quality of many vegetables due to PGPR application have substantially been increased (Table 5.1). In the following section an attempt is made to highlight some of the serious diseases of most commonly grown and consumed vegetables and their management through the use of PGPR inoculations.

Table 5.1 Diseases of some common vegetables and their management by PGPR

Disease	Affected host plant	Causative agent	Principle antagonist	Active biomolecules	Reference
Fusarium wilt	Tomato, brinjal	*Fusarium oxysporum* f. sp. *lycopersici*; *F. oxysporum* f. sp. *melongenae*	*Bacillus subtilis*; *Trichoderma* sp.	Enzymes; secretion of extracellular cell wall-degrading enzymes	Loganathan et al. (2014), Abdel-Monaim et al. (2014)
Bacterial wilt	Tomato, brinjal	*Ralstonia solanacearum*	*Bacillus amyloliquefaciens*; *Pseudomonas fluorescens*	Antibiotics and secondary metabolites; rhizosphere colonization	Singh et al. (2016), Chakravarty and Kalita (2012)
Root rot	Okra	*Rhizoctonia solani*	*Pseudomonas fluorescens*	Siderophores, HCN and indole acetic acid	Adhikari et al. (2013)
Damping-off	Cucumber	*Pythium ultimum*	*Pseudomonas fluorescens*; *Pseudomonas* sp.; *Bacillus subtilis*	Antibiotics and metabolites	Khabbaz and Abbasi (2014)
Bacterial spot	Pepper	*Xanthomonas campestris* pv. *vesicatoria*	Lactic acid bacteria	Siderophores	Shrestha et al. (2014)
Black rot	Crucifers	*Xanthomonas campestris* pv. *campestris*	*Bacillus* sp.	Antibiosis	Luna et al. (2002)
Downy mildew	Cucumber	*Pseudoperenospora cubensis*	Consortium of *Achromobacter* sp.; *Streptomyces* sp. and *Bacillus licheniformis*	Induced systemic resistance	Sen et al. (2014)
Late blight	Potato; pepper	*Phytophthora infestans*; *Phytophthora capsici*	*Chaetomium globosum*; *Burkholderia cepacia*	Endo and exoglucanases; antimicrobial activity of organic acids	Shanthiyaa et al. (2013), Sopheareth et al. (2013)
Early blight	Potato	*Alternaria solani*	*Trichoderma harzianum* + *Pseudomonas fluorescens*	ND	Mane et al. (2014)

ND Not determined

5.3   How Plant Growth-Promoting Rhizobacteria Combat Phytopathogen Attack: A General Perspective

Plant growth-promoting rhizobacteria (Kloepper and Schroth 1978) are certain beneficial bacteria that colonize plant roots and improve the performance of crop plants through enhanced nutrient uptake from soil and several other mechanisms. They are known to antagonize several plant pathogenic microorganisms by releasing antimicrobial metabolites (George et al. 2015) and also by chelating the iron present in the soil, thus creating a competition for iron requirement by plant pathogens (Haas and Défago 2005; Haas and Keel 2003; Raaijmakers et al. 2002). Plant growth-promoting rhizobacteria are effective antagonists toward various bacterial (Liu et al. 2016), fungal (Kumari and Khanna 2014) and viral diseases (Li et al. 2016) attacking the crops. Some PGPR secretes antibiotics, for example, pyrrolnitrin, pyoluteorin, 2,4-DAPG, etc. and inhibit the growth of plant pathogens (Beneduzi et al. 2012). The biocontrol activity of many disease-suppressive microorganisms is also attributed to stimulation of defence-related mechanisms within the host plants, what is better known as induced systemic resistance (ISR). Some PGPR combine different mechanisms of antagonism and plant growth promotion and are therefore able to suppress a wide range of plant diseases while simultaneously enhancing plant growth and development (Vassilev et al. 2006). For instance, *Pseudomonas fluorescens* CHA0 has been reported to synthesize antifungal compounds like 2,4-diacetylphloroglucinol (DAPG) (Keel et al. 1992; Keel et al. 1990) and pyoluteorin (PLT) (Maurhofer et al. 1994; Maurhofer et al. 1992). These compounds in turn have been found to suppress various soilborne plant diseases (Haas and Keel 2003).

Although several strains of PGPR have been reported as suitable candidates for plant disease suppression, PGPR belonging to the genus *Pseudomonas* have received considerable attention as potential biocontrol agent (Cabanás et al. 2014). The process of plant growth promotion and disease control by *Pseudomonas* sp. are interlinked involving various direct and indirect mechanisms that include synthesis of some metabolites like auxins, cytokinins, gibberellins, etc., ACC deaminase activity, production of iron-chelating compounds (siderophores), antibiotics and numerous cyanogenic and volatile compounds. Other mechanisms may include mineral phosphate solubilization, competition for nutrients and induced systemic resistance (Lucy et al. 2004; Adesemoye et al. 2008). These beneficial bacteria are able to improve the yield of vegetable crops, thereby reducing economic losses with minimum cost inputs involved (Dias et al. 2013). In addition to *Pseudomonas* sp. acting as effective biocontrol agent in the agricultural system, some strains of *Bacillus subtilis* are also known to inhibit the growth of phytopathogenic fungi by producing certain wide-spectrum antibiotics and thermostable metabolites as a disease control measure (Mercado-Flores et al. 2014). To understand the importance of PGPR in vegetable disease suppression and eventually plant growth promotion, the present section highlights some of the active biomolecules secreted by PGPR which are involved in combating the attack of phytopathogens.

5.3.1 Release of Siderophores

Siderophores are low molecular weight (200–2000 Daltons) compounds released by PGPR (Gupta et al. 2015) which chelate iron present within the soil system and transport it through the bacterial cells. Siderophores are secreted by many bacterial genera, for example, *Bacillus* (Bharucha et al. 2013), *Pseudomonas* (Luján et al. 2015), etc. to solubilize iron from the surrounding environment, thus forming a ferric-siderophore complex that can diffuse through the cell and be returned to the cell surface (Andrews et al. 2003). Thus, siderophores play an important role in the control of some soilborne plant pathogens through competition for iron nutrition (Loper and Buyer 1991). Since siderophores are known to sequester iron (III) present within the surroundings, they limit its availability to the pathogens and ultimately suppress their growth and disease-causing ability (Schroth et al. 1984). Among most of the siderophores released by the bacteria, those produced by pseudomonads, for example, pyoverdin (Peek et al. 2012), can inhibit the growth of plant pathogenic bacteria and fungi (Ruiz et al. 2015). Moreover, a pseudobactin siderophore produced by *P. putida* strain B10 has been found to suppress *Fusarium oxysporum* in the soil and the diseases caused by this pathogenic fungus by limiting the supply of iron. Also, recent studies have demonstrated the suppression of soilborne fungal pathogens with the help of iron-chelating siderophores by fluorescent pseudomonads, thus making it unavailable to other pathogenic microorganisms (Vanitha and Ramjegathesh 2014; Dwivedi and Johri 2003). Production of siderophores is therefore considered as one of the most potent mechanisms of disease suppression and an indirect means of growth promotion employed by numerous PGPR. Besides, iron-chelating siderophores (Beneduzi et al. 2012), various antibiotics (Sivasakthi et al. 2014) and cyanogenic compounds (Sureshbabu et al. 2016) are also produced by PGPR strains that aid in combating the phytopathogens attack and promoting plant growth and development by alleviating the disease severity. Some of the other biomolecules involved in disease suppression are discussed in the following sections.

5.3.2 Production of Cyanogenic Compounds

Production of cyanogenic compounds like hydrogen cyanide (HCN) by PGPR (Lukkani and Reddy 2014) is yet another active biomolecules that aid in successfully controlling various plant diseases by inhibiting the growth and proliferation of plant pathogenic bacteria and fungi, thereby assisting in plant growth promotion. Interestingly, the phenomenon of cyanogenesis by PGPR was predominantly thought to be associated with pseudomonads, and it enhanced in the presence of glycine added as an additional supplement to the culture media (Lakshmi et al. 2015). Cyanide, a highly toxic secondary metabolite is produced by most microorganisms including PGPR (Fouzia et al. 2015) and fungi (Ng et al. 2015) as a means of defence mechanism to safeguard the crops from the pathogens and, therefore, indirectly promotes the growth of plants. Mechanistically, hydrogen cyanide synthesized mostly by *Pseudomonas* (Reetha et al. 2014) and *Bacillus* species inhibits the electron transport chain and the energy supply to the bacterial cell and eventually thus cause the death of the pathogenic microbes. For instance, certain rhizobacterial strains have been reported to have the ability to synthesize HCN by which they restrict the growth of phytopathogens and, hence, exert positive effects on seedling root growth of various plants (Kremer and Souissi 2001).

5.3.3 Production of Antibiotics

Antibiotic production is an important mechanism of antagonism associated with PGPR to fight the target phytopathogens (Glick et al. 2007). Plant growth-promoting rhizobacteria are known to synthesize a vast array of antibiotics, as yet another major defence tool that provides protection to plants from nuisance of phytopathogens (Ulloa-Ogaz et al. 2015). And hence one or more antibiotics produced by the PGPR (Wang et al. 2015) play a prime role in disease suppression. The mechanism of antibiosis is to produce low molecular weight compounds that may pose deleterious impacts on the metabolism of pathogenic microorganisms and thus retards their growth. Several studies have shown that the production of certain antibiotics like pyrrolnitrin, phycocyanin, 2,4-diacetylphloroglucinol (DAPG) (Meyer et al. 2016), etc. by various microbial genera belonging to PGPR can cause suppression of phytopathogens (Subba Rao 1993; Glick 1995). Since then a variety of antibiotics have been isolated from various bacterial strains that could eventually inhibit the synthesis of cell walls of the pathogenic microflora (Dilantha et al. 2005). Also, the antibiotics damage the membrane integrity of the cells and the formation of initiation complexes on the small subunit of the ribosome (Maksimov et al. 2011). For example, 2,4-diacetylphloroglucinol (DAPG) is an effective and extensively studied antibiotic produced by pseudomonads that has been reported to damage the membrane of *Pythium* sp. and causes inhibition of zoospore formation (De Souza et al. 2003). Pseudomonads also produce some other antibiotics like phenazine that possesses redox activity and is capable of suppressing *F. oxysporum* and *Gaeumannomyces graminis* (Chin-A-Woeng et al. 2003). Besides *Pseudomonas* sp., several strains of *Bacillus* also produce antibiotics like polymyxin, circulin and colistin that are active against numerous Gram-positive and Gram-negative bacteria, as well as many plant pathogenic fungi (Maksimov et al. 2011).

5.3.4 Secretion of Lytic Enzymes

Several lytic enzymes are released by PGPR (Gupta et al. 2015) that are able to destruct/lyse the cell walls of fungal pathogens. Secretion of lytic enzymes, e.g. chitinase (Shrivastava et al. 2016), glucanase (Figueroa-Lopez et al. 2016), β-1,3-glucanase (El-Gamal et al. 2016), cellulases (Ashwini and Srividya 2014), proteases (Illakiam et al. 2013), lipases (Tiru et al. 2013), etc. is yet another mode of defence adopted by PGPR to protect plants from damage caused by phytopathogens. These lytic enzymes can degrade the cell wall of the pathogenic fungi and ultimately cause their death. Since the fungal cell walls are mainly composed of chitin and beta-glucans, the beneficial antagonistic PGPR could inhibit the growth of pathogenic fungi by degrading their cell walls through these lytic enzymes. Symbiotic nitrogen-fixing plant growth-promoting rhizobacteria, *Sinorhizobium fredii* strain KCC5, and free-living PGPR, *P. fluorescens* strain LPK2, have been reported to produce lytic enzymes such as chitinase and beta-glucanases, which have been found to inhibit the growth of *Fusarium udum* leading consequently to manage the fusarium wilt disease caused by the fungus (Kumar et al. 2010).

5.3.5 Induced Systemic Resistance (ISR)

Some PGPR do not directly inhibit the pathogens, instead they activate the host plants to develop resistance against specific attacking pathogen, through a mechanism commonly known as induced systemic resistance (ISR). Principally, ISR is defined as the mechanism of enhanced resistance at specific sites of plant tissue at which disease induction has occurred. Only when a potent pathogen attacks the host plant, the defence mechanism of ISR is activated in its response. In other words, ISR is a condition of enhanced defence developed by a plant when appropriately stimulated by an attacking pathogen (Van Loon et al. 1998). There are numerous biotic and abiotic agents that can protect crops from pathogenic microorganisms by eliciting ISR (Da Rocha and Hammerschmidt 2005; Reglinski and Walters 2009; De Vleesschauwer and Höfte 2009). Of these, the biotic agents include a varied range of plant growth promoters including *Bacillus* sp. (Jourdan et al. 2009; Kloepper et al. 2004), *Pseudomonas* sp. (Bakker et al. 2007), *Serratia* sp. (Press et al. 1997; Schuhegger et al. 2006), *Trichoderma* sp. (Koike et al. 2001; Segarra et al. 2009), *Piriformospora indica* (Shoresh et al. 2010), *Penicillium simplicissimum* (Elsharkawy et al. 2012), *Phoma* sp. (Sultana et al. 2009), non-pathogenic *F. oxysporum* (Fravel et al. 2003) and arbuscular mycorrhizal fungi (Pozo et al. 2009). However, the ISR is not specific against particular pathogen but may play a major role in controlling plant diseases. The major role in providing systemic resistance by plants to various plant pathogens is primarily due to plant hormones jasmonic acid and ethylene. The crosstalk between these two molecules leads to enhanced resistance to pathogens.

5.3.6 Competition

The ability to compete for limited space and scarcely available nutrients within the rhizosphere is another defence mechanism that has evolved within PGPR strains. The plant growth-promoting rhizobacteria sometimes compete with the plant pathogenic microbes for various nutrients present in trace amounts which can limit the growth of the disease-causing pathogens. The beneficial microflora of the rhizosphere, especially the pseudomonads, are efficient colonizers (Zhao et al. 2013) which very efficiently colonize the surface of plant roots and in turn limit the growth of pathogenic microbes. Moreover, the growth-promoting rhizobacteria, when inoculated onto seeds or soils, compete for the available nutrients. Through active uptake of essential nutrients, the PGPR inhibits the growth of pathogenic fungi and bacteria by limiting the availability of nutrients to competing microbiota. Summarily, various beneficial soilborne PGPR such as *Pseudomonas* sp. and *Bacillus* sp. endowed with massive potential of protecting plants against pathogenic microorganisms involving a wide range of mechanisms, such as competition for space and nutrients, production of secondary metabolites, release of antibiotics and bacteriocins, production of iron-chelating siderophores, secretion of lytic enzymes and elicitation of induced systemic resistance (ISR) (Pieterse et al. 2014), could be used to protect crops including vegetables from negative impact of phytopathogens.

5.4   Some Examples of Growth Promotion and Vegetable Disease Management by PGPR Wilt Disease: A General Perspective

Bacterial wilt is a common disease among vegetables and affects mainly tomato, eggplant, potato, tobacco and pepper. The causal organism of bacterial wilt is *Ralstonia solanacearum* which is highly devastating for the crops (Hayward 1991). Moreover, nearly 450 different species of other crops serve as suitable hosts for this bacterial pathogen (Swanson et al. 2005). *Ralstonia solanacearum* thrives mainly in the tropical and subtropical regions of the world (Kelman 1998) and is known to cause enormous yield losses of the vegetable crops. Attempts have been made to control the menace caused by bacterial wilt using PGPR formulations having antagonistic abilities against *R. solanacearum* (Nguyen and Ranamukhaarachchi 2010).

5.4.1   Diseases of Tomato and their Management

5.4.1.1 Bacterial Wilt of Tomato

Tomato (*Solanum lycopersicum* L.) is an important vegetable crop grown and consumed worldwide. It is a rich source of vitamin A and C and is most popular among vegetables because of high nutritive value. Among various diseases, bacterial wilt is the most common and destructive disease of tomato caused by *R. solanacearum* (Tahat and Kamaruzaman 2010). The yield loss of the crops due to this pathogen ranges from 2 to 90% in various agro-climatic conditions (Mishra et al. 1995). To overcome the losses caused due to bacterial wilt, various strategies including the use of agrochemicals have been adopted to control the disease (Singh et al. 2012). However, application of these chemicals has not been found effective enough to control the disease; rather such chemicals following deposition in soils have resulted in deleterious impact on soil fertility and plant health. Thus, growers, in order to avoid chemicals threat, rely on biological control measures for the management of bacterial wilt disease (Singh et al. 2013). In this regard, several antagonistic bacteria, such as *P. fluorescens*, *P. putida*, *Bacillus* sp., etc., have been used to control wilt disease in tomato (Singh et al. 2016; Toua et al. 2013). Among various bacterial antagonists, *Bacillus* spp. including *B. amyloliquefaciens*, *B. coagulans*, *B. cereus*, *B. licheniformis*, *B. pumilus*, *B. subtilis* and *B. vallismortis* have been used extensively for controlling the disease effectively (Tan et al. 2013). In a study various strains of *Bacillus* including *B. amyloliquefaciens* DSBA-11 and DSBA-12, *B. cereus* JHTBS-7, *B. pumilus* MTCC-7092 and *B. subtilis* DTBS-5 were selected to test their comparative antagonistic ability to control wilt disease as well as growth promotion of tomato. The results revealed minimum disease intensity (17.95%) and maximum biocontrol efficacy (68.19%) in tomato plants inoculated with *B. amyloliquefaciens* DSBA-11. The intensity of the disease was, however, a little higher in case of other treatments, for example, *B. amyloliquefaciens* strain DSBA-12 which showed the disease intensity up to 20.81% while *B. subtilis* strain DTBS-5 could reduce the intensity of the disease up to 21.63% after 30 days of initiation of infection by *R. solanacearum*. Furthermore, the population of *R. solanacearum* decreased in *Bacillus*-treated plants. Also, *Bacillus* strains improved other growth parameters of tomato plants. For instance, maximum shoot length (39.50 cm) was recorded in *B. subtilis* DTBS-5-inoculated plants which was followed by *B. amyloliquefaciens* DSBA-11 (38.50 cm) and *B. amyloliquefaciens* DSBA-12 (38.40 cm). Likewise, root length was maximum in plants inoculated with *B. amyloliquefaciens* strain DSBA-11, followed by *B. amyloliquefaciens* DSBA-12 after 30 days of inoculation. Similarly, the dry matter accumulation in root and shoots also enhanced correspondingly (Singh et al. 2016).

5.4.1.2 Fusarium Wilt of Tomato

Fusarium wilt caused by *Fusarium* sp. causes severe tomato yield losses. Yellowing and wilting of the lower leaves are the initial symptoms of the disease that could be visible on the plant (Khan and Khan 2002). The fungus invades the host tissue and the microconidia and grows intercellularly within the xylem of the stem and root of the host plant. The xylem tissue is then infected by the fungus resulting in severe damage to the xylem. The damage caused to xylem leads to disruption of water transportation within the plant, which results in death of the infected tomato plant (Burgess et al. 2008). On the other hand, the conidia forms chlamydospores that fall back into the soils (Jones 2000) which germinates under amenable environmental conditions, and thus the reproductive cycle of the fungus continues. The management of *Fusarium* wilt is however a big challenge for tomato growers (Srinon et al. 2006). The use of fungicides and other chemicals has not been a practical method for controlling the disease. Rather, disease management through biocontrol mechanisms involving PGPR is considered an effective and suitable approach. For controlling the disease, several microorganisms like species of *Pseudomonas* and *Bacillus* have been used as successful antagonists against this disease. Of all the antagonists, *Bacillus* sp. has been found very effective in plant disease management (Jacobsen et al. 2004). To substantiate this further, a study conducted by Ajilogba et al. (2013) revealed a significant growth inhibition of *Fusarium solani* by four *Bacillus* strains, namely, *B. amyloliquefaciens*, *B. cereus*, *B. pumilus* and *B. subtilis*. A 95.2% reduction in the growth of *F. solani* was observed when tomato plants were inoculated with *B. Amyloliquefaciens*. Despite the variation in effectiveness of each bacteria strain, all four strains of *Bacillus* served as potential antagonists and successfully protected tomato plants from fusarium wilt disease. Mechanistically, the antagonistic potential of *B. amyloliquefaciens* strain was attributed to the release of various metabolites and antifungal compounds by the test bacterial strains used in this study (Dihazi et al. 2012). Several other studies have also revealed the production of a variety of antibiotics like as zwittermicin, bacillomycin, fengycin, bacilysin and difficidin by *B. amyloliquefaciens* strains which explains the possible mechanism of resistance to fusarium wilt of tomato, thereby leading to improved growth and yield (Athukorala et al. 2009; Chen et al. 2009).

5.4.1.3 Bacterial Wilt of Brinjal

Bacterial wilt of brinjal (*Solanum melongena*) is yet another important disease caused by plant pathogenic bacterium *R. solanacearum* and is a major challenge to brinjal production causing severe losses in crop yield. Several strategies like crop rotation and introduction of resistant cultivars, etc. have been employed for the management of wilt disease, but complete control of the disease has not been achieved so far, since the survivability of the pathogen in soil is longer, and therefore, the same pathogen can reinfect the healthy plants under favourable environmental conditions. Moreover, the strain exists in diverse forms due to which the development of resistant cultivars has become difficult and ineffective (Wang et al. 1998). To minimize the yield losses caused by *R. solanacearum*, application of hazardous chemicals to soil, modification of soil pH, soil solarization, and the use of plant essential oils (e.g. thymol) or phosphoric acid (Norman et al. 2006) have been practised over the years. However, these methods have not been found successful due to one or other reasons (Champoiseau Patrice et al. 2009). Thus, there is an urgent need to overcome this disease so as to safeguard the vegetables and minimize the adverse impact on the environment. In this regard, biological strategies to control plant diseases have been suggested (Lwin and Ranamukhaarachchi 2006). Among various PGPR, strains of *P. fluorescens* are well-known for suppressing soilborne diseases caused by phytopathogens (O'Sullivan and O'Gara 1992). To assess the potential of *P. fluorescens* as a biocontrol agent against bacterial wilt, a study was conducted and the efficacy of *P. fluorescens*-based bioformulations in disease suppression was determined under pot and field trials. During the experiment, the population density of *P. fluorescens* at 30 days after transplanting increased significantly up to 60 days. Besides reducing the disease severity, *P. fluorescens*-based bioformulation also improved the growth and yield attributes of brinjal. Various biological parameters like leaf area, average fruit weight, yield/plant, no. of fruits/plant, no. of branches/plant and plant height were enhanced in the presence of *P. fluorescens* (Chakravarty and Kalita 2011). The formulations when applied to seed, root and soil were more effective in reducing the incidence and severity of bacterial wilt disease in brinjal which could possibly be due to the correct placement of the antagonist *P. fluorescens* on the seed, from where it migrated to the elongating roots (Burr et al. 1978), on the roots which is the best location for colonization by microbes (Anuratha and Gnanamanickam 1990) and on the soil, the collection of both beneficial and pathogenic microorganisms (Dupler and Baker 1984). Thus, the strategy adopted by *P. fluorescens* for disease management including both its colonization on the root surface of brinjal plants and its ability to survive and establish within the soil provides a competitive advantage to the antagonists over the native soil/rhizosphere microflora (Loper et al. 1985).

5.4.1.4 Fusarium Wilt of Brinjal

Fusarium wilt of eggplant is one of the most destructive diseases caused by *F. oxysporum* f. sp. *melongenae*. The pathogenic fungus is soilborne and causes disease in healthy eggplants by invading the vascular bundles. The invasion of vascular bundles ultimately results in severe wilting and finally the death of the plants which occur due to blocking of the xylem tissue and collapsing of the water transport system within the plant (Altinok 2005). Since the spores of *Fusarium* are resistant to environmental stress and can survive in the soil for many years, it becomes difficult to control the fungal growth and spread of the disease through conventional disease management strategies. Thus, the application of beneficial PGPR as biocontrol agents has become important, since they are endowed with multiple disease resistance mechanisms. Realizing the importance of PGPR, a study was conducted to assess the biocontrol potential of certain PGPR isolates against Fusarium wilt disease in brinjal. Among the PGPR isolates, *Pseudomonas aeruginosa* (P07-1), *P. putida* (P11-4), *P. aeruginosa* (85A-2), *Bacillus amyloliquefaciens* (76A-1) and *B. cereus* (B10a) could significantly reduce the incidence of the disease by up to 85%. Interestingly, the PGPR strains exhibited some traits of disease suppression that ultimately led to the inhibition of the mycelial growth of the pathogenic fungus. The percentage of inhibition varied from 38 to 72% depending upon the potentiality of each PGPR strain. Moreover, of all the PGPR strains, *P. aeruginosa* (P07-1) and *P. putida* (P11-4) successfully colonized within the seedlings of eggplant and eliminated the chances of entry of the fungal mycelium within the host tissue and thus prevented the disease incidence. The experiment further revealed that the PGPR isolates could suppress the disease more efficiently when applied singly, rather than when used in combination. Also, the eggplants exhibited the property of induced systemic resistance which was triggered by the PGPR strains in response to *F. oxysporum* f. sp. *melongenae*. The brinjal plants could synthesize several enzymes like peroxidase (POX, EC 1.11.1.7), polyphenol oxidase (PPO, EC 1.14.18.1) catalase (CAT, 1.11.1.6) along with several lytic enzymes capable of degrading the fungal cell wall. The production of enzymes could be a possible mechanism of resistance against Fusarium wilt in brinjal. The study, thus, demonstrated the use of beneficial PGPR that could serve as antagonists and enhance disease resistance for sustainable production of brinjal (Altinok et al. 2013).

5.4.1.5 Diseases of Okra

Root Rot Disease
Okra (*Abelmoschus esculentus*) is one of the important summer vegetables of India with a high average productivity. Field-grown okra is attacked largely by a number of phytopathogens including bacteria, fungi, viruses, nematodes and various insect pests which adversely affect the production, and if the crop is not cured off the pathogens at the right time, it may lead to serious destruction resulting in heavy yield losses that may reach up to 80–90% (Hamer and Thompson 1957). Among various diseases of vegetables, root rot of okra incited by *Rhizoctonia solani* is one of the most serious and devastating diseases of okra and is a menace for its cultivation on a large scale. To highlight the potential of *Pseudomonas* strains as a biocontrol agent against root rot of okra, two isolates of *Pseudomonas flourescens* PF-7 and PF-8 were used in a study where they inhibited the mycelial growth of *R. solani* by 72.05 and 68.25%, respectively. On the other hand, the vigour index of okra was recorded maximum for isolate PF-8 (2415.7) followed by PF-7 (2063.25) (Adhikari et al. 2013). The strains of *P. fluorescens* produced secondary metabolites responsible for the inhibition of fungal growth and proliferation, as a major mechanism of biocontrol of *R. solani*. The other antagonistic attributes of *P. fluorescens* strains included production of pigments, iron-chelating siderophores, cyanogenic compounds like HCN, etc. Besides exhibiting biocontrol properties, *P. fluorescens* strains PF-7 and PF-8 also released certain plant growth-promoting substances like indole acetic acid and salicylic acid and could solubilize inorganic P. All these growth-promoting properties of *P. fluorescens* make this organism a suitable choice for the enhancement of okra production while limiting the root rot disease of okra.

5.4.1.6 Blight Diseases

Early Blight of Potato
Among the most important food crops of the world, potato (*Solanum tuberosum* L.) ranks third after rice and wheat (Anonymous 2012). Globally, India ranks fourth in terms of area under production and fifth overall in the world (Shailbala and Pathak 2008). Potato, popularly known as the king of vegetables, is cultivated mainly in the tropics and in subtropics during the cool and dry seasons. Cultivation of potato suffers heavily from attack of pathogenic microorganisms leading to enormous yield losses. Among various potato diseases, early blight is one of the most common foliar diseases of potato occurring worldwide (Christ 1990; Van der Walls et al. 2001) caused by *Alternaria solani*. In recent past, a constant increase in disease incidence on potato foliage caused by *A. solani* has been reported in various potato-growing areas (Vloutoglou and Kalogerakis 2000). Initial symptoms of the disease begin with premature defoliation of the potato plants, leading to reduction in the yield of potato tubers. The symptoms first occur on the lower senescing leaves, which later on become chlorotic and abscise prematurely. The disease appears as brown spots that enlarge slowly to completely destroy the leaves. The pathogenic fungus infects young seedlings to cause stem canker or collar rot. Sunken spots or

cankers on older stems, dark leathery fruit spots, etc. are some of the other symptoms that appear on the potato plants simultaneously. Sometimes, lesions appear on upper stems and petioles, indicating the severity of the disease (Raziq and Ishtiaq 2010). The loss in yield of potato following infection by *A. solani* depends mainly on season of cropping, location of planting, type of cultivars and the stage of potato at which infection starts. Early blight may also result in other infections including dry rot of tubers, which reduces the quality and quantity of the tubers to be sold in the market (Nnodu et al. 1982). Rotem (2004) reported that high water content in the surrounding atmosphere is favourable for germination of conidia leading to augmentation of infection. Moreover, alternating low and high humidity in the environment also favours disease development (Van der Walls et al. 2001). The incidence of this disease is also enhanced through repeated and continuous production of potato (Olanya et al. 2009). Management of such lethal diseases is a challenge for potato growers. Even though fungicides can be used to circumvent such diseases, the adverse effects of fungicides and chemicals on plants have warranted to search for a safer and inexpensive method to control early blight disease while simultaneously enhancing the potato growth and productivity. Apart from the sole application of some fungi, for example, *Trichoderma* (Chet et al. 1981; Kumar and Mukerji 1996), a bioformulation comprising of *Trichoderma harzianum* and *Pseudomonas fluorescens* has been applied to potato plants along with the fungicide mancozeb to ward off the pathogenic fungus *A. solani*. The severity and incidence of the disease were greatly reduced in the presence of biocontrol agents. Also, the growth and yield of potato were enhanced significantly (Mane et al. 2014). Although the exact mechanism of control of early blight disease by composite culture of *T. harzianum* and *P. fluorescens* is not determined, these combinations were found effective against *A. solani* and, hence, could be developed as a substitute to chemical treatments.

Late Blight of Potato
Late blight disease of potato is another highly destructive disease and is one of the major constraints in potato cultivation (Chycoski and Punja 1996; Fry and Goodwin 1997; Song et al. 2003). In the mid 1800, the disease resulted in severe crop losses throughout Northern Europe including Ireland where it was responsible for the Irish famine (Elansky et al. 2001). Since then, it has spread very rapidly and, in the present time, attacks potatoes on a large scale wherever potatoes are cultivated. The annual losses of potato caused due to *Phytophthora infestans* have been estimated to € 12 billion worldwide, out of which a productivity and yield loss of approximately € 10 billion per annum has been estimated for the developing nations (Haverkort et al. 2009). The causal organism of this disease (*P. infestans*) produces lesions on potato plants which is small and chlorotic initially, but enlarge in size when the climatic conditions are humid, thereby destroying almost the entire plant. The most prominent disease symptom is the appearance of irregular pale green lesions around the tip and margins of the leaves that enlarges to form brown to purplish black necrotic spots. Also, a white mildew, consisting of sporangia and viable spores of the pathogen can be seen on the ventral side of the infected leaves. The stems of the potato plant also get affected by this disease and exhibit light to dark

brown lesions. The entire affected crop appears blackened and may be destroyed within a week if the conditions are favourable for the growth and survival of the pathogen. The sporangia from the diseased foliage fall to the ground and reach the tubers to infect them. Irregular reddish brown to purple coloured spots appear as disease symptoms on the infected potato tubers. As a consequence, rotting of the potato tubers occurs when the favourable conditions arrive and results in heavy yield losses of potato (Flier et al. 2001), thereby leading to a reduction in global production of potato by approximately 15% (Anonymous 1997). The infected tubers may consequently be attacked by soft rot-causing bacteria upon storage. In conventional farming systems, late blight disease is controlled mainly through repeated and injudicious applications of various chemical protectants like fungicides that, after a long term usage, may pose serious threats to plant and soil health (Cooke et al. 2011; Axel et al. 2012). To overcome the losses caused by late blight disease, biocontrol measures have been introduced and employed nowadays as an effective alternate strategy for protection against such devastating diseases (Velivelli et al. 2014).

Considering these, a study was conducted where three *Pseudomonas* strains were tested for their protective ability against late blight disease of potato. The green house experiment revealed that *P. chlororaphis* strain R47 was the most active protectant PGPR. This strain possessed biocontrol potential against *P. infestans* when tested in vitro. However, the protective effect provided by *P. chlororaphis* strain R47 against *P. infestans*, its survival in the phyllosphere and its ability to colonize the potato rhizosphere in a very high number suggest that this strain could be used as a suitable antagonist to late blight of potato under field conditions. *P. chlororaphis* R47 responded to the pathogen most efficiently and showed the highest level of inhibition of *P. infestans* in vitro, followed by *P. fluorescens* R76 and *P. marginalis* S35. The prime mechanism of management of late blight of potato by *Pseudomonas* strains is through the secretion of some antifungal compounds that could probably inhibit the growth of *P. infestans*, thereby leading to a better potato production with highly minimized yield losses (Guyer et al. 2015). *Pseudomonas* strains, in general, have also been reported as the best producers of various antifungal metabolites (Hunziker et al. 2015). Together, these studies suggest that *Pseudomonas* isolates could be used as a potent biocontrol agent against *P. infestans* for potato cultivation on a large scale in different production systems.

Blight Disease of Pepper

Blight of pepper (*Capsicum annuum* L.) is caused by *Phytophthora capsici* and results in severe yield losses. The disease is soilborne in origin and affects the pepper plants cultivated worldwide across major pepper-growing countries like China (Ma et al. 2008), Mexico (Robles-Yerena et al. 2010), Turkey (Akgül and Mirik 2008), Spain (Silvar et al. 2006), The United States of America (Hausbeck and Lamour 2004) and Nigeria (Alegbejo et al. 2006). Although, the disease is difficult to control, yet there are numerous reports where disease has been controlled employing various chemical (Hausbeck and Lamour 2004) and microbial (Kim et al. 2010) fungicides. For example, some *Pseudomonas* isolates from various crops have been used to inhibit the growth of *P. capsici* in vitro and for the production of biosurfactant. Also, the efficacy of selected *Pseudomonas* strains against *P. capsici* was determined in two experiments where the antagonistic bacteria were applied to infected pepper plants along with fungicide acibenzolar-S-methyl (ASM) and mefenoxam, either singly or in combination. Bacterial strains were applied by soil drenching method whereas the fungicides were applied as foliar sprays. The application of four *Pseudomonas* strains resulted in significant reduction in the severity of pepper blight ranging from 48.4 to 61.3% in infected pepper. In another experiment, when *P. fluorescens* was applied along with olive oil, the biocontrol efficiency of the *Pseudomonas* isolates enhanced significantly, resulting in a significant decrease in the level of disease severity from 56.8 to 81.1%. The reduction in severity of disease and consequently the inhibition of germination of zoospores and hyphal growth of *P. capsici* was attributed to the synthesis of rhamnolipid-type biosurfactants by *Pseudomonas* sp. (D'aes et al. 2010). Besides this, other molecules that could be involved in disease management by *P. fluorescens* include the production of a vast array of antibiotics like phenazines, pyrrolnitrin, pyoluteorin and 2,4-diacetylphloroglucinol (Cui and Harling 2006), HCN, indolic compounds and siderophores, etc. Thus, it is established that the use of *P. fluorescens* strains possessing biosurfactant producing properties can be a successful and effective method of blight disease management and plant growth promotion in pepper plants while reducing the use of chemicals and fungicides in pepper production to a great extent (Özyilmaz and Benlioglu 2013).

5.4.1.7 Diseases of Crucifers

Bacterial Soft Rot of Cabbage
Bacterial soft rot is another detrimental disease of vegetables occurring worldwide and affecting several economically important crop plants including crucifers (Pérombelon and Kelman 1980). The disease is caused by *Pectobacterium carotovorum* subsp. *carotovorum* (*Pcc*), one of the most hazardous plant pathogenic bacterium (Kyeremeh et al. 2000) which hinders the production of Chinese cabbage (Kikumoto 2000). Several methods including biological approaches have been attempted to control/minimize the severity of soft rot diseases (Hayward 1991; Bernal et al. 2002). There are few reports available on the control measures of soft rot disease either by using microbial pesticide formulations (Takahara 1994), avirulent mutant strains of *Erwinia* (Takahara et al. 1993; Kyeremeh et al. 2000) or through fluorescent antagonistic bacterium (Togashi et al. 2000) as biocontrol agents. Moreover, disease-resistant transgenic cultivars of Chinese cabbage (Vanjildorj et al. 2009) showing resistance to soft rot have been developed by the growers in an attempt to eradicate this disease to avoid the yield losses. Among microbiological preparations for use against soft rot of cabbage, few bacterial formulations comprising of *Lactobacillus*, *Lactococcus* and *Paenibacillus* strains have been tried against the same disease. Biocontrol efficacies of these bacterial strains were tested against soft rot of cabbage and were found significantly effective as antagonists to the disease. The disease severity for the strains KLF01, KLC02 and

KPB3 was reported as 23, 20 and 20%, respectively, whereas the biocontrol efficacy of KLF01, KLC02 and KPB3 was 55, 60 and 62%, respectively, when tested in field trials. Among various strains used in the study, strain KPB3 proved to be the best biocontrol agent with the highest biocontrol efficacy (Shrestha et al. 2009). The factors affecting growth promotion and disease suppression by *Lactobacillus* and *Lactococcus* strains were suggested as the production of various antibacterial substances like acetic acid, lactic acid (Ariyapitipun et al. 1999), hydrogen peroxide (Chang et al. 1997) and bacteriocins (Klaenhammer 1982); furthermore, these bacterial strains could exhibit antagonistic effect (Visser et al. 1986) and antifungal activity (Laitila et al. 2002) against phytopathogens most probably due to the release of biomolecules mentioned earlier.

5.4.1.8 Diseases of Cucumber

Damping-Off and Root Rot of Cucumber

Damping-off and root rot diseases are mainly caused by an oomycete plant pathogen *Pythium* sp. and damage young seedlings of several horticultural and vegetable crops both under greenhouse and field conditions (Howard et al. 1994; Paulitz and Bélanger 2001). The causal organism of root rot of cucumber is *Pythium ultimum*. The oomycete pathogen generally attacks the juvenile tissues of bedding plants (Gravel et al. 2009), greenhouse transplants and floral crops (Moorman et al. 2002) and direct seeded field crops (Paulitz 2006; Leisso et al. 2009). The most favourable conditions for the growth of damping-off and root rot pathogen are cool and wet environment when it can cause infection of the seedlings in poorly drained soils and eventually kill the young seedlings either before or soon after emergence. Also, it has been reported that various young emerging plant organs like the radicle, hypocotyl, cotyledons, seed coat, endosperm and embryo are highly prone to attack by the pathogen-causing damping-off and root rot diseases (Paulitz et al. 1992). The severity of the disease caused by damping-off and root rot pathogens, however, can be reduced considerably provided some measures are taken to check or slow down the initial attacks by the phytopathogen. In this context, several fungicides such as captan, thiram, iprodione, fenaminosulf, fosetyl-Al and metalaxyl have been applied as seed treatments to control the disease (Leisso et al. 2009). But the biological control has been considered as a good and safe option for the management of damping-off and root rot diseases in both conventional and organic farming practices with least destruction to the environment (Jacobsen and Backman 1993; Georgakopoulos et al. 2002; Nagarajkumar et al. 2004). To further promote and popularize the use of biocontrol agents to eradicate/reduce this disease, several species of non-pathogenic bacteria belonging to the genera *Pseudomonas* and *Bacillus* have been used as potential antagonists to damping-off and root rot pathogen *P. ultimum*. In a study, the biocontrol potential of three most effective antagonistic bacteria was evaluated against seedling damping-off and root rot of cucumber caused by *P. ultimum*. Based on phenotypic characteristics, biochemical characterization and 16S rDNA gene sequence analysis, the three antagonistic bacteria were identified as *P. fluorescens* (9A-14), *Pseudomonas* sp. (8D-45) and *Bacillus subtilis* (8B-1). All of the three bacteria could promote plant growth and simultaneously suppress the effects of damping-off and root rot caused by *P. ultimum* on cucumber seedlings when tested in growth chamber trials. Interestingly, both pre- and post-planting application of bacterial treatment led to a decrease in damping-off and root rot severity in cucumber by 27–50%, thereby resulting in an improved growth (Khabbaz and Abbasi 2014). All the strains could successfully reduce the disease incidence when applied as seed treatment either singly or in combination. The production of antibiotics and some specific metabolites could probably be a possible reason of disease suppression by PGPR isolates. Additionally, the ISR may also be involved in providing protection to cucumber against damping-off and root rot disease (Van Loon et al. 1998; Powell et al. 2000; Van Loon 2007). This study thus suggests that various formulations of PGPR can be used to develop biofungicides to minimize the crop losses caused by seedling damping-off and root rot disease in cucumber and other vegetables of economic importance.

5.5   Conclusion and Future Prospects

Vegetables are grown on a large scale worldwide to fulfil human food demands. But unfortunately, most of the vegetable crops are lost due to bacterial and fungal phytopathogens that cause major diseases leading eventually to enormous yield losses. To minimize the yield loss in vegetables, several conventional approaches for plant disease management like developing resistant cultivars, crop rotation, field sanitization, spraying of fungicides, etc. have been practised over the years. But these methods have not been found fully effective in controlling plant diseases, and more so such strategies are expensive and labour intensive. Also, the use of fungicides and other chemicals adversely affects the quality and productivity of the vegetables. Thus, production of disease-free vegetables becomes a challenging task for the growers. In this context, biological control measures could be an effective alternate approach for containing vegetable diseases. Several plant growth-promoting rhizobacteria are known to suppress various diseases of vegetables by employing one or a combination of mechanisms leading eventually to enhancement in production. Application of such beneficial microbes is likely to reduce the use of chemicals in vegetable production practices in different production systems.

References

Abdel-Monaim MF, Abdel-Gaid MA, Zayan SA et al (2014) Enhancement of growth parameters and yield components in eggplant using antagonism of *Trichoderma* spp. against fusarium wilt disease. Int J Phytopathol 3(1):33–40

Adesemoye AO, Obini M, Ugoji EO (2008) Comparison of plant growth-promotion with *Pseudomonas aeruginosa* and *Bacillus subtilis* in three vegetables. Braz J Microbiol 39:423–426

Adhikari A, Dutta S, Nandi S et al (2013) Antagonistic potentiality of native rhizobacterial isolates against root rot disease of okra incited by *Rhizoctonia solani*. Afr J Agric Res 8(4):405–412

Ajilogba CF, Babalola OO, Ahmad F (2013) Antagonistic effects of *Bacillus* species in biocontrol of tomato *Fusarium* wilt. Ethno Med 7(3):205–216

Akgül SD, Mirik M (2008) Biocontrol of *Phytophthora capsici* on pepper plants by *Bacillus megaterium* strains. J Plant Pathol 90:29–34

Alegbejo MD, Lawal AB, Chindo PS (2006) Outbreak of basal stem rot and wilt disease of pepper in Katsina, Nigeria. Arch Phytopathol PFL 39:93–98

Altinok HH (2005) First report of Fusarium wilt of eggplant caused by *Fusarium oxysporum* f. sp. *melongenae* in Turkey. Plant Pathol 54:577

Altinok HH, Dikilitas M, Yildiz HN (2013) Potential of *Pseudomonas* and *Bacillus* isolates as biocontrol agents against fusarium wilt of eggplant. Biotechnol Biotechnol Eq 27(4):3952–3958

Andrews SC, Robinson AK, Rodríguez-Quiñones F (2003) Bacterial iron homeostasis. FEMS Microbiol Rev 27:215–237

Anonymous (1997) The International Potato Centre annual report. International Potato Centre, Lima, p 59

Anonymous (2012) Small farmer's agriculture consortium and Indian Agriculture Systems Pvt. Ltd. http://sfacindia.com

Anuratha CS, Gnanamanickam SS (1990) Biological control of bacterial wilt caused by *Pseudomonas solanacearum* in India with antagonistic bacteria. Plant Soil 124:109–116

Ariyapitipun T, Mustapha A, Clarke AD (1999) Microbial shelf life determination of vacuum packaged fresh beef treated with polylacetic acid and nisin solutions. J Food Protect 62:913–920

Ashwini N, Srividya S (2014) Potentiality of *Bacillus subtilis* as biocontrol agent for management of anthracnose disease of chilli caused by *Colletotrichum gloeosporioides* OGC1. 3 Biotech 4:127–136

Athukorala SNP, Fernando WGD, Rashid KY (2009) Identification of antifungal antibiotics of *Bacillus* species isolated from different microhabitats using polymerase chain reaction and MALDI-TOF mass spectrometry. Can J Microbiol 55:1021–1032

Axel C, Zannini E, Coffey A et al (2012) Ecofriendly control of potato late blight causative agent and the potential role of lactic acid bacteria: a review. Appl Microbiol Biotechnol 96:37–48. doi:10.1007/s00253-012-4282-y

Bakker PAHM, Pieterse CMJ, Van Loon LC (2007) Induced systemic resistance by fluorescent *Pseudomonas* sp. Phytopathology 97:239–243

Beneduzi A, Ambrosini A, Luciane MP et al (2012) Plant growth-promoting rhizobacteria (PGPR): their potential as antagonists and biocontrol agents. Gen Mol Biol 35(4, Suppl):1044–1051

Bernal G, Illanes A, Ciampi L (2002) Isolation and partial purification of a metabolite from a mutant strain of *Bacillus* sp. with antibiotic activity against plant pathogenic agents. Electron J Biotechnol 5:12–20

Bharucha UD, Patel KC, Trivedi UB (2013) Antifungal activity of catecholate type siderophore produced by *Bacillus* sp. Int J Res Pharm Sci 4(4):528–531

Burgess LW, Knight TE, Tesoriero L et al (2008) Diagnostic manual for plant diseases in Vietnam. ACIAR, Canberra

Burr TJ, Schroth MN, Suslow T (1978) Increased potato yields by treatment of seed pieces with specific strains of *Pseudomonas fluorescens* and *Pseudomonas putida*. Phytopathology 68:1377–1383

Cabanás CGL, Schilirò E, Valverde-Corredor A et al (2014) The biocontrol endophytic bacterium *Pseudomonas fluorescens* PICF7 induces systemic defence responses in aerial tissues upon colonization of olive roots. Front Microbiol. doi:10.3389/fmicb.2014.00427

Chakravarty G, Kalita MC (2011) Management of bacterial wilt of brinjal by *P. fluorescens* based bioformulation. J Agric Biol Sci 6(3):1–11

Chakravarty G, Kalita MC (2012) Biocontrol potential of *Pseudomonas fluorescens* against bacterial wilt of brinjal and its possible plant growth promoting effects. Ann Biol Res 3(11):5083–5094

Champoiseau PG, Jones JB, Allen C (2009) *Ralstonia solanacearum* Race 3 Biovar 2 causes tropical losses and temperate anxieties. Plant Health Progress. doi: 10.1094/PHP-2009-0313-01-RV

Chang IS, Kim BH, Shin PK (1997) Use of sulphite and hydrogen peroxide to control bacterial contamination in ethanol production. Appl Environ Microbiol 63:1–6

Chen X, Scholz R, Borriss M et al (2009) Difficidin and bacilysin produced by plant-associated *Bacillus amyloliquefaciens* dare efficient in controlling fire blight disease. J Biotechnol 140:38–44

Chet I, Harman GE, Baker R (1981) *Trichoderma hamatum*: it's hyphal interactions with *Rhizoctoniasolani* & *Pythium* sp. Microb Ecol 7:29–38

Chin-A-Woeng TF, Bloemberg GV, Lugtenberg BJ (2003) Phenazines and their role in biocontrol by *Pseudomonas* bacteria. New Phytol 157:503–523

Christ BJ (1990) Influence of potato cultivars on the effectiveness of fungicide control of early blight. Am Potato J 67:3–11

Chycoski CI, Punja ZK (1996) Characteristics of populations of *Phytophthora infestans* from potato in British Columbia and other regions of Canada during 1993 to 1995. Plant Dis 80:579–589

Cooke LR, Schepers HTAM, Hermanse A et al (2011) Epidemiology and integrated control of potato late blight in Europe. Potato Res 54:183–222. doi:10.1007/s11540-011-9187-0

Cui X, Harling R (2006) Evaluation of bacterial antagonists for biological control of broccoli head rot caused by *Pseudomonas fluorescens*. Phytopathology 96:408–416

D'aes J, De Maeyer K, Pauwelyn E et al (2010) Biosurfactants in plant–*Pseudomonas* interactions and their importance to biocontrol. Environ Microbiol Rep 2:359–372

Da Rocha AB, Hammerschmidt R (2005) History and perspectives on the use of disease resistance inducers in horticultural crops. Hortic Technol 15:518–529

De Souza JT, Arnould C, Deulvot C et al (2003) Effect of 2,4-diacetylphloroglucinol on *Pythium*: cellular responses and variation in sensitivity among propagules and species. Phytopathology 93:966–975

De Vleesschauwer D, Höfte M (2009) Rhizobacteria-induced systemic resistance. Adv Bot Res 51:223–281

DESA, Department of Economic and Social Affairs (2000) World population prospects: the 2000 revision. United nation population division, department of economic and social affairs in Badaurakis. Int J Agribus 18(4):543–558

Dias A, Santos SG, Vasconcelos VGS et al (2013) Screening of plant growth promoting rhizobacteria for the development of vegetable crops inoculants. Afr J Microbiol Res 7(19):2087–2092

Dihazi A, Jaiti FW, Wafataktak et al (2012) Use of two bacteria for biological control of bayoud disease caused by *Fusarium oxysporum* in date palm (*Phoenix dactylifera* L) seedlings. Plant Physiol Biochem 55:7–15

Dilantha WG, Nakkeeran S, Zhang Y (2005) Biosynthesis of antibiotics by PGPR and its relation in biocontrol of plant diseases. Biocont Biofertil:67–109

Dupler M, Baker R (1984) Survival of *Pseudomonas putida*, a biological control agent in soil. Phytopathology 74:195–200

Dwivedi D, Johri BN (2003) Antifungals from fluorescent pseudomonads: biosynthesis and regulation. Curr Sci 12:1693–1703

Elansky SN, Smirnov AN, Dyakov Y et al (2001) Genotypic analysis of Russian isolates of *Phytophthora infestans* from the Moscow region, Siberia, and Far East. J Phytopathol 149:605–611

El-Gamal NG, Shehata AN, Hamed ER et al (2016) Improvement of lytic enzymes producing *Pseudomonas fluorescens* and *Bacillus subtilis* isolates for enhancing their biocontrol potential against root rot disease in tomato plants. Res J Pharm Biol Chem Sci 7(1):1394–1400

Elsharkawy MM, Shimizu M, Takahashi H et al (2012) Induction of systemic resistance against Cucumber mosaic virus by *Penicillium simplicissimum* GP17-2 in *Arabidopsis* and tobacco. Plant Pathol. doi:10.1111/j.1365-3059.2011.02573.x

Figueroa-Lopez AM, Cordero-Ramirez JD, Martinez-Alvarez JC et al (2016) Rhizospheric bacteria of maize with potential for biocontrol of *Fusarium verticillioides*. SpringerPlus 5:330

Flier WG, Turkensteen LJ, Van Den Bosch GBM et al (2001) Differential interaction of *Phytophthora infestans* on tubers of potato cultivars with different levels of blight resistance. Plant Pathol 50(3):292–301

Fouzia A, Allaoua S, Hafsa CS et al (2015) Plant growth promoting and antagonistic traits of indigenous fluorescent *pseudomonas* spp. isolated from wheat rhizosphere and *A. halimus* endosphere. Eur Sci J 11(24):130–148

Fravel D, Olivain C, Alabouvette C (2003) *Fusarium oxysporum* and its biocontrol. New Phytol 157:493–502

Fry WE, Goodwin SB (1997) Re emergence of potato and tomato late blight in the United States. Plant Dis 81:1349–1357

Gafar MO, Elhag AZ, Abdelgader MA (2013) Impact of pesticides malathion and sevin on growth of snake cucumber (*Cucumis melo* L. var. Flexuosus) and soil. Univ J Agric Res 1(3):81–84

Georgakopoulos DG, Fiddaman P, Leifert C et al (2002) Biological control of cucumber and sugar beet damping-off caused by *Pythium ultimum* with bacterial and fungal antagonists. J Appl Microbiol 92:1078–1086

George E, Kumar SN, Jacob J et al (2015) Characterization of the bioactive metabolites from a plant growth-promoting rhizobacteria and their exploitation as antimicrobial and plant growth-promoting agents. Appl Biochem Biotechnol 176(2):529–546

Glick BR (1995) The enhancement of plant growth by free-living bacteria. Can J Microbiol 41:109–117

Glick BR, Cheng Z, Czarny J et al (2007) Promotion of plant growth by ACC deaminase-producing soil bacteria. Eur J Plant Pathol 119:329–339

Gravel V, Ménard C, Dorais M (2009) Pythium root rot and growth responses of organically grown geranium plants to beneficial microorganism. Hortic Sci 44:1622–1627

Gupta G, Parihar SS, Ahirwar NK et al (2015) Plant growth promoting rhizobacteria (PGPR): current and future prospects for development of sustainable agriculture. J Microb Biochem Technol 7:2

Guyer A, DeVrieze M, Bönisch D et al (2015) The Anti *Phytophthora* effect of selected potato associated *Pseudomonas* strains: from the laboratory to the field. Front Microbiol 6:1309

Haas D, Défago G (2005) Biological control of soil-borne pathogens by fluorescent pseudomonads. Nat Rev Microbiol 3:307–319

Haas D, Keel C (2003) Regulation of antibiotic production in root colonizing *Pseudomonas* sp. and relevance for biological control of plant disease. Annu Rev Phytopathol 41:117–153

Hamer C, Thompson T (1957) Vegetable crops. McGraw Hill Co., Inc., N. X. Toronto, London

Hausbeck MK, Lamour KH (2004) *Phytophthora capsici* on vegetable crops: research progress and management challenges. Plant Dis 88:1292–1303

Haverkort AJ, Struik PC, Visser RGF et al (2009) Applied biotechnology to control late blight in potato caused by *Phytophthora infestans*. Potato Res 52:249–264

Hayward AC (1991) Biology and epidemiology of bacterial wilt caused by *Pseudomonas solanacearum*. Annu Rev Phytopathol 29:65–87

Howard RJ, Garland JA, Seaman WL (1994) Diseases and pests of vegetable crops in Canada: an illustrated compendium. Canadian Phytopathological Society and Entomological Society of Canada, Ottawa, ON

Hunziker L, Bonisch D, Groenhagen U et al (2015) *Pseudomonas* strains naturally associated with potato plants produce volatiles with high potential for inhibition of *Phytophthora infestans*. Appl Environ Microbiol 81:821–830

Ikeda K, Banno S, Furusawa A et al (2015) Crop rotation with broccoli suppresses Verticillium wilt of eggplant. J Gen Plant Pathol 81(1):77–82

Illakiam D, Anuj NL, Ponraj P et al (2013) Proteolytic enzyme mediated antagonistic potential of *Pseudomonas aeruginosa* against *Macrophomina phaseolina*. Indian J Exp Biol 51:1024–1031

Jacobsen BJ, Backman PA (1993) Biological and cultural plant disease controls: alternatives and supplements of chemicals in IPM systems. Plant Dis 77:311–315

Jacobsen BJ, Zidack NK, Larson BJ (2004) The role of *Bacillus*-based biological control agents in integrated pest management systems: plant diseases. In: Symposium—the nature and application of biocontrol microbes: *Bacillus* sp. Phytopathology 94:1272–1275

Jones DR (2000) History of banana breeding. In: Jones D (ed) Diseases of banana, abaca and enset. CAB International, Wallingford, UK, pp 425–449

Jourdan E, Henry G, Duby F et al (2009) Insights into the defence related events occurring in plant cells following perception of surfactin-type lipopeptide from Bacillus subtilis. Mol Plant-Microbe Interact 22:456–468

Keel C, Schnider U, Maurhofer M et al (1992) Suppression of root diseases by *Pseudomonas fluorescens* CHA0: importance of the bacterial secondary metabolite 2,4-diacetylphloroglucinol. Mol Plant-Microbe Interact 5:4–13

Keel C, Wirthner P, Oberhänsli T et al (1990) Pseudomonads as antagonists of plant pathogens in the rhizosphere: role of the antibiotic 2,4-diacetylphloroglucinol in the suppression of black root-rot of tobacco. Symbiosis 9:327–341

Kelman A (1998) One hundred and one years of research on bacterial wilt. In: Prior P, Allen C, Elphinstone J (eds) Bacterial wilt: molecular and ecological aspects. INRA Editions, Paris, France, pp 1–5

Khabbaz SE, Abbasi PA (2014) Isolation, characterization, and formulation of antagonistic bacteria for the management of seedlings damping-off and root rot disease of cucumber. Can J Microbiol 60:25–33

Khan MR, Khan SM (2002) Effects of root-dip treatment with certain phosphate solubilizing microorganisms on the fusariam wilt of tomato. Bioresour Technol 85:213–215

Kikumoto T (2000) Ecology and biocontrol of soft rot of Chinese cabbage. J Gen Plant Pathol 66:275–277

Kim SG, Jang Y, Kim HY et al (2010) Comparison of microbial fungicides in antagonistic activities related to the biological control of phytophthora blight in chilli pepper caused by *Phytophthora capsici*. Plant Pathol J 26:340–345

Klaenhammer TR (1982) Bacteriocins of lactic acid bacteria. Biochimie 70:337–349

Kloepper JW, Schroth MN (1978) Plant growth-promoting rhizobacteria on radishes. In: Proceedings of the 4th international conference on plant pathogenic bacteria, vol II. Gilbert-Clay, Tours, France, pp 879–882

Kloepper JW, Ryu CM, Zhang S (2004) Induced systemic resistance and promotion of plant growth by *Bacillus* sp. Phytopathology 94:1259–1266

Koike N, Hyakumachi M, Kageyama K et al (2001) Induction of systemic resistance in cucumber against several diseases by plant growth-promoting fungi: lignifications and superoxide generation. Eur J Plant Pathol 108:187–196

Kremer RJ, Souissi T (2001) Cyanide production by rhizobacteria and potential for suppression of weed seedling growth. Curr Microbiol 43(3):182–186

Kumar RN, Mukerji KG (1996) Integrated disease management future perspectives. In: Mukerji KG, Mathur B, Chamala BP, Chitralekha C (eds) Advances in botany. APH Publishing Corporation, New Delhi, India, pp 335–347

Kumar H, Bajpai VK, Dubey RC (2010) Wilt disease management and enhancement of growth and yield of *Cajanus cajan* (L) var. Manak by bacterial combinations amended with chemical fertilizer. Crop Protect 29:591–598

Kumari S, Khanna V (2014) Effect of antagonistic rhizobacteria inoculated with *Mesorhizobium ciceri* on control of fusarium wilt in chickpea (*Cicer arietinum* L.) Afr J Microbiol Res 8(12):1255–1265

Kyeremeh GAT, Kikumoto D, Chuang Y et al (2000) Biological control of soft rot of Chinese cabbage using single and mixed treatments of bacteriocin producing avirulent mutants of *Erwinia carotovora* subsp. *carotovora*. J Gen Plant Pathol 66:264–268

Laitila AH, Alakomi L, Raaska L et al (2002) Antifungal activities of two *Lactobacillus plantarum* strains against *Fusarium* moulds in vitro and inmalting of barley. J Appl Microbiol 93:556–576

Lakshmi V, Kumari S, Singh A et al (2015) Isolation and characterization of deleterious *Pseudomonas aeruginosa* KCl from rhizospheric soils and its interaction with weed seedlings. J King Saud Univ Sci 27:113–119

Leisso RS, Miller PR, Burrows ME (2009) The influence of biological and fungicidal seed treatments on chickpea (*Cicer arietinum*) damping off. Can J Plant Pathol 31:38–46

Li H, Ding X, Wang C et al (2016) Control of tomato yellow leaf curl virus disease by *Enterobacter asburiae* BQ9 as a result of priming plant resistance in tomatoes. Turk J Biol 40:150–159

Liu K, Garrett C, Fadamiro H et al (2016) Antagonism of black rot in cabbage by mixtures of plant growth-promoting rhizobacteria (PGPR). BioControl 61:1–9

Loganathan M, Garg R, Venkataravanappa V et al (2014) Plant growth promoting rhizobacteria (PGPR) induces resistance against *Fusarium* wilt and improves lycopene content and texture in tomato. Afr J Microbiol Res 8(11):1105–1111

Loper JE, Buyer JW (1991) Siderophores in microbial interactions on plant surfaces. Mol Plant Microbe Int 4:5–13

Loper JE, Haack C, Schroth MN (1985) Population dynamics of soil pseudomonads in the rhizosphere of potato (*Solanum tuberosum* L.) Appl Environ Microbiol 49:416–422

Lucy M, Reed E, Glick BR (2004) Applications of free living plant growth-promoting rhizobacteria. Antonie Van Leeuwenhoek 86:1–25

Luján AM, Gómez P, Buckling A (2015) Siderophore cooperation of the bacterium *Pseudomonas fluorescens* in soil. Biol Lett 11:20140934

Lukkani NJ, Reddy ECS (2014) Evaluation of plant growth promoting attributes and biocontrol potential of native fluorescent *pseudomonas* spp. against *Aspergillus niger* causing collar rot of ground nut. Int J Plant Animal Environ Sci 4(4):256–262

Luna CL, Mariano RLR, Souto-Maior AM (2002) Production of a biocontrol agent for Crucifers black rot disease. Braz J Chem Eng 19(2):133–140

Lwin M, Ranamukhaarachchi SL (2006) Development of biological control of *Ralstonia solanacearum* through antagonistic microbial populations. Int J Agric Biol 8(5):657–660

Ma Y, Chang Z, Zhao J et al (2008) Antifungal activity of *Penicillium striatisporum* Pst10 and its biocontrol effect on *Phytophthora* root rot of chilli pepper. Biol Control 44:24–31

Maksimov IV, Abizgil'dina RR, Pusenkova LI (2011) Plant growth promoting rhizobacteria as alternative to chemical crop protectors from pathogens (Review). Appl Biochem Microbiol 47:333–345

Mane MM, Lal A, Zghair QN et al (2014) Efficacy of certain bio agents and fungicides against early blight of potato (*Solanum tuberosum* L.) Int J Plant Protect 7(2):433–436

Maurhofer M, Keel C, Haas D et al (1994) Pyoluteorin production by *Pseudomonas fluorescens* strain CHA0 is involved in the suppression of *Pythium* damping-off of cress but not of cucumber. Eur J Plant Pathol 100:221–232

Maurhofer M, Keel C, Schnider U et al (1992) Influence of enhanced antibiotic production in *Pseudomonas fluorescens* strain CHA0 on its disease suppressive capacity. Phytopathology 82:190–195

Mercado-Flores Y, Cárdenas-Álvarez IO, Rojas-Olvera AV et al (2014) Application of *Bacillus subtilis* in the biological control of the phytopathogenic fungus *Sporisorium reilianum*. Biol Control 76:36–40

Meyer SLF, Everts KL, McSpadden Gardener B et al (2016) Assessment of DAPG-producing *Pseudomonas fluorescens* for management of *Meloidogyne incognita* and *Fusarium oxysporum* on watermelon. J Nematol 48(1):43–53

Mishra A, Mishra SK, Karmakar SK et al (1995) Assessment of yield loss due to wilting and some popular tomato cultivars. Environ Ecol 28:287–290

Moorman GW, Kang S, Geiser DM et al (2002) Identification and characterization of *Pythium* species associated with greenhouse floral crops in Pennsylvania. Plant Dis 86:1227–1231. doi:10.1094/PDIS.2002.86.11.1227

Nagarajkumar M, Bhaskaran R, Velazhahan R (2004) Involvement of secondary metabolites and extracellular lytic enzymes produced by *Pseudomonas fluorescens* in inhibition of *Rhizoctonia solani*, the rice sheath blight pathogen. Microbiol Res 159:73–81

Ng LC, Ngadin A, Azhari M et al (2015) Potential of *Trichoderma* spp. as biological control agents against bakanae pathogen (*Fusarium fujikuroi*) in Rice. Asian J Plant Pathol 9(2):46–58

Nguyen MT, Ranamukhaarachchi SL (2010) Soil-borne antagonists for biological control of bacterial wilt disease caused by *Ralstonia solanacearum* in tomato and pepper. J Plant Pathol 92:395–406

Nnodu EC, Harrison MD, Parke RV (1982) The effect of temperature and relative humidity on wound healing and infection of potato tubers by *Alternaria solani*. Am Pot J 59:297–311

Norman DJ, Chen J, Yuen JMF et al (2006) Control of bacterial wilt of *Geranium* with phosphorous acid. Plant Dis 90:798–802

O'Sullivan DJ, O'Gara F (1992) Traits of fluorescent *Pseudomonas* sp. involved in suppression of plant root pathogens. Microbiol Rev 56:662–672

Olanya GM, Moneycutt CW, Larkin RP et al (2009) The effect of cropping systems and irrigation management on development of potato early blight. J Gen Plant Pathol 75:267–275

Özyilmaz U, Benlioglu K (2013) Enhanced biological control of *Phytophthora* blight of pepper by biosurfactant-producing *Pseudomonas*. Plant Pathol J 29(4):418–426

Paulitz TC (2006) Low input no-till cereal production in the Pacific Northwest of the U.S.: the challenges of root diseases. Eur J Plant Pathol 115:271–281. doi:10.1007/s10658-006-9023-6

Paulitz TC, Bélanger RR (2001) Biological control in greenhouse systems. Annu Rev Phytopathol 39:103–133. doi:10.1146/annurev.phyto.39.1.103. PMID: 11701861

Paulitz TC, Anas O, Fernando DG (1992) Biological control of *Pythium* damping-off by seed-treatment with *Pseudomonas putida*: relationship with ethanol production by pea and soybean seeds. Biocontrol Sci Tech 2:193–201. doi:10.1080/09583159209355233

Peek ME, Bhatnagar A, McCarty NA et al (2012) Pyoverdine, the major siderophore in *Pseudomonas aeruginosa*, evades NGAL recognition. Interdiscip Perspect Infect Dis . doi:10.1155/2012/843509Article ID: 843509

Pérombelon MCM, Kelman A (1980) Ecology of the soft rot *Erwinias*. Annu Rev Phytopathol 18:361–387

Pieterse CMJ, Zamioudis C, Berendsen RL et al (2014) Induced systemic resistance by beneficial microbes. Annu Rev Phytopathol 52:347–375. doi:10.1146/annurev-phyto-082712-102340. PMID: 24906124

Powell JF, Vargas JM Jr, Nair MG et al (2000) Management of dollar spot on creeping bentgrass with metabolites of *Pseudomonas aureofaciens* (TX-1). Plant Dis 84:19–28. doi:10.1094/PDIS.2000.84.1.19

Pozo MJ, Verhage A, García-Andrade J et al (2009) Priming plant defences against pathogens by arbuscular mycorrhizal fungi. In: Aguilar CA, Barea JM, Gianinazzi S, Gianinazzi-Pearson V (eds) Mycorrhizas: functional processes and ecological impact. Springer-Verlag, Heidelberg, pp 137–149

Press CM, Wilson M, Tuzun S et al (1997) Salicylic acid produced by *Serratia marcescens* 90–166 is not the primary determinant of induced systemic resistance in cucumber or tobacco. Mol Plant-Microbe Interact 10:761–768

Raaijmakers JM, Vlami M, De Souza JT (2002) Antibiotic production by bacterial biocontrol agents. Antonie Van Leeuwenhoek 81:537–547

Raziq F, Ishtiaq S (2010) Integrated control of *Alternaria solani* with *Trichoderma* sp. and fungicides under *in vitro* conditions. Sarhad J Agric 26(4):613–619

Reetha AK, Pavani SL, Mohan S (2014) Hydrogen cyanide production ability by bacterial antagonist and their antibiotics inhibition potential on *Macrophomina phaseolina* (Tassi.) Goid. Int J Curr Microbiol Appl Sci 3(5):172–1783

Reglinski T, Walters D (2009) Induced resistance for plant disease control. In: Walters D (ed) Disease control in crops. Wiley-Blackwell, Oxford, UK, pp 62–92

Robles-Yerena L, Rodríguez-Villarreal RA, Ortega-Amaro MA et al (2010) Characterization of a new fungal antagonist of *Phytophthora capsici*. Sci Hortic Amsterdam 125:248–255

Rotem J (2004) The genus *Alternaria*: biology, epidemiology and pathogenicity. American Phytopathological Society Press, Saint Paul, MN

Ruiz JA, Bernar EM, Jung K (2015) Production of siderophores increases resistance to fusaric acid in *Pseudomonas protegens* Pf-5. PLoS One 10(1):0117040. doi:10.1371/journal.pone.0117040

Schroth MN, Loper JE, Hildebrand DC (1984) Bacteria as biocontrol agents of plant disease. In: Klug MJ, Reddy CA (eds) Current perspectives in microbial ecology. American Society for Microbiology, Washington, DC, pp 362–369

Schuhegger R, Ihring A, Gantner S et al (2006) Induction of systemic resistance in tomato by N-acyl-L-homoserine lactone producing rhizosphere bacteria. Plant Cell Environ 29:909–918

Segarra G, Van der Ent S, Trillas I et al (2009) MYB72, a node of convergence in induced systemic resistance triggered by a fungal and a bacterial beneficial microbe. Plant Biol 11:90–96

Sen K, Sengupta C, Saha J (2014) PGPR consortium in alleviating downy mildew of cucumber. Int J Plant Animal Environ Sci 4(4):150–159

Shailbala, Pathak C (2008) Harnessing the potential of potato to meet increasing food demand. Kurukshetra 56(3):45–48

Shanthiyaa V, Saravanakumar D, Rajendran L et al (2013) Use of *Chaetomium globosum* for biocontrol of potato late blight disease. Crop Protect 52:33–38

Shoresh M, Harman GE, Mastouri F (2010) Induced systemic resistance and plant responses to fungal biocontrol agents. Annu Rev Phytopathol 48:21–43

Shrestha A, Kim EC, Lim CK, Cho S, Hur JH, Park DH (2009) Biological control of soft rot on chinese cabbage using beneficial bacterial agents in greenhouse and field. Korean J Pestic Sci 13(4):325–331

Shrestha A, Kim BS, Park DH (2014) Biological control of bacterial spot disease and plant growth-promoting effects of lactic acid bacteria on pepper. Biocontrol Sci Tech 24(7):763–779

Shrivastava P, Kumar R, Yandigeri MS (2016) *In vitro* biocontrol activity of halotolerant *Streptomyces aureofaciens* K20: a potent antagonist against *Macrophomina phaseolina* (Tassi) Goid. Saudi J Biol Sci. doi:10.1016/j.sjbs.2015.12.004

Silvar C, Merino F, Díaz J (2006) Diversity of *Phytophthora capsici* in northwest Spain: analysis of virulence, metalaxyl response, and molecular characterization. Plant Dis 90:1135–1142

Singh D, Yadav DK, Chaudhary G et al (2016) Potential of *Bacillus amyloliquefaciens* for biocontrol of bacterial wilt of tomato incited by *Ralstonia solanacearum*. J Plant Pathol Microbiol 7:327

Singh D, Yadav DK, Shweta S et al (2013) Genetic diversity of iturin producing strains of Bacillus species antagonistic to *Ralstonia solanacerarum* causing bacterial wilt disease in tomato. Afr J Microbiol Res 7:5459–5470

Singh D, Yadav DK, Sinha S et al (2012) Utilization of plant growth promoting *Bacillus subtilis* isolates for the management of bacterial wilt incidence in tomato caused by *Ralstonia solanacearum* race 1 biovar 3. Indian Phytopathol 65:18–24

Sivasakthi S, Usharani G, Saranraj P (2014) Biocontrol potentiality of plant growth promoting bacteria (PGPR)—*Pseudomonas fluorescens* and *Bacillus subtilis*: a review. Afr J Agric Res 9(16):1265–1277

Song J, Bradeen JM, Naess SK et al (2003) Gene AB cloned from *Solanum tuberosum* L. confers broad spectrum resistance to potato late blight. Proc Natl Acad Sci U S A 100:9128–9133

Sopheareth M, Chan S, Naing KW et al (2013) Biocontrol of late blight (*Phytophthora capsici*) disease and growth promotion of pepper by *Burkholderia cepacia* MPC-7. Plant Pathol J 29(1):67–76

Srinon W, Chuncheen K, Jirattiwarutkul K et al (2006) Efficacies of antagonistic fungi against *Fusarium* wilt disease of cucumber and tomato and the assay of its enzyme activity. J Agric Technol 2(2):191–201

Srivastava MP, Sharma S (2014) Potential of PGPR bacteria in plant disease management. In: Sharma N (ed) Biological controls for preventing food deterioration: strategies for pre- and post-harvest management. John Wiley & Sons Ltd, Chichester, UK. doi:10.1002/9781118533024.ch5

Subba Rao NS (1993) Biofertilizers in agriculture and forestry. Oxford and IBH Publishing Co. Pvt. Ltd, New Delhi, p 242

Sultana S, Hossian MM, Kubota M et al (2009) Induction of systemic resistance in Arabidopsis thaliana in response to a culture filtrate from a plant growth-promoting fungus, *Phoma* sp. GS8-3. Plant Biol 11:97–104

Sureshbabu K, Amaresan N, Kumar K (2016) Amazing multiple function properties of plant growth promoting rhizobacteria in the rhizosphere Soil. Int J Curr Microbiol Appl Sci 5(2):661–683

Swanson JK, Yao J, Tans-Kersten J et al (2005) Behavior of *Ralstonia solanacearum* race 3 biovar 2 during latent and active infection of geranium. Phytopathology 95:136–143

Tahat MM, Kamaruzaman S (2010) *Ralstonia solanacearum*: the bacterial wilt causal agent. Asian J Plant Sci 9:385–393

Takahara Y (1994) Development of the microbial pesticide for the soft rot disease. PSJ Biocont Rept 4:1–7

Takahara Y, Iwabuchi T, Shiota T et al (1993) Suppression of soft-rot lesion development by avirulent strains of *Erwinia carotovora* subsp. *carotovora*. Ann Phytopathol Soc Jpn 59:581–586

Tan S, Jiyang Y, Song S et al (2013) Two *Bacillus amyloliquefaciens* strains isolated using the competitive tomato root enrichment method and their effects on suppressing *Ralstonia solanacearum* and promoting tomato plant growth. Crop Protect 43:134–140

Tiru M, Muleta D, Berecha G et al (2013) Antagonistic effects of rhizobacteria against coffee wilt disease caused by *Gibberella xylarioides*. Asian J Plant Pathol 7:109–122

Togashi J, Uehara D, Namai T (2000) Biological control of the soft rot of Chinese cabbages by fluorescent antagonistic bacterium. Bull Yamagata Univ Agric Sci 13(3):225–232

Toua D, Benchabane M, Bensaid F et al (2013) Evaluation of *Pseudomonas fluorescens* for the biocontrol of fusarium wilt in tomato and flax. Afr J Microbiol Res 7(48):5449–5458

Ulloa-Ogaz AL, Muñoz-Castellanos LN, Nevárez-Moorillón GV (2015) Biocontrol of phytopathogens: antibiotic production as mechanism of control. In: Mendez-Vilas A (ed) The battle against microbial pathogens: basic science, technological advances and educational programs, Formatex Research Center, Spain, pp 305–309

Van der Walls JE, Korsen L, Aveling TAS (2001) A review of early blight of potato. Afr Plant Protect 70:91–102

Van Loon LC (2007) Plant responses to plant growth-promoting rhizobacteria. Eur J Plant Pathol 119:243–254. doi:10.1007/s10658-007-9165-1

Van Loon LC, Bakker PAHM, Pieterse CMJ (1998) Systemic resistance induced by rhizosphere bacteria. Annu Rev Phytopathol 36:453–483

Vanitha S, Ramjegathesh R (2014) Bio control potential of *Pseudomonas fluorescens* against Coleus root rot disease. J Plant Pathol Microb 5:1

Vanjildorj E, Song SY, Yang ZH et al (2009) Enhancement of tolerance to soft rot disease in the transgenic Chinese cabbage (*Brassica rapa* L. ssp. *pekinensis*) inbred line, Kenshin. Plant Cell Rep 28:1581–1591

Vassilev N, Vassileva M, Nikolaeva I (2006) Simultaneous P-solubilizing and biocontrol activity of microorganisms: potentials and future trends. Appl Microbiol Biotechnol 71:137–144

Velivelli SLS, DeVos P, Kromann P et al (2014) Biological control agents: from field to market, problems and challenges. Trends Biotechnol 32:493–496. doi:10.1016/j.tibtech.2014.07.002

Visser R, Holzapfel WH, Bezuidenhout JJ et al (1986) Antagonism of lactic acid bacteria against phytopathogenic bacteria. Appl Environ 52:552–555

Vloutoglou I, Kalogerakis SN (2000) Effects of inoculum concentration, wetness duration and plant age on development of early blight (*Alternaria solani*) and on shedding of leaves in tomato plants. Plant Pathol 49:339–345

Wang JF, Hanson P, Barnes JA (1998) Worldwide evaluation of an international set of resistance sources of bacterial wilt in tomato. In: Prior P, Allen C, Elphinstone J (eds) Bacterial wilt disease: molecular and ecological aspects. Springer Verlag, Berlin, Germany, pp 269–275

Wang X, Mavrodi DV, Ke L et al (2015) Biocontrol and plant growth-promoting activity of rhizobacteria from Chinese fields with contaminated soils. Microb Biotechnol 8(3):404–418

Witek K, Jupe F, Witek AI et al (2016) Accelerated cloning of a potato late blight–resistance gene using RenSeq and SMRT sequencing. Nat Biotechnol. doi:10.1038/nbt.3540

Zaidi A, Ahmad E, Khan MS et al (2015) Role of plant growth promoting rhizobacteria in sustainable production of vegetables: current perspective. Sci Hortic 193:231–239

Zhao LF, Xu YJ, Ma ZQ et al (2013) Colonization and plant growth promoting characterization of endophytic *Pseudomonas chlororaphis* strain Zong1 isolated from *Sophora alopecuroides* root nodules. Braz J Microbiol 44(2):623–631

Perspectives of Plant Growth Promoting Rhizobacteria in Growth Enhancement and Sustainable Production of Tomato

6

Bilal Ahmed, Almas Zaidi, Mohd. Saghir Khan, Asfa Rizvi, Saima Saif, and Mohammad Shahid

Abstract

Tomato is an important horticultural product with a high content of bioactive compounds such as folate, ascorbate, polyphenols, and carotenoids and many other essential nutrients. Due to these, tomatoes are considered extremely valuable to human health. To optimize tomato production, chemical fertilizers and pesticides are frequently used. These chemicals are however, destructive for both crops and soil ecosystems. A reduction of these detrimental practices is therefore urgently required to protect both tomato and environments from damaging effects of agrochemicals. In this context, microbial inoculation especially those consisting of plant growth-promoting rhizobacteria (PGPR) could be used to replace chemical fertilizers/pesticides. Also, PGPR can be integrated with such chemical practices to reduce their application in tomato cultivation. Plant growth-promoting rhizobacteria that naturally inhabit the rhizosphere stimulate the growth and development of tomato plants directly or indirectly via availability of many essential plant nutrients, phytohormones, or through suppression/destruction of plant diseases. A better understanding of the plant growth-promotion activity of these bacterial strains is likely to enhance the production of safe, fresh, and high-quality tomatoes while reducing chemical inputs in different agronomic setups.

B. Ahmed (✉) • A. Zaidi • M.S. Khan • A. Rizvi • S. Saif • M. Shahid
Faculty of Agricultural Sciences, Department of Agricultural Microbiology, Aligarh Muslim University, Aligarh 202002, India
e-mail: bilalahmed.amu@gmail.com

© Springer International Publishing AG 2017
A. Zaidi, M.S. Khan (eds.), *Microbial Strategies for Vegetable Production*,
DOI 10.1007/978-3-319-54401-4_6

Table 6.1 Nutritional composition of edible portion of tomato

Nutrient	Unit	Value per 100 g	Data points	Std. error	Cup cherry tomatoes 149 g	1 cup, chopped or sliced 180 g	1 Italian tomato 62 g	1 cherry 17 g	1 large whole (3″ dia) 182 g	1 medium whole (2–3/5″ dia) 123 g	1 slice, medium (1/4″ thick) 20 g	1 plum tomato 62 g	1 small whole (2–2/5″ dia) 91 g	1 slice, thick/large (1/2″ thick) 27 g	1 wedge (1/4 of medium tomato) 31 g	1 slice, thin/small 15 g	1 NLEA serving 148 g
Water	g	94.52	390	0.139	140.83	170.14	58.6	16.07	172.03	116.26	18.9	58.6	86.01	25.52	29.3	14.18	139.89
Protein	g	0.88	19	0.039	1.31	1.58	0.55	0.15	1.6	1.08	0.18	0.55	0.8	0.24	0.27	0.13	1.3
Total lipid (fat)	g	0.2	26	0.034	0.3	0.36	0.12	0.03	0.36	0.25	0.04	0.12	0.18	0.05	0.06	0.03	0.3
Carbohydrate	g	3.89	—	—	5.8	7	2.41	0.66	7.08	4.78	0.78	2.41	3.54	1.05	1.21	0.58	5.76
Fiber, total dietary	g	1.2	5	0.234	1.8	2.2	0.7	0.2	2.2	1.5	0.2	0.7	1.1	0.3	0.4	0.2	1.8
Sugars, total	g	2.63	—	—	3.92	4.73	1.63	0.45	4.79	3.23	0.53	1.63	2.39	0.71	0.82	0.39	3.89
Calcium, Ca	mg	10	69	0.291	15	18	6	2	18	12	2	6	9	3	3	2	15
Iron, Fe	mg	0.27	70	0.015	0.4	0.49	0.17	0.05	0.49	0.33	0.05	0.17	0.25	0.07	0.08	0.04	0.4
Magnesium, Mg	mg	11	70	0.198	16	20	7	2	20	14	2	7	10	3	3	2	16
Phosphorus, P	mg	24	69	0.534	36	43	15	4	44	30	5	15	22	6	7	4	36
Potassium, K	mg	237	67	5.318	353	427	147	40	431	292	47	147	216	64	73	36	351
Sodium, Na	mg	5	66	0.565	7	9	3	1	9	6	1	3	5	1	2	1	7
Zinc, Zn	mg	0.17	70	0.036	0.25	0.31	0.11	0.03	0.31	0.21	0.03	0.11	0.15	0.05	0.05	0.03	0.25
Vitamin C	mg	13.7	25	0.860	20.4	24.7	8.5	2.3	24.9	16.9	2.7	8.5	12.5	3.7	4.2	2.1	20.3
Thiamin	mg	0.037	19	0.004	0.055	0.067	0.023	0.006	0.067	0.046	0.007	0.023	0.034	0.01	0.011	0.006	0.055
Riboflavin	mg	0.019	19	0.001	0.028	0.034	0.012	0.003	0.035	0.023	0.004	0.012	0.017	0.005	0.006	0.003	0.028
Niacin	mg	0.594	19	0.032	0.885	1.069	0.368	0.101	1.081	0.731	0.119	0.368	0.541	0.16	0.184	0.089	0.879
Vitamin B6	mg	0.08	19	0.003	0.119	0.144	0.05	0.014	0.146	0.098	0.016	0.05	0.073	0.022	0.025	0.012	0.118
Vitamin E	mg	0.54	16	0.070	0.8	0.97	0.33	0.09	0.98	0.66	0.11	0.33	0.49	0.15	0.17	0.08	0.8
Vitamin K	µg	7.9	15	4.743	11.8	14.2	4.9	1.3	14.4	9.7	1.6	4.9	7.2	2.1	2.4	1.2	11.7
Myricetin	mg	0.1	22	0.03	0.2	0.2	0.1	0.0	0.2	0.2	0.0	0.1	0.1	0.0	0.0	0.0	0.2
Quercetin	mg	0.6	96	0.01	0.9	1.0	0.4	0.1	1.1	0.7	0.1	0.4	0.5	0.2	0.2	0.1	0.9

Modified from USDA National Nutrient Database for Standard Reference; April 21, 2016

In other study, Kumar et al. (2007) reported that lycopene should be used as a first line of therapy in the management of oral submucous fibrosis. Additionally, scientists have reported that lycopene can kill oral cancer cells when supplemented in culture medium. This killing effect has been attributed to be due to the ability of lycopene to restore gap junction communication, which is destroyed in oral malignancies (Livny et al. 2002). Lycopene has also been found effective against several other cancerous cells such as breast, endometrium, lung, etc. Lycopene also reduces the occurrence of oxidative DNA damage in lymphocytes, deactivates the nitric oxide (NO) and hydrogen peroxide (H_2O_2), and protects cell ROS-induced membrane damage.

6.4   Rhizosphere and PGPR

The limited zone of soil in close proximity of plant root system or surrounding the root system of plants is generally referred to as rhizosphere (Hiltner 1904). The bacterial populations inhabiting rhizosphere, generally termed as "rhizobacteria," are competent in colonizing the roots and can interact well with other soil microflora both positively or negatively (Kloepper et al. 1991). Broadly, there are three distinct zones in the rhizosphere: (i) the endo-rhizosphere, (ii) the rhizoplane, and (iii) the ecto-rhizosphere (Lynch 1987). In general, rhizosphere acts as nutrient (macro and micro) pool for many organisms including microbes. Due to this, the rhizospheric zone is considered the zone of maximum microbial diversity and activity compared to non-rhizosphere region (Walker et al. 2003). The concentration of such compounds added by plants to rhizosphere however varies from plant genotype to genotype and acts as chemoattractants for a variety of heterogeneously growing soil microorganism. About 5–30% of photosynthates exuded by plants are sugars, amino acids, and other secondary metabolites which are taken up as nutrients by microbes (Glick 2014). Also, the nutrients can modify the physicochemical properties of the rhizosphere and in effect regulate the structure and composition of soil microbial community (Dakora and Phillips 2002). Among variously distributed heterotrophic microflora, bacterial populations belonging to different species form about 15% of the total microbial populations (Jha et al. 2010; Govindasamy et al. 2011). Among numerous species of soil bacteria, the bacteria that are beneficent for plant growth and metabolism are collectively termed as "plant growth-promoting rhizobacteria" (PGPR) (Kloepper and Schroth 1978). The PGPR has the ability to (1) colonize root surface, (2) survive and multiply, and (3) compete with other microorganisms. Once established, they play a pivotal role in geochemical nutrient cycling and enhance plant growth by several mechanisms such as (1) fixing the atmospheric N, (2) solubilizing insoluble form of P, (3) producing phytohormones (auxin, cytokinins, etc.), (4) secreting ACC deaminase, etc. Besides such beneficial activities, PGPR also play some role in the management of disease by producing siderophores, hydrogen cyanide (HCN), and antibiotics (Kloepper et al. 1980; Son et al. 2014). The notable PGPR fostering plant growth belongs to genera *Rhizobium, Bradyrhizobium, Azotobacter, Bacillus, Thiobacillus, Pseudomonads, Azospirillum, Burkholderia, Arthrobacter,*

Acinetobacter, *Agrobacterium*, *Serratia*, *Frankia*, etc. (Gopalakrishnan et al. 2015; Kasa et al. 2015; Anandham et al. 2014; Manzanera et al. 2015; Majeed et al. 2015; Ahmad et al. 2016; Seneviratne et al. 2016). Despite such a varied group, only 2–5% of rhizosphere bacteria have been found as potent PGPR (Jha et al. 2010; Siddikee et al. 2010).

6.4.1 Availability of Nutrients in Tomato Rhizosphere and Their Interaction with PGPR

Root exudates form the basis for communications between plants and microorganisms inhabiting rhizosphere (Badri et al. 2013; Chaparro et al. 2013). This chemical communication is influenced largely by carbohydrates, amino acids, hormones, and other plant secondary metabolites. Of these, carbohydrates and amino acids are considered prominent chemoattractants for PGPR which in effect facilitate plant growth promotion/growth protection through several direct and indirect mechanisms (Neal et al. 2012). The quantitative and qualitative compositions of root exudates vary among plant species and developmental stage of plant, cultivar, type and pH of soil, temperature, and composition of microbial flora (Bulgarelli et al. 2012; Lundberg et al. 2012; Chaparro et al. 2013; Turner et al. 2013). Like other plants, tomato plants also secrete different compounds like amino acids, for example, aspartic acid, glutamic acid, lysine, leucine, and isoleucine (Simons et al. 1997), carbohydrates such as xylose and glucose (Lugtenberg et al. 1999), and organic acids, for instance, citric acid, succinic acid, and malic acid (Kamilova et al. 2006). Even though sugars and organic acids are important for root colonization in tomato (Lugtenberg et al. 2002) yet without the bioavailability of amino acids, there cannot be the efficient root tip colonization (Simons et al. 1997; Shishido and Chanway 1998).

6.4.2 Tomato Growth Enhancement by PGPR: A General Perspective

The PGPR have been utilized as an eco-friendly option to restore and/or increase the nutrient availability to numerous vegetable species including tomato (Table 6.2). A wide range of PGPR, for example, *Pseudomonas*, *Bacillus*, *Acinetobacter*, *Streptomyces*, *Micrococcus*, *Azotobacter*, *Flavobacterium*, and *Streptococcus*, are associated with tomato rhizosphere (Prashar et al. 2014) which either directly (through IAA production and P solubilization) or indirectly enhance tomato growth (Fig. 6.1). Inoculation of tomato seeds with selected PGPR strains has also been found to circumvent stresses by modifying the structure of microbial community (Kloepper and Schroth 1981). Some of the mechanisms/active biomolecules (Table 6.3) by which PGPR promote plant growth are discussed briefly in the following section.

Table 6.2 Examples of vegetable growth promotion by some common plant growth-promoting rhizobacteria

Plant growth-promoting activities	PGPR	Crops	References
Nitrogen fixation	*Acetobacter* and *Azospirillum*	Tomato	Ordookhani et al. (2010)
	Acinetobacter, *Pseudomonas*		Indiragandhi et al. (2008)
Siderophore production	*Rhizobium*	Tomato, pepper, carrot, lettuce	Garcia-Fraile et al. (2012), Flores-Félix et al. (2013)
	Pseudomonas	Potato	Beneduzi et al. (2012)
	Chryseobacterium	Tomato	Radzki et al. (2013)
	Bacillus	Pepper	Beneduzi et al. (2012)
Phytohormone production			
Cytokinin	*Azotobacter*	Cucumber	Aloni et al. (2006)
	Bacillus	Cucumber	Sokolova et al. (2011)
Auxin	*Bacillus*	Potato	Ahmed and Hasnain (2010)
	Achromobacter, *Stenotrophomonas*	Pumpkin, tomato	Ngoma et al. (2013)
	Rhizobium	Pepper, tomato, lettuce, carrot	Garcia-Fraile et al. (2012), Flores-Félix et al. (2013)
	Bacillus	Sweet potato	Dawwam et al. 2013
	Bacillus	Brassica	Turan et al. (2014)
Gibberellin	*Bacillus*	Pepper	Joo et al. (2005)
	Sphingomonas	Tomato	Khan et al. (2014)
	Pseudomonas	Mentha	Sivasakthi et al. (2013)
	Rhizobium	Pepper, tomato, mung beans	Ahmad et al. (2013), Garcia-Fraile et al. (2012)
Hydrogen cyanide production	*Rhizobia*	Legumes	Thamer et al. (2011)
	Bacillus and *Pseudomonas*	Tomato	Vaikuntapu et al. (2014)
	Pseudomonas, *Achromobacter*, *Stenotrophomonas*	Green amaranth, pumpkin, tomato	Ngoma et al. (2013)
Phosphate solubilization	*Bacillus*, *Achromobacter*	Sweet potato	Dawwam et al. 2013
	Pseudomonas, *Achromobacter*, *Stenotrophomonas*	Pumpkin, tomato	Ngoma et al. 2013

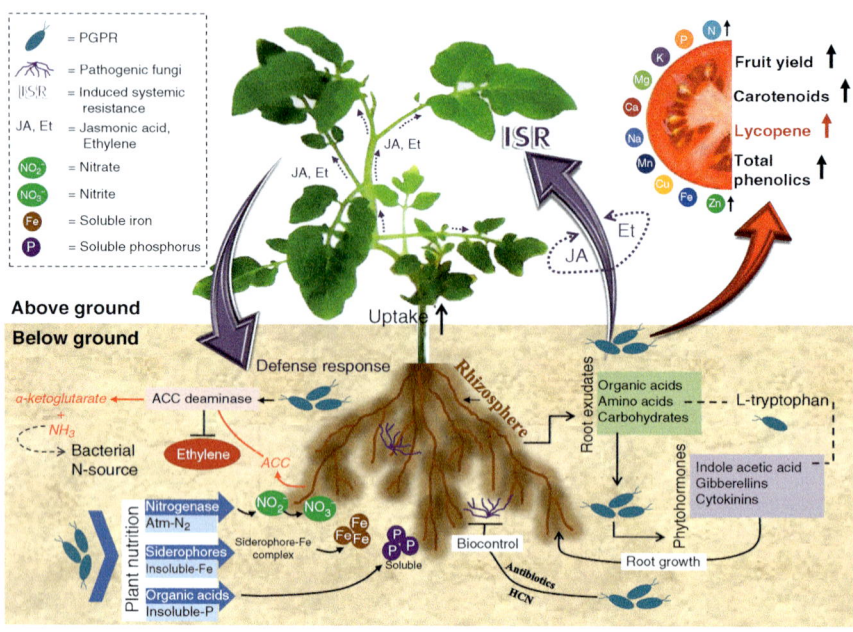

Fig. 6.1 Schematic overview of rhizospheric bacteria mediated enhancement of the overall growth, fruit yield, and nutrient content of tomato

Table 6.3 An array of plant growth-promoting rhizobacteria associated with tomato root system/rhizosphere and their plant growth-promoting traits

PGPR	Growth-promoting traits	References
Fluorescent *Pseudomonas* spp.	Biocontrol	Botelho and Mendonça-Hagler (2006)
P. fluorescens, P. marginalis, P. putida, and *P. syringae*	IAA and ACC deaminase production	Gravel et al. (2007)
Sphingomonas sp. LK11	IAA and gibberellin production, increase in shoot length, chlorophyll contents, shoot, and root dry weight	Khan et al. (2014)
Bacillus subtilis 101 and *Azospirillum brasilense* Sp245	Colonization of roots, longer primary roots, significant increase in biomass of plant, auxin biosynthesis	Felici et al. (2008)
Burkholderia graminis M12	Increase the shoot height and neck diameter, salt tolerance (more than 95%)	Barriuso et al. (2008)
Consortium of *Pseudomonas* sp. and *Trichoderma harzianum*	shoot length, root length, fresh and dry weight	Singh et al. (2013)

(continued)

Table 6.3 (continued)

PGPR	Growth-promoting traits	References
B. subtilis and *P. fluorescens*	Siderophore production	Selvakumar et al. (2013)
P. fluorescens strains Psf 4, Pp1, and Pa2	Siderophore production	Hammami et al. (2013)
Pseudomonas, Bacillus, Acinetobacter, Streptomyces, Micrococcus, Azotobacter, Flavobacterium and *Streptococcus*	Phosphate solubilization and IAA production	Prashar et al. (2014)
Aeromonas, Pseudomonas, Bacillus, and *Enterobacter*	Positive for biofilm formation, production of ACC deaminase, antagonism to *Fusarium solani, F. moniliforme,* and *Macrophomina phaseolina*	Vaikuntapu et al. (2014)
P. fluorescens	Higher seed germination and enhanced seedling vigor index	Murthy et al. (2014)
Bacillus and *Pseudomonas* sp.	Phosphate solubilization, IAA production	Bellishree et al. (2014)
Bacillus fortis IAGS162 and *B. subtilis* IAGS174	IAA, siderophore production, phosphate solubilization, increase in carotenoids of tomato, total sugar contents, and total chlorophyll	Akram et al. (2015)
Bacillus spp. and *Pseudomonas* spp.	HCN production, phosphate solubilization, IAA production, and starch hydrolysis	Lachisa and Dabassa (2015)
Chryseobacterium sp., *Bacillus* sp., *Microbacterium* sp., *Pseudomonas* sp., *Rhizobium* sp., *Rhodococcus* sp., and *Agrobacterium* sp.	Organic acid, IAA, ACC deaminase, and siderophore production	Abbamondi et al. (2016)
B. subtilis, B. amyloliquefaciens, and *P. fluorescens*	Antibiotic, siderophore, ammonia, IAA production, and phosphorus solubilization	Singh et al. (2016)

6.4.2.1 Direct Mechanisms: Biological Nitrogen Fixation

Nitrogen, the most vital element for growth and productivity of plants, is converted into plant utilizable forms by nitrogen-fixing organisms to ammonia using a complex enzyme system termed as nitrogenase (Khan et al. 2009). The symbiotic relationship as observed between rhizobia and legumes, even though it does not exist in tomato but the nonsymbiotic nitrogen fixation carried out by either free living, associative, or endophytic bacteria, occurs in tomato root system. Examples of such nitrogen-fixing PGPR include *Azotobacter, Azospirillum, Herbaspirillum, Bacillus, Burkholderia,* and *Paenibacillus* (Heulin et al. 2002; Goswami et al. 2015). Without penetrating the plant tissues, free-living nitrogen-fixing bacteria form a very close association with plants where they live in a sufficient number to provide the fixed form of N to plants.

6.4.2.2 Phosphate Solubilization

Phosphorus is yet another important element affecting growth and development of plants including tomato. It plays a key role in almost all metabolic processes of plant such as photosynthesis, respiration, signal transduction, and energy transfer (Khan et al. 2010). Despite the abundance of P in soil, it is not available in the soluble forms for plants (Jha et al. 2012; Jha and Saraf 2015). Approximately 95–99% P is rendered unavailable for plant uptake due to its rapid complexation with cations such as Ca, Fe, or Al. However, numerous PGPR involved especially in transforming insoluble P to soluble forms and quite often referred to phosphate solubilizing microorganisms (PSM) provides P to plants. Such PSM involving bacteria, fungi, and actinomycetes employ several mechanisms for making P available to plants. These are (1) production and release of organic acids, OH^- ions, CO_2, and protons (phosphate solubilization), (2) release of extracellular enzymes (biochemical phosphate mineralization), and (3) release of available phosphate during degradation of substrate (biological phosphate mineralization) (Sharma et al. 2013). Examples of some phosphate-solubilizing PGPR are *Bacillus, Beijerinckia, Enterobacter, Flavobacterium, Microbacterium, Arthrobacter, Burkholderia, Erwinia, Pseudomonas, Rhizobium*, and *Serratia* (Bhattacharyya and Jha 2012). Like other conventional PGPR, P-solubilizing microbes have also been used in isolation (El-Tantawy and Mohamed 2009) and in combination with other microorganisms such as *Azotobacter, Azospirillum*, and *Actinomycetes* to enhance yield of tomato (Meena et al. 2015). Among all phosphate solubilizing bacteria (PSB), *Pseudomonas* and *Bacillus* species are abundant in tomato rhizosphere and can provide tomato with available P (Lachisa and Dabassa 2015).

6.4.2.3 Production of Phytohormones

The PGPR secretes phytohormones such as auxins (Talboys et al. 2014; Majeed et al. 2015), cytokinins (Goswami et al. 2015), gibberellins (Pandya and Desai 2014), abscisic acid (Porcel et al. 2014), and ethylene (Patten and Glick 1996). Generally, phytohormones affect cell enlargement, cell division, and cell extension in roots (Glick 2014). Some of the common phytohormones involved in growth enhancement is discussed briefly in the following section.

Indole-3-acetic Acid

Indole-3-acetic acid (IAA), one of the most physiologically active auxins, is secreted by 80% of rhizosphere microbial populations (Patten and Glick 1996). The IAA controls organogenesis; tropic responses; cellular responses such as cell expansion, division, and differentiation; gene regulation; and responses to light and gravity (Teale et al. 2006; Lambrecht et al. 2000). Moreover, the plants exposed to IAA for long times have extremely developed root system, which in turn provides the plants greater access to nutrients. This in turn allows the plant to absorb more nutrients and, hence, aid in overall growth of the plants (Aeron et al. 2011). Also, IAA loosens cell walls of roots and as a result alleviates an increasing amount of root exudates that ultimately supports PGPR growth by providing additional nutrients (Glick 2012).

In a study carried out by Felici et al. (2008) on tomato plant, in vitro auxin biosynthesis assay showed that *Azospirillum brasilense* Sp245 could produce auxin at the rate of 7.95 ± 0.002 mg/l after 72 h of incubation, in the absence of L-trp, which, however, increased by 40.2 ± 0.015 mg/l in the presence of L-trp. Similarly, other bacterial populations such as *Rhizobium* sp., *Pseudomonas* sp., *Bacillus* sp., *Agrobacterium*, etc. isolated from rhizosphere of tomato cultivars demonstrated highest potential for IAA production (89%) as compared to other PGP activities (Abbamondi et al. 2016). Of these, strains DSBA-11, DSBA-12, and FZB42 of *B. amyloliquefaciens* produced substantial amounts of IAA and following inoculation promoted the growth of tomato plant (Singh et al. 2016). In another study, Bellishree et al. (2014) quantitatively estimated the IAA production by strains of *Bacillus* spp. isolated from tomato rhizosphere. The results exhibited that IAA production was higher in *B. subtilis* strain BCA-6 (11.75 µg/ml) followed by *B. amyloliquefaciens* strain BCA-17 (11.57 µg/ml) and *B. megaterium* strain BCA-5 (10.95 µg/ml). The minimum amount of IAA synthesized was seen in *P. stutzeri* strain BCA-20 (7.91 µg/ml). In a follow-up study, *Burkholderia seminalis* showed the production of IAA in nutrient broth supplemented with tryptophan. Also, when used as inoculant against tomato seed, it had a positive impact on germination of seeds (Tallapragada et al. 2015). These and other similar studies therefore confirm that IAA-positive PGPR could be used to augment the growth of tomato in different agronomic setups.

Cytokinins

Cytokinins are adenine derivative phytohormones that are secreted by PGPR (Goswami et al. 2015) and are similar in functions to IAA. Cytokinins when applied exogenously to plants are known to control cell division and cell cycle and stimulate developmental processes in plants (Srivastava 2002; Jha and Saraf 2015). Stimulatory or inhibitory function of cytokinins in different developmental processes such as regulation of root and shoot growth as well as branching, control of apical dominance in the shoot, chloroplast development, and leaf senescence has also been described (Werner et al. 2001; Oldroyd 2007; Amara et al. 2015). Additionally, cytokinin affects cell division activity in embryonic and mature plants by altering the size and activity of meristems (Werner et al. 2001). In a study, *Azotobacter chroococcum*, when inoculated with tomato plants grown in P-deficient soil, significantly increased the dry weight of inoculated plants compared to that of uninoculated plants (Puertas and Gonzales 1999).

Gibberellins

Gibberellins (GAs) are yet another group of important plant hormones which influence many developmental processes such as seed germination, flowering, stem elongation, and fruit setting (Hedden and Phillips 2000). The production of gibberellins by bacterial strains is rare. Among *Bacillus* spp. only two strains have been documented that are capable of producing gibberellins: *B. licheniformis* and *B. pumilus* (Gutierrez-Manero et al. 2001). The GAs are also synthesized by *Azotobacter* sp. The first report on the growth promotion of tomato plant by GA-like substances synthesized by asymbiotic nitrogen-fixing bacterium *Azotobacter* sp. was by Azcorn and Barea (1975). Being first of its nature, study on gibberellin characterization in bacteria using physicochemical methods was documented by Atzorn et al. (1988), who established the presence of GA-1, GA-4, GA-9, and GA-20 in gnotobiotic cultures of *Rhizobium meliloti*. Synthesis of GAs has also been confirmed in *Herbaspirillum seropedicae* and *Acetobacter diazotrophicus* (Bastián et al. 1998).

6.4.2.4 1-Aminocyclopropane-1-carboxylic Acid (ACC) Deaminase (EC 4.1.99.4)

Ethylene, also known as stress hormone, is a key phytohormone having a vast range of biological functions including plant growth and development. It promotes root initiation, inhibits root elongation, reduces wilting, enhances fruit ripening, stimulates seed germination, and activates the production of other plant hormones (Glick et al. 2007). Ethylene on the other hand also has detrimental effect on plants, where it causes leaf senescence, leaf abscission, chlorosis, flower wilting, etc. The enzyme ACC deaminase is a prerequisite for ethylene production, catalyzed by ACC oxidase. A variety of PGPR produce ACC deaminase that cleaves ACC in to ammonia and α-ketobutyrate inhibiting its transition to ethylene. The ACC is produced in larger volumes by plants surviving under abiotic stresses such as water flooding, ultraviolet radiations, temperature, and heavy metals (Ali et al. 2012; Glick 2014). In this way, production of ACC deaminase by PGPR defends plants against detrimental effects of ethylene surviving under abiotic stresses (Glick 2014). Bacterial synthesizing ACC deaminase belongs to genera *Pseudomonas*, *Bacillus*, *Acinetobacter*, *Azospirillum*, *Achromobacter*, *Enterobacter*, *Burkholderia*, *Agrobacterium*, *Alcaligenes*, *Rhizobium*, *Serratia*, etc. (Gupta et al. 2015). Vaikuntapu et al. (2014) evaluated the plant growth-promoting potential of selected strains of *Pseudomonas*, *Bacillus*, *Aeromonas*, and *Enterobacter* which revealed their PGP activities including the production of ACC deaminase, in addition to in vitro growth inhibition of fungal pathogens of tomato rhizosphere.

In a study, Hassan et al. (2014) examined the response of tomato to various PGPR and observed that inoculation with rhizobacteria increased all the measured physical, chemical, and enzymatic growth parameters compared to control. However, among PGPR, the TAN1 isolate had the highest effect and significantly ($P < 0.05$) increased the root length (8.25-fold); root fresh (8.36-fold) and dry (12.6-fold) weight; shoot length (6.92-fold); shoot fresh (7.18-fold) and dry (6.9-fold) weight; number of leaves (11-fold); chlorophyll a (6.25-fold); chlorophyll b (10.7-fold); carotenoid contents (8.8-fold); seedlings fresh (9-fold) and dry (8.71-fold) weight; plant macronutrient uptake such as N (7.7- and 8.9-fold), P (10.5- and 11.4-fold), K (7.8- and 8.8-fold), Ca (12.7- and 8.2-fold), and Mg (12.6- and 9-fold) in shoot and root; plant micronutrient uptake, i.e., Zn (6.6- and 10.2-fold), Cu (9.3- and 10.3-fold), Fe (7.7- and 10.7-fold), and Mn (4.7- and 5.7-fold) in shoot and root; and plant antioxidant enzymes, i.e., glutathione S-transferase (10.7-fold), peroxidase (8.1-fold), and catalase (10.5-fold). This study concluded that the rhizobacteria having both ACC deaminase and N_2-fixing activity together could be more effective than rhizobacteria possessing either of the two activities for growth promotion of crops.

Apart from their role in crop improvement under conventional soil, numerous ACC deaminase positive PGPR have also been reported to promote tomato growth under derelict/stressed soils (Khaliq et al. 2013). For example, Yan et al. (2014) conducted an experiment to evaluate whether the ACC deaminase producing PGPR *Pseudomonas putida* UW4 can maintain and promote plant growth in saline environments and modulate the expression of chloroplast import apparatus genes in salt-treated tomato plants. For this, tomatoes were grown in the presence and absence of the *P. putida*, and shoot length, fresh and dry mass, and chlorophyll concentration were measured after 6 weeks. The expression levels of the Toc GTPases of the chloroplast protein import apparatus were measured using quantitative real-time PCR. The results revealed that the *P. putida* UW4 significantly increased length and fresh and dry biomass of shoots, and the chlorophyll concentration of tomato seedlings grown in the presence of up to 90 mmol/l NaCl. Furthermore, the expression of most of the Toc GTPase genes was upregulated in tomato seedlings after 6 weeks of exposure to NaCl, which may help facilitate the import into chloroplasts of proteins that are involved in the stress response. In a similar study, the effect of salt-tolerant bacterium containing ACC deaminase (*Bacillus licheniformis* B2r) on the germination and growth of tomato was investigated under saline conditions (Chookietwattana and Maneewan 2012). Tomato plants inoculated with the selected bacterium under various saline conditions (0, 30, 60, 90, and 120 mM NaCl) revealed a significant increase in the germination percentage, germination index, root length, and seedling dry weight especially at salinity levels ranging from 30 to 90 mM NaCl. These and other similar studies therefore suggest that the salt-tolerant ACC deaminase-containing bacteria could be advantageous over others to thrive in a new saline environment in sufficient numbers to deliver beneficial effects on tomato plants.

6.4.3 Indirect Mechanisms: Biomanagement of Phytopathogens

6.4.3.1 Siderophore Production

Siderophores are low-molecular-weight (usually below 1 kDa) iron-chelating compounds which are synthesized by many PGPR in order to prevent the proliferation of pathogenic microorganisms by sequestering Fe^{3+} in root surroundings (Mehnaz 2013). Plants take up iron produced by a large number of PGPR including *Pseudomonas* (Sulochana et al. 2014), *Bacillus* (Jikare and Chavan 2013), *Aeromonas* (Sayyed et al. 2013), *Azotobacter* (Ahmad et al. 2008), *Burkholderia* (Jiang et al. 2008), *Rhizobium* (Bano and Musarrat 2003), and *Serratia* (Selvakumar et al. 2008) as confirmed by using radiolabeled ferric siderophores as a sole source of Fe (Gupta et al. 2015). Siderophore-producing pseudomonads, specifically *P. fluorescens* and *P. aeruginosa*, release pyoverdine and pyochelin type of siderophores (Goswami et al. 2015). Regarding the growth promotion of tomato plant by PGPR, the strains of *Aeromonas*, *Pseudomonas*, *Bacillus*, and *Enterobacter* were evident in production of siderophores. Additionally, all strains positive for siderophore production were also capable of antagonizing the population of fungal pathogens, for example, *Fusarium solani*, *F. moniliforme*, and *Macrophomina phaseolina* (Vaikuntapu et al. 2014). A wide range of cultivable PGPR strains associated with different tomato cultivars and new tomato hybrids such as Plus Licopene, Indigo Perù, San Marzano-X, Super San Marzano, and Black Tomato were isolated and identified as *Ensifer* sp., *Chryseobacterium* sp., *Bacillus* sp., *Microbacterium* sp., *Pseudomonas* sp., *Rhizobium* sp., *Rhodococcus* sp., and *Agrobacterium* sp. and assessed for their PGP traits. Among them, 87% isolates were found positive for siderophores (Abbamondi et al. 2016). *Bacillus subtilis, P. fluorescens* (Selvakumar et al. 2013), and four strains of *P. fluorescens* isolated from tomato plant also produced siderophore on CAS agar medium (Hammami et al. 2013).

6.4.3.2 Antibiosis

Many of the PGPR strains also produce substances that are inhibitory to pathogenic microbial populations and suppress their growth (Beneduzi et al. 2012). This phenomenon is called antibiosis. Bacteria belonging to *Bacillus* spp. produce a wide range of antibacterial and antifungal antibiotics (Chowdhury et al. 2015). Some of these antibiotics have ribosomal origin such as subtilin, subtilosin A, TasA, and sublancin, but synthesis of others is mediated by non-ribosomal peptide synthetases (NRPSs) and/or polyketide synthases (PKS) such as rhizocticins, chlorotetain, bacilysin, difficidin, mycobacillin, bacillaene, and lipopeptides (Leclere et al. 2005). On the contrary, PGPR *Pseudomonas* spp. produce mainly 2,4-diacetylphloroglucinol (DAPG), phenazine-1-carboxamide (PCN), phenazine-1-carboxylic acid (PCA), oomycin A, pyoluteorin (Plt), pyrrolnitrin (Prn), butyrolactones, viscosinamide, zwittermicin A, kanosamine, aerugine, cepaciamide A, rhamnolipids, pseudomonic acid, ecomycins, antitumor antibiotics, azomycin, cepafungins, and antibiotic karalicin (reference). These antibiotics are known to possess antiviral, antimicrobial, insect, and PGP activities (Goswami et al. 2015). Lanteigne et al. (2012) reported the production of DAPG by *Pseudomonas* contributing to the biocontrol of bacterial canker of tomato. Apart from antibiotic production, some PGPR also produce a popular volatile chemical compound, HCN. Strains of *Aeromonas, Pseudomonas, Bacillus*, and *Enterobacter* isolated from tomato rhizosphere have been found positive for HCN production (Vaikuntapu et al. 2014). Isolates of *Bacillus* spp. and *Pseudomonas* spp. recovered from tomato rhizosphere were positive for HCN production which very effectively controlled *Fusarium* wilt of tomato caused by *Fusarium* spp. (Lachisa and Dabassa 2015). Results obtained in the study of Alvarez et al. (1995) showed that the addition of compost to rhizospheric soil increased the incidence of antagonistic activities of bacteria in tomato rhizosphere exhibiting antagonism toward *F. oxysporum* f. sp. *radicis-lycopersici, Pythium ultimum, Rhizoctonia solani*, and *Pyrenochaeta lycopersici*. A variety of plant growth-promoting and antagonistic bacterial population belonging to genera *Bacillus* (*B. amyloliquefaciens, B. subtilis, B. thuringiensis, B. pumilus, B. polymyxa*, and *B. cereus*) and *Pseudomonas* (*P. fluorescens, P. putida, P. chlororaphis*, and *P. aeruginosa)* have been reported to provide protection against severe diseases of tomato plants (Table 6.4) resulting in major crop enhancement (Singh et al. 2016; Loganathan et al. 2014; Myresiotis et al. 2012; Murthy et al. 2014). The pathogen *Ralstonia solanacearum* of bacterial wilt of tomato causing very heavy losses,

Table 6.4 Characteristic diseases of tomato crop, their causative agent, and suppression of disease symptoms by antagonistic PGPR

Disease/Stress	Causal organism	Antagonistic PGPR	References
Fungal leaf disease	*Alternaria alternata*	N-Acyl-L-homoserine lactone producing *Serratia liquefaciens* MG1 and *Pseudomonas putida* IsoF	Schuhegger et al. (2006)
Several diseases	*Pseudomonas syringae* pv. Tomato	*Bacillus cereus* (a phylloplane resident)	Bernardo et al. (2006)
Late blight	*Phytophthora infestans*	*Pseudomonas fluorescens* SS101	Tran et al. (2007)
Yield losses symptoms, i.e., root galling	*Meloidogyne incognita*	*Azotobacter chroococcum*, *Bacillus subtilis*, and *Pseudomonas putida*	Siddiqui and Futai (2009)
Soilborne diseases	*Meloidogyne incognita*	*Bacillus polymyxa* and *Bacillus* sp. in combination with AM fungi	Liu et al. (2012)
	Botrytis cinerea	*Pseudomonas putida* BTP1	Mariutto et al. (2011)
Fusarium crown and root rot (FCRR)	*Fusarium oxysporum* f. sp. *radicis-lycopersici* (Forl)	*B. subtilis* GB03 and FZB24, *B. amyloliquefaciens* IN937a, and *B. pumilus* SE34	Myresiotis et al. (2012)
Bacterial wilt	*Ralstonia solanacearum*	Commercial formulation of PGPR: equity and Tricho-Shield	Dairo and Akintunde (2012)
Fusarium wilt	*F. oxysporum lycopersici*	*Bacillus thuringiensis* strain 199	Akram et al. (2013)
Damping-off	*Sclerotinia sclerotiorum*	Fluorescent *Pseudomonas* spp.	Hammami et al. (2013)
Root rot	*F. solani*		
Stem canker and leaf blight	*Alternaria alternate*		
Fusarium wilt	*F. oxysporum*	*Bacillus subtilis* and *Pseudomonas fluorescens*	Selvakumar et al. (2013)
Wilting, blight, basal stem rot, and fruit rot	*Sclerotium rolfsii*	Consortium of *Pseudomonas* and *Trichoderma harzianum*	Singh et al. (2013)
Early blight	*A. alternate*	*Bacillus* spp.	Abdalla et al. (2014)
Bacterial wilt	*R. solanacearum*	*Pseudomonas fluorescens*	Murthy et al. (2014)
Fusarium wilt	*F. oxysporum* f. sp. lycopersici	*Bacillus subtilis* (BS2)	Loganathan et al. (2014)
Fusarium wilt	*Fusarium* spp.	*Bacillus* spp. and *Pseudomonas* spp.	Lachisa and Dabassa (2015)
Bacterial wilt	*R. solanacearum*	*Bacillus amyloliquefaciens* DSBA-11 and DSBA-12	Singh et al. (2016)

varying from 2 to 90% in different agroclimatic conditions, can survive in soil for days to years. This disease can be managed by antagonist PGPR such as *Bacillus* spp. through induction of systemic resistance. In addition to this, the antagonist PGPR offer some other advantages such as they are easy to apply, harmless to human beings and environment, and simultaneously enhance plant growth and improve yield of the crops (Singh et al. 2016).

6.4.4   Impact of PGPR Inoculation on Growth and Yield of Tomato

Many studies employing the use of PGPR in tomato cultivation have confirmed the remarkable increase in overall growth and yield of tomato crop. For instance, the average weight of tomato fruit per plant inoculated with *Rhodopseudomonas* sp. KL9 (82.7 g) was found greater than uninoculated control. Also, the content of lycopene in the ripe tomato was increased by 48.3% following strain KL9 application (Lee et al. 2008). Integrated nutrient management (INM) using *Azotobacter* sp. along with farmyard manure (FYM), NPK, and micronutrient showed an overall improvement in growth and yield of tomato. The application of *Azotobacter* sp. in INM enhanced the supply of N to plant resulting in crop improvement (Pandey and Chandra 2013). Various measurable growth parameters of tomato plant such as plant height, fresh and dry weight of root and shoot have also been reported to be positively affected as a result of inoculation with *Bacillus pumilus* alone, mixed inoculation of *Piriformospora indica* (a beneficial root endophytic fungus) + *B. pumilus*, and co-cultured *P. indica* and *B. pumilus* (Anith et al. 2015). Likewise, two PGPR strains *P. aeruginosa* and *Stenotrophomonas rhizophila* significantly increased the plant growth and fruit yield of tomato cultivars compared to the healthy controls and the plants infected with the CMV strains (Dashti et al. 2014). The highest fruit weight was obtained for plants treated with PGPR alone, followed closely by the healthy controls. The other pomological parameters such as fruit diameters, fruit weight, fruit volume, and specific gravity were found to be proportional with individual fruit weight of tomatoes in cultivars UC82B and supermarmande (Dashti et al. 2014). In the fall experiments, the plants were grown under suboptimal environmental conditions in a green house where *P. fluorescens* strain 63-28 increased significantly the marketable fruit yield by 13.3% and Grade No. 1 fruit weight by 18.2% (Gagne et al. 1993). The highest average fruit weight, fruit weight per plant, and plant length were obtained from *Pantoea agglomerans* FF inoculations. Fruit number per plant was found maximum in *Acinetobacter baumannii* CD-1 application. Fruit width, fruit length, and dry matter were highest in *B. megaterium*-GC subgroup A., MFD-2 application, than that of the other application in tomato (Dursun et al. 2010). Like the impact of PGPR on overall growth of tomato under conventional soil, there are reports suggesting that PGPR can also enhance the production of tomato under stressed soils. As an example, the effects of three PGPR on plant growth, yield, and quality of tomato under simulated seawater irrigation varied considerably (Shen et al. 2012). Of the three PGPR used in this study, *Erwinia persicinus* RA2 containing ACC deaminase showed maximum increase in marketable yields of fresh and dried fruits in tomato, grown under simulated seawater irrigation especially under high saline condition compared to those recorded for ACC deaminase-negative strains of *B. pumilus* (WP8) and *P. putida* (RBP1). Moreover, strain WP8 of *B. pumilus* had significant positive effects on tomato fruit quality, irrigated with 1% and 2% NaCl solution. Irrespective of the salt concentration, there were more Na (+) accumulated in mid-shoot leaves of tomato plants than in fruits. Also, sodium concentration was greater in leaves of *E. persicinus* RA2 and *B. pumilus* WP8 inoculated tomato. This study thus suggested that *E. persicinus* and *B. pumilus* could be the promising PGPR strains for enhancing the production of tomato under saline environment.

6.4.4.1 PGPR Mediated Augmentation in Lycopene Production and Other Components of Tomato

Lycopene is one of the most important components that determines the nutritional and marketable quality of tomato (George et al. 2004). Following PGPR inoculation, increase in lycopene and other valuable components has been reported. For example, increase in carotenoids contents measured through colorimetric assay in three tomato varieties inoculated with *B. subtilis* IAGS174 and *B. fortis* IAGS162 is reported (Akram et al. 2015). Also, the total sugar contents increased in all three varieties when bacterized with *B. subtilis* IAGS174 as compared to uninoculated control. Studies have also revealed a considerable increase in lycopene content following mixed application of *Trichoderma harzianum* and *Glomus mosseae* (AMF) (Nzanza et al. 2012). Significant effects of *T. harzianum* inoculation on Ca and Mg fruit contents were also evident. A noticeable increase in fruit lycopene content and antioxidant activity was observed in both the PGPR and PGPR × AMF application. Maximum lycopene content was found when three cultures of *Pseudomonas*, *Azotobacter*, and *Azospirillum* were used together (80.42 mg/kg fresh weight), and *Azotobacter* was applied with *Azospirillum* (78.51 mg/kg fresh weight). Synergistic interactions between AMF and asymbiotic N_2-fixing bacteria such as *A. chroococcum*, *Acetobacter diazotrophicus*, and *Azospirillum* spp. have also been reported elsewhere (Ordookhani et al. 2010). In a follow-up study, PGPR in combination with AM fungi resulted in maximum increase in lycopene content whereas antioxidant activity was observed maximum due to combined inoculation of *Pseudomonas* and *Azotobacter* (71.66 mg/kg fresh weight and 50.73%, respectively). When PGPR was co-inoculated with AMF, lycopene content was maximum in the presence of *Pseudomonas* + *Azotobacter* + AMF (76.43 mg/kg fresh weight), which did not differ significantly from *Pseudomonas* + AMF (73.65 mg/kg fresh weight) treatment (Ordookhani and Zare 2011). In a similar investigation, Loganathan et al. (2014) reported that inoculation of tomato with *Bacillus* spp. resulted in better lycopene content. Of the different *Bacillus* species used, *B. subtilis* BS2 recorded the highest content in all the harvest (71.28 mg/kg) when compared with *B. amyloliquefaciens* BA1 (46.21 mg/kg) and control (35.21 mg/kg). However, the influence of strain BS2 on lycopene content of tomato fruit was remarkable. The presence of PGPR increased the total protein content in tomatoes. Nevertheless, it had a negative impact on the ascorbic acid content in the fruits. Also, the total phenolic content and the lycopene content of the tomato fruits increased in the presence of PGPR (Dashti et al. 2014). The PGPR inoculations particularly affected the N, P, Mg, Ca, Na, K, Cu, Mn, Fe, and Zn contents of tomato fruit (Dursun et al. 2010).

Conclusion

The PGPR expressing multiple activities of growth promotion by supplying essential nutrients to tomato crop have shown impressive results under green house and field conditions. Moreover, PGPR application has been found to enhance the nutritional value and overall yield of tomato. The PGPR have also been found to reduce the excessive usage of mineral fertilizers in tomato production and, hence, are considered as an alternative to agrochemicals. Considering these, there is a need to identify tomato-specific PGPR and to explore their growth-enhancing potential in promoting the yield of tomato in different agronomic regions.

References

Abbamondi GR, Tommonaro G, Weyens N (2016) Plant growth-promoting effects of rhizospheric and endophytic bacteria associated with different tomato cultivars and new tomato hybrids. Chem Biol Technol Agric 3:1

Abdalla SA, Soad AAA, Elshiekh AI, Ahmed MEN (2014) *In vitro* screening of *Bacillus* isolates for biological control of early blight disease of tomato in Shambat soil. W J Agric Res 2:47–50

Adewuyi GO, Ademoyegun OT (2008) Analysis of vitamin C and major carotenoids in different fractions of tomatoes. Proc Int Conf Sci Tech Africa 2:65–73

Aeron A, Pandey P, Kumar S, Maheshwari DK (2011) Emerging role of plant growth promoting rhizobacteria. In: Maheshwari DK (ed) Bacteria in agrobiology: crop ecosystem. Springer Verlag, Berlin/Heidelberg, pp 1–26

Ahmad F, Ahmad I, Khan MS (2008) Screening of free-living rhizospheric bacteria for their multiple plant growth promoting activities. Microbiol Res 163:173–181

Ahmad M, Zahir ZA, Khalid M (2013) Efficacy of *Rhizobium* and *Pseudomonas* strains to improve physiology, ionic balance and quality of mung bean under salt-affected conditions on farmer's fields. Plant Physiol Biochem 63:170–176

Ahmad E, Zaidi A, Khan MS (2016) Effects of plant growth promoting rhizobacteria on the performance of greengram under field conditions. JJBS 9(2):79–88

Ahmed A, Hasnain S (2010) Auxin producing *Bacillus* sp.: auxin quantification and effect on the growth *Solanum tuberosum*. Pure Appl Chem 82:313–319

Ahmed L, Martin-Diana AB, Rico D, Barry-Ryan C (2011) The antioxidant properties of whey permeate treated fresh-cut tomatoes. Food Chem 24:1451–1457

Akram W, Anjum T, Ali B (2015) Co-cultivation of tomato with two *Bacillus* strains: effects on growth and yield. J Anim Plant Sci 25:1644–1651

Akram W, Mahboob A, Javed AA (2013) *Bacillus thuringiensis* strain 199 can induce systemic resistance in tomato against *Fusarium* wilt. Eur J Microbiol Immunol 3:275–280

Ali S, Charles TC, Glick BR (2012) Delay of flower senescence by bacterial endophytes expressing 1-aminocyclopropane-1-carboxylate deaminase. J Appl Microbiol 113:1139–1144

Almaghrabi OA, Massoud S, Abdelmoneim TS (2013) Influence of inoculation with plant growth promoting rhizobacteria (PGPR) on tomato plant growth and nematode reproduction under greenhouse conditions. Saudi J Biol Sci 20:57–61

Aloni R, Aloni E, Langhans M (2006) Role of cytokinin and auxin in shaping root architecture: regulating vascular differentiation, lateral root initiation, root apical dominance and root gravitropism. Ann Bot 97:883–889

Alvarez MADB, Gagne S, Antoun H (1995) Effect of compost on rhizosphere microflora of the tomato and on the incidence of plant growth-promoting rhizobacteria. Appl Environ Microbiol:194–199

Amara U, Khalid R, Hayat R (2015) Soil bacteria and phytohormones for sustainable crop production. In: Maheshwari DK (ed) Bacterial metabolites in sustainable agroecosystem. Springer International, Switzerland, pp 87–103

Anandham R, Janahiraman V, Gandhi PI, Kwon SW, Chung KY, Han GH, Choi JH, Sa TM (2014) Early plant growth promotion of maize by various sulfur oxidizing bacteria that uses different thiosulfate oxidation pathway. Afr J Microbiol Res 8:19–27

Anith KN, Sreekumar A, Sreekumar J (2015) The growth of tomato seedlings inoculated with co-cultivated *Piriformospora indica* and *Bacillus pumilus*. Symbiosis 65:9–16

Atzorn R, Crozier A, Wheeler CT, Sandberg G (1988) Production of gibberellins and indole-3-acetic acid by *Rhizobium phaseoli* in relation to nodulation of *Phaseolus vulgaris* roots. Planta 175:532–538

Azcorn R, Barea JM (1975) Synthesis of auxins, gibberellins and cytokinins by *Azotobacter vinelandi* and *Azotobacter beijerinckii* related to effects produced on tomato plants. Plant Soil 43:609–619

Badri DV, Chaparro JM, Zhang R, Shen Q, Vivanco JM (2013) Application of natural blends of phytochemicals derived from the root exudates of *Arabidopsis* to the soil reveal that phenolic-related compounds predominantly modulate the soil microbiome. J Biol Chem 288:4502–4512

Bano N, Musarrat J (2003) Isolation and characterization of phorate degrading soil bacteria of environmental and agronomic significance. Lett Appl Microbiol 36:349–353

Barriuso J, Solano BR, Fray RG, Cámara M, Hartmann A, Gutiérrez Mañero FJ (2008) Transgenic tomato plants alter quorum sensing in plant growth-promoting rhizobacteria. Plant Biotechnol J 6:442–452

Bastián F, Cohen A, Piccoli P, Luna V, Bottini R, Baraldi R, Bottini R (1998) Production of indole-3-acetic acid and gibberellins A1 and A3 by *Acetobacter diazotrophicus* and *Herbaspirillum seropedicae* in chemically-defined culture media. Plant Grow Reg 24:7–11

Bellishree K, Ganeshan G, Ramachandra YL, Rao AS, Chethana BS (2014) Effect of plant growth promoting rhizobacteria (pgpr) on germination, seedling growth and yield of tomato. Int J Rec Sci Res 5:1437–1443

Beneduzi A, Ambrosini A, Passaglia LMP (2012) Plant growth-promoting rhizobacteria (PGPR): their potential as antagonists and biocontrol agents. Genet Mol Biol 35(4 Suppl):1044–1051

Benner M, Linnemann AR, Jongen WMF et al (2007) An explorative study on the systematic development of tomato ketchup with potential health benefits using the chain information model. Trends Food Sci Tech 18:150–158

Bernardo AH, José RVJ, Reginaldo SR, Harllen SAS, Maria CB (2006) Induction of systemic resistance in tomato by the autochthonous phylloplane resident *Bacillus cereus*. Pesq Agropec Bras, Brasília 41:1247–1252

Beutner S, Bloedorn B, Frixel S et al (2001) Quantitative assessment of antioxidant properties of natural colorants and phytochemicals: carotenoids, flavonoids, phenols and indigoids. The role of β-carotene in antioxidant functions. J Sci Food Agric 81:559–568

Bhattacharyya PN, Jha DK (2012) Plant growth-promoting rhizobacteria (PGPR): emergence in agriculture. World J Microbiol Biotechnol 28:1327–1350

Bhowmik D, Kumar KPS, Paswan S, Srivastava S (2012) Tomato—a natural medicine and its health benefits. J Pharmacog Phytochem 1:33

Botelho GR, Mendonça-Hagler LC (2006) Fluorescent pseudomonads associated with the rhizosphere of crops-an overview. Braz J Microbiol 37:401–416

Bulgarelli D, Rott M, Schlaeppi K, Ver Loren van Themaat E, Ahmadinejad N, Assenza F, Rauf P, Huettel B, Reinhardt R, Schmelzer E, Peplies J, Gloeckner FO, Amann R, Eickhorst T, Schulze-

Lefert P (2012) Revealing structure and assembly cues for *Arabidopsis* root-inhabiting bacterial microbiota. Nature 488:91–95

Chandra RV, Prabhuji V, Roopa DA, Ravirajan S, Kishore HC (2007) Efficacy of lycopene in treatment of gingivitis: a randomized placebo controlled clinical trial. Oral Health Prev Dent 5:327–336

Chaparro JM, Badri DV, Vivanco JM (2013) Rhizosphere microbiome assemblage is affected by plant development. ISME J 8:790–803

Chookietwattana K, Maneewan K (2012) Selection of efficient salt-tolerant bacteria containing ACC deaminase for promotion of tomato growth under salinity stress. Soil Environ 31:30–36

Chowdhury SP, Hartmann A, Gao X, Borriss R (2015) Biocontrol mechanism by root-associated *Bacillus amyloliquefaciens* FZB42—a review. Front Microbiol 6:780

Dairo KP, Akintunde JK (2012) Evaluation of plant growth-promoting rhizobacteria for the control of bacterial wilt disease of tomato. GJBB 1:253–256

Dakora FD, Phillips DA (2002) Root exudates as mediators of mineral acquisition in low-nutrient environments. Plant Soil 245:35–47

Dashti NH, Montasser MS, Ali NYA, Cherian VM (2014) Influence of plant growth promoting rhizobacteria on fruit yield, pomological characteristics and chemical contents in cucumber mosaic virus-infected tomato plants. Kuwait J Sci 41:205–220

Dawwam GE, Elbeltagy Emara AHM, Abbas IH, Hassan MM (2013) Beneficial effect of plant growth promoting bacteria isolated from the roots of potato plant. Ann Agric Sci 58:195–201

Dorais M, Ehret DL, Papadopoulos AP (2008) Tomato (*Solanum lycopersicum*) health components: from the seed to the consumer. Phytochem Rev 7:231–250

Dursun A, Ekinci M, Dönmez MF (2010) Effects of foliar application of plant growth promoting bacterium on chemical contents, yield and growth of tomato (*Lycopersicon esculentum* L.) and cucumber (*Cucumis sativus* L.) Pak J Bot 42:3349–3356

El-Tantawy ME, Mohamed MAN (2009) Effect of inoculation with phosphate solubilizing bacteria on the tomato rhizosphere colonization process, plant growth and yield under organic and inorganic fertilization. J Appl Soil Res 5:1117–1131

FAOSTAT (2007) http://faostat.fao.org/cgi-bin/nph-db.pl?Subset=agriculture

Felici C, Vettori L, Giraldi E, Forino LMC, Toffanin A, Tagliasacchi AM, Nuti M (2008) Single and co-inoculation of *Bacillus subtilis* and *Azospirillum brasilense* on *Lycopersicon esculentum*: effect on plant growth and rhizosphere microbial community. Appl Soil Ecol 40:260–270

Flores-Félix JD, Menéndez E, Rivera LP, Marcos-García M, Martínez-Hidalgo P, Mateos PF, Martínez-Molina E, Velázquez ME, García-Fraile P, Rivas R (2013) Use of *Rhizobium leguminosarum* as a potential biofertilizer for *Lactuca sativa* and *Daucus carota* crops. J Plant Nutr Soil Sci 176:876–882

Frohlich K, Kaufmann K, Bitsch R, Bohm V (2006) Effects of ingestion of tomatoes, tomato juice and tomato puree on contents of lycopene isomers, tocopherols and ascorbic acid in human plasma as well as on lycopene isomer pattern. Br J Nutr 95:734–741

Gagne S, Dehbi L, Quéré DL, Fournier N (1993) Increase of greenhouse tomato fruit yields by plant growth-promoting rhizobacteria (PGPR) inoculated into the peat-based growing media. Soil Biol Biochem 25:269–272

Garcia-Fraile P, Carro L, Robledo M (2012) *Rhizobium* promotes non-legumes growth and quality in several production steps: towards a biofertilization of edible raw vegetables healthy for humans. PLoS One 7(5):e38122. doi:10.1371/journal.pone.0038122

George B, Kaur C, Khurdiya DS, Kapoor HC (2004) Antioxidants in tomato (*Lycopersium esculentum*) as a function of genotype. Food Chem 84:45–51

Glick BR (2012) Plant growth-promoting bacteria: mechanisms and applications. Scientifica. Article ID: 963401. doi:10.6064/2012/963401

Glick BR (2014) Bacteria with ACC deaminase can promote plant growth and help to feed the world. Microbiol Res 169:30–39

Glick BR, Todorovic B, Czarny J, Cheng Z, Duan J et al (2007) Promotion of plant growth by bacterial ACC deaminase. Crit Rev Plant Sci 26:227–242

Gopalakrishnan S, Sathya A, Vijayabharathi R, Varshney RK, Gowda CLL, Krishnamurthy L (2015) Plant growth promoting rhizobia: challenges and opportunities. 3 Biotech 5:355–377

Goswami D, Parmar S, Vaghela H, Dhandhukia P, Thakker J (2015) Describing *Paenibacillus mucilaginosus* strain N3 as an efficient plant growth promoting rhizobacteria (PGPR). Cogent Food Agric 1:1

Govindasamy V, Senthilkumar M, Magheshwaran V, Kumar U, Bose P, Sharma V, Annapurna K (2011) *Bacillus* and *Paenibacillus* spp.: potential PGPR for sustainable agriculture. In: Maheshwari DK (ed) Plant growth and health promoting bacteria. Springer-Verlag, Berlin, pp 333–364

Gravel V, Antoun H, Tweddell RJ (2007) Growth stimulation and fruit yield improvement of greenhouse tomato plants by inoculation with *Pseudomonas putida* or *Trichoderma atroviride*: possible role of indole acetic acid (IAA). Soil Biol Biochem 39:1968–1977

Gupta G, Parihar SS, Ahirwar NK, Snehi SK, Singh V (2015) Plant growth promoting rhizobacteria (PGPR): current and future prospects for development of sustainable agriculture. J Microb Biochem Technol 7:2

Gutierrez-Manero FJ, Ramos-Solano B, Probanza A, Mehouachi JR, Tadeo F, Talon M (2001) The plant-growth-promoting rhizobacteria *Bacillus pumilus* and *Bacillus licheniformis* produce high amounts of physiologically active gibberellins. Physiol Plantarum 111:206–211

Hammami I, Hsouna AB, Hamdi N, Gdoura R, Triki MA (2013) Isolation and characterization of rhizosphere bacteria for the biocontrol of the damping-off disease of tomatoes in Tunisia. C R Biol 336:557–564

Hassan W, David J, Bashir F (2014) ACC-deaminase and/or nitrogen-fixing rhizobacteria and growth response of tomato (*Lycopersicon pimpinellfolium* Mill.) J Plant Inter 9:869–882

Hedden P, Phillips AL (2000) Gibberellin metabolism: new insights revealed by the genes. Trends Plant Sci 5:523–530

Heeb A, (2005) Organic or mineral fertilization. Effects on tomato plant growth and fruit quality. Doctoral Thesis, Swedish University of Agricultural Sciences, Uppsala

Heulin T, Achouak W, Berge O, Normand P, Guinebretière MH (2002) *Paenibacillus graminis* sp. nov. and *Paenibacillus odorifer* sp. nov., isolated from plant roots, soil and food. Int J Syst Evol Microbiol 52:607–616

Hiltner L (1904) About recent experiences and problems the field of soil bacteriology with special consideration of green manure and fallow. Arbeiten der Deutschen Landwirtschaftlichen Gesellschaft 98:59–78

Hortencia GM, Olalde V, Violante P (2007) Alteration of tomato fruit quality by root inoculation with plant growth-promoting rhizobacteria (PGPR): *Bacillus subtilis* BEB-13bs. Sci Hortic 113:103–106

Indiragandhi P, Anandham R, Madhaiyan M, Sa TM (2008) Characterization of plant growth-promoting traits of bacteria isolated from larval guts of diamondback moth *Plutella xylostella* (Lepidoptera: Plutellidae). Curr Microbiol 56:327–333

Jha CK, Saraf M (2015) Plant growth promoting rhizobacteria (PGPR): a review. E3 J Agric Res Dev 5:108–119

Jha CK, Patel D, Rajendran N, Saraf M (2010) Combinatorial assessment on dominance and informative diversity of PGPR from rhizosphere of *Jatropha curcas* L. J Basic Microbiol 50:211–217

Jha CK, Patel B, Saraf M (2012) Stimulation of the growth of *Jatropha curcas* by the plant growth promoting bacterium *Enterobacter cancerogenus* MSA2. World J Microbiol Biotechnol 28:891–899

Jiang C, Sheng X, Qian M, Wang Q (2008) Isolation and characterization of a heavy metal resistant *Burkholderia* sp. from heavy metal-contaminated paddy field soil and its potential in promoting plant growth and heavy metal accumulation in metal-polluted soil. Chemosphere 72:157–164

Jikare AM, Chavan MD (2013) Siderophore produced by *Bacillus shackletonii* GN-09 and its plant growth promoting activity. IJPBS 3:198–202

Joo GJ, Kim YM, Kim JT (2005) Gibberellins-producing rhizobacteria increase endogenous gibberellins content and promote growth of red peppers. J Microbiol 43:510–515

Kalloo G (1991) Introduction. In: Kalloo G (ed) Monographs on theoretical and applied genetics 14. Genetic improvement of tomato. Springer-Verlag, Berlin, pp 1–9

Kamilova F, Kravchenko LV, Shaposhnikov AI, Azarova A, Makarova N, Lugtenberg B (2006) Organic acids, sugars, and L-tryptophane in exudates of vegetables growing on stone wool and their effects on activities of rhizosphere bacteria. Mol Plant-Microbe Interact 19:250–256

Kasa P, Modugapalem H, Battini K (2015) Isolation, screening, and molecular characterization of plant growth promoting rhizobacteria isolates of *Azotobacter* and *Trichoderma* and their beneficial activities. J Nat Sci Biol Med 6:360–363

Khaliq S, Khalid A, Saba B, Mahmood S, Siddique MT, Aziz I (2013) Effect of ACC deaminase bacteria on tomato plants containing azo dye wastewater. Pak J Bot 45:529–534

Khan AL, Waqas M, Kang S, Al-Harrasi A, Hussain J, Al-Rawahi A, Al-Khiziri S, Ullah I, Ali L, Jung H, Lee I (2014) Bacterial endophyte *Sphingomonas* sp. LK11 produces gibberellins and IAA and promotes tomato plant growth. J Microbiol. doi:10.1007/s12275-014-4002-7

Khan MS, Zaidi A, Ahemad M, Oves M, Wani PA (2010) Plant growth promotion by phosphate solubilizing fungi-current perspective. Arch Agron Soil Sci 56:73–98

Khan MS, Zaidi A, Wani PA, Oves M (2009) Role of plant growth promoting rhizobacteria in the remediation of metal contaminated soils. Environ Chem Lett 7:1–19

Kloepper JW, Schroth MN (1978) Plant growth-promoting rhizobacteria on radishes. In: Station de Pathologie (ed) Proceedings of the 4th international conference on plant pathogenic bacteria, Végétale et Phyto-Bactériologie, pp 879–882

Kloepper JW, Schroth MN (1981) Relationship of *in vitro* antibiosis of plant growth promoting rhizobacteria to plant growth and the displacement of root microflora. Phytopathology 71:1020–1024

Kloepper JW, Leong J, Teintze M, Schroth MN (1980) Enhanced plant growth by siderophores produced by plant growth promoting rhizobacteria. Nature 286:885–886

Kloepper JW, Zablotowick RM, Tipping EM, Lifshitz R (1991) Plant growth promotion mediated by bacterial rhizosphere colonizers. In: Keister DL, Cregan PB (eds) The rhizosphere and plant growth. Kluwer Academic Publishers, Dordrecht, Netherlands, pp 315–326

Kumar A, Bagewadi A, Keluskar V, Singh M (2007) Efficacy of lycopene in the management of oral submucous fibrosis. Oral Surg Oral Med Oral Pathol Oral Radiol Endod 103:207–213

Lachisa L, Dabassa A (2015) Synergetic effect of rhizosphere bacteria isolates and composted manure on fusarium wilt disease of tomato plants. Res J Microbiol 11:20–27

Lambrecht M, Okon Y, Broek AV, Vanderleyden J (2000) Indole-3-acetic acid: a reciprocal signaling molecule in bacteria plant interactions. Trends Microbiol 8:298–300

Lanteigne C, Gadkar VJ, Wallon T, Novinscak A, Filion M (2012) Production of DAPG and HCN by *Pseudomonas* sp. LBUM300 contributes to the biological control of bacterial canker of tomato. Phytopathology 102:967–973

Leclere V, Bechet M, Adam A, Guez JS, Wathelet B, Ongena M, Thonart P (2005) Mycosubtilin overproduction by *Bacillus subtilis* BBG100 enhances the organism's antagonistic and biocontrol activities. Appl Environ Microbiol 71:4577–4584

Lee KH, Koh RH, Song HG (2008) Enhancement of growth and yield of tomato by *Rhodopseudomonas* sp. under green house conditions. J Microbiol 46:641–646

Liu R, Dai M, Wu X, Li M, Liu X (2012) Suppression of the root-knot nematode [*Meloidogyne incognita* (Kofoid & White) Chitwood] on tomato by dual inoculation with arbuscular mycorrhizal fungi and plant growth-promoting rhizobacteria. Mycorrhiza 22:289–296

Livny O, Kaplan I, Reifen R, Charcon SP, Madat Z, Schwartz B (2002) Lycopene inhibits proliferation and enhances gap junction communication of KB-1 human oral tumor cells. J Nutr 132:3754–3759

Loganathan MR, Garg V, Saha VS, Rai AB (2014) Plant growth promoting rhizobacteria (PGPR) induces resistance against *Fusarium* wilt and improves lycopene content and texture in tomato. Afr J Microbiol Res 8:1105–1111

Lugtenberg B, Chin-A-Woeng T, Bloemberg G (2002) Microbe–plant interactions: principles and mechanisms. Antonie Van Leeuwenhoek 81:373–383

Lugtenberg BJ, Kravchenko LV, Simons M (1999) Tomato seed and root exudate sugars: composition, utilization by *Pseudomonas* biocontrol strains and role in rhizosphere colonization. Environ Microbiol 1:439–446

Lundberg DS, Lebeis SL, Paredes SH, Yourstone S, Gehring J, Malfatti S, Tremblay J, Engelbrektson A, Kunin V, del Rio TG, Edgar RC, Eickhorst T, Ley RE, Hugenholtz P, Tringe SG, Dangl JL (2012) Defining the core *Arabidopsis thaliana* root microbiome. Nature 488:86–90

Lynch JM (1987) The rhizosphere. Wiley Interscience, Chichester, UK

Majeed A, Abbasi MK, Hameed S, Imran A, Rahim N (2015) Isolation and characterization of plant growth-promoting rhizobacteria from wheat rhizosphere and their effect on plant growth promotion. Front Microbiol 6:198

Manzanera M, Narváez-Reinaldo JJ, García-Fontana C, Vílchez JI, González-López J (2015) Genome sequence of *Arthrobacter koreensis* 5J12A, a plant growth promoting and desiccation-tolerant strain. Genome Announc 3(3):e00648-15. doi:10.1128/genomeA.00648-15

Mariutto M, Duby F, Adam A, Bureau C, Fauconnier M, Ongena M, Thonart P, Dommes J (2011) The elicitation of a systemic resistance by *Pseudomonas putida* BTP1 in tomato involves the stimulation of two lipoxygenase isoforms. BMC Plant Biol 11:29

Meena G, Borkar SG, Nisha ML (2015) Population dynamics of plant growth promoting microbes on root surface and rhizosphere of tomato crop and their beneficial effect as bioinoculants on tomato and chilli crop. Int J Adv Res 3:990–996

Mehnaz S (2013) Secondary metabolites of *Pseudomonas aurantiaca* and their role in plant growth promotion. In: Arora NK (ed) Plant microbe symbiosis: fundamentals and advances. Springer, India, pp 373–394

Murthy KN, Uzma F, Chitrashree SC (2014) Induction of systemic resistance in tomato against *Ralstonia solanacearum* by *Pseudomonas fluorescens*. Am J Plant Sci 5:1799–1811

Myresiotis CK, Karaoglanidis GS, Vryzas Z, Papadopoulou-Mourkidou E (2012) Evaluation of plant-growth-promoting rhizobacteria, acibenzolar-S-methyl and hymexazol for integrated control of Fusarium crown and root rot on tomato. Pest Manage Sci 68:404–411

Neal AL, Ahmad S, Gordon-Weeks R, Ton J (2012) Benzoxazinoids in root exudates of maize attracts *Pseudomonas putida* to the rhizosphere. PLoS One 7(4):e35498. doi:10.1371/journal.pone.0035498

Ngoma L, Esau B, Babalola OO (2013) Isolation and characterization of beneficial indigenous endophytic bacteria for plant growth promoting activity in Molelwane Farm Mafikeng, South Africa. Afr J Biotechnol 12:4105–4114

Nzanza B, Marais D, Soundy P (2012) Yield and nutrient content of tomato (*Solanum lycopersicum* L.) as influenced by *Trichoderma harzianum* and *Glomus mosseae* inoculation. Sci Hortic 144:55–59

Odriozola-Serrano I, Soliva-Fortuny R, Hernandez-Jover T, Martin-Belloso O (2009) Carotenoid and phenolics profile of tomato juices processed by high intensity pulse electric fields compared with conventional thermal treatments. Food Chem 112:258–266

Oldroyd GED (2007) Nodules and hormones. Science 315(5808):52–53

Ordookhani K, Zare M (2011) Effect of *Pseudomonas, Azotobacter* and Arbuscular Mycorrhiza Fungi on lycopene, antioxidant activity and total soluble solid in tomato (*Lycopersicon esculentum* F1 Hybrid, Delba). Adv Environ Biol 5:1290–1294

Ordookhani K, Khavazi K, Moezzi A, Rejali F (2010) Influence of PGPR and AMF on antioxidant activity, lycopene and potassium contents in tomato. Afr J Agric Res 5:1108–1116

Pandey SK, Chandra KK (2013) Impact of integrated nutrient management on tomato yield under farmers field conditions. J Environ Biol 34:1047–1051

Pandya ND, Desai PV (2014) Screening and characterization of GA3 producing *Pseudomonas monteilii* and its impact on plant growth promotion. Int J Curr Microbiol Appl Sci 3:110–115

Patten CL, Glick BR (1996) Bacterial biosynthesis of indole-3-acetic acid. Can J Microbiol 42:207–220

Peralta IE, Spooner DM (2007) History, origin and early cultivation of tomato (Solanaceae). In: Razdan MK, Mattoo AK (ed) Genetic improvement of solanaceous crops, Vol 2: Tomato, Science Publishers, New Hampshire, USA, p 1–24

Porcel R, Zamarreño ÁM, García-Mina JM, Aroca R (2014) Involvement of plant endogenous ABA in *Bacillus megaterium* PGPR activity in tomato plants. BMC Plant Biol 14:36

Prashar P, Kapoor N, Sachdeva S (2014) Plant growth promoting activities of rhizobacteria associated with tomato in semi-arid region. Adv Life Sci Health 1:1

Puertas A, Gonzales LM (1999) Aislamiento de cepasnativas de *Azotobacter chroococcum* en la provincial Granmayevaluacion de suactividadestimuladora en plantulas de tomate. Cell Mol Life Sci 20:5–7

Radzki W, Manero FJG, Algar E (2013) Bacterial siderophores efficiently provide iron to iron-starved tomato plants in hydroponics culture. Antonie Van Leeuwenhoek 104:321–330

Sayyed RZ, Chincholkar SB, Reddy MS, Gangurde NS, Patel PR (2013) Siderophore producing PGPR for crop nutrition and phytopathogen suppression. In: Maheshwari DK (ed) Bacteria in agrobiology: disease management. Springer-Verlag, Berlin, Heidelberg

Schuhegger R, Ihring A, Gantner S, Knappe GBC, Vogg G, Hutzler P, Schmid M, Eberl FVBL, Hartmann A, Langebartels C (2006) Induction of systemic resistance in tomato by N-acyl-L homoserine lactone-producing rhizosphere bacteria. Plant Cell Environ 29:909–918

Selvakumar G, Mohan M, Kundu S, Gupta AD, Joshi P, Nazim S, Gupta HS (2008) Cold tolerance and plant growth promotion potential of *Serratia marcescens* strain SRM (MTCC 8708) isolated from flowers of summer squash (*Cucurbita pepo*). Lett Appl Microbiol 46:171–175

Selvakumar GR, Reetha R, Thamizhiniyan P (2013) The PGPR as elicitors of plant defence mechanisms and growth stimulants on tomato (*Lycopersicum esculentum* Mill.) Bot Res Int 6:47–55

Seneviratne M, Gunaratne S, Bandara T, Weerasundara L, Rajakaruna N, Seneviratne G, Vithanage M (2016) Plant growth promotion by *Bradyrhizobium japonicum* under heavy metal stress. South Afr J Bot 105:19–24

Sharma SB, Sayyed RZ, Trivedi MH, Gobi TA (2013) Phosphate solubilizing microbes: sustainable approach for managing phosphorus deficiency in agricultural soils. SpringerPlus 2:587

Shen M, Jun Kang Y, Li Wang H, Sheng Zhang X, Xin Zhao Q (2012) Effect of plant growth-promoting rhizobacteria (PGPRs) on plant growth, yield, and quality of tomato (*Lycopersicon esculentum* Mill.) under simulated seawater irrigation. J Gen Appl Microbiol 58:253–262

Shi J, Le Maguer M (2000) Lycopene in tomatoes: chemical and physical properties affected by food processing. Crit Rev Food Sci Nutr 40:1–42

Shishido M, Chanway CP (1998) Storage effects on indigenous soil microbial communities and PGPR efficacy. Soil Biol Biochem 30:939–947

Siddikee MA, Chauhan PS, Anandham R, Han GH, Sa T (2010) Isolation, characterization, and use for plant growth promotion under salt stress, of ACC deaminase producing halo tolerant bacteria derived from coastal soil. J Microbiol Biotechnol 20:1577–1584

Siddiqui ZA, Futai K (2009) Biocontrol of *Meloidogyne incognita* on tomato using antagonistic fungi, plant-growth-promoting rhizobacteria and cattle manure. Pest Manage Sci 65: 943–948

Simons M, Permentier HP, de Weger LA, Wijffelman CA, Lugtenberg BJJ (1997) Amino acid sysnthesis is necessary for tomato root colonization by *Pseudomonas fluorescens* strain WCS365. Mol Plant-Microbe Interact 10:102–106

Singh SP, Singh HB, Singh DK (2013) *Trichoderma harzianum* and *Pseudomonas* sp. mediated management of *Sclerotium rolfsii* rot in tomato (*Lycopersicon esculentum* Mill.) Bioscan 8:801–804

Singh D, Yadav DK, Chaudhary G, Rana VS, Sharma RK (2016) Potential of *Bacillus amyloliquefaciens* for biocontrol of bacterial wilt of tomato incited by *Ralstonia solanacearum*. J Plant Pathol Microbiol 7:327

Sivasakthi S, Kanchana D, Usharani G, Saranraj P (2013) Production of plant growth promoting substance by *Pseudomonas fluorescens* and *Bacillus subtilis* isolates from paddy rhizosphere soil of Cuddalore district Tamil Nadu, India. Int J Microbiol Res 4:227–233

Sokolova MG, Akimova GP, Vaishlia OB (2011) Effect of phytohormones synthesized by rhizosphere bacteria on plants. Prikl Biokhim Mikrobiol 47:302–307

Son J, Sumayo M, Hwang Y, Kim B, Ghim S (2014) Screening of plant growth-promoting rhizobacteria as elicitor of systemic resistance against gray leaf spot disease in pepper. Appl Soil Ecol 73:1–8

Srivastava LM (2002) Plant growth and development: hormones and environment. Academic Press, San Diego, CA

Súarez MH, Rodriguez EMR, Romero CD (2008) Chemical composition of tomato (*Lycopersicon esculentum*) from Tenerife, the Canary Islands. Food Chem 106:1046–1056

Sulochana MB, Jayachandra SY, Kumar SA, Parameshwar AB, Reddy KM, Dayanand A (2014) Siderophore as a potential plant growth-promoting agent produced by *Pseudomonas aeruginosa* JAS-25. Appl Biochem Biotechnol 174:297–308

Talboys PJ, Owen DW, Healey JR, Withers PJA, Jones DL (2014) Auxin secretion by *Bacillus amyloliquefaciens* FZB42 both stimulates root exudation and limits phosphorus uptake in *Triticum aestivum*. BMC Plant Biol 14:51

Tallapragada P, Dikshit R, Seshagir S (2015) Isolation and optimization of IAA producing *Burkholderia seminalis* and its effect on seedlings of tomato. Songklanakarin J Sci Technol 37:553–559

Teale WD, Paponov IA, Palme K (2006) Auxin in action: signaling, transport and the control of plant growth and development. Mol Cell Biol 7:847–859

Thamer S, Schädler M, Bonte D (2011) Dual benefit from a belowground symbiosis: nitrogen fixing rhizobia promote growth and defense against a specialist herbivore in a cyanogenic plant. Plant Soil 341:209–219

Thompson KA, Marshall MR, Sims CA, Wei CI, Sargent SA, Scott JW (2000) Cultivar, maturity and heat treatment on lycopene content in tomatoes. J Food Sci 65:791–795

Tran H, Ficke A, Asiimwe T, Höfte M, Raaijmakers JM (2007) Role of the cyclic lipopeptide massetolide A in biological control of *Phytophthora infestans* and in colonization of tomato plants by *Pseudomonas fluorescens*. New Phytol. doi:10.1111/j.1469-8137.2007.02138.x

Turan M, Ekinci M, Yildirim E, Günes A, Karagöz K, Kotan R, Dursun A (2014) Plant growth-promoting rhizobacteria improved growth, nutrient, and hormone content of cabbage (*Brassica oleracea*) seedlings. Turk J Agric For 38:327–333

Turner TR, Ramakrishnan K, Walshaw J, Heavens D, Alston M, Swarbreck D, Osbourn A, Grant A, Poole PS (2013) Comparative meta transcriptomics reveals kingdom level changes in the rhizosphere microbiome of plants. ISME J 7:2248–2258

Vaikuntapu PR, Dutta S, Samudrala RB, Rao VRVN, Kalam S, Podile AR (2014) Preferential promotion of *Lycopersicon esculentum* (Tomato) growth by plant growth promoting bacteria associated with tomato. Indian J Microbiol 54:403–412

Vessey JK (2003) Plant growth promoting rhizobacteria as biofertilizers. Plant Soil 255:571–586

Walker TS, Bais HP, Grotewold E, Vivanco JM (2003) Root exudation and rhizosphere biology. Plant Physiol 132:44–51

Werner T, Motyka V, Strnad M, Schmulling T (2001) Regulation of plant growth by cytokinin. Proc Natl Acad Sci U S A 98:10487–11049

Yan J, Smith MD, Glick BR, Liang Y (2014) Effects of ACC deaminase containing rhizobacteria on plant growth and expression of Toc GTPases in tomato (*Solanum lycopersicum*) under salt stress. Botany 92:775–781

Beneficial Role of Plant Growth-Promoting Bacteria in Vegetable Production Under Abiotic Stress

7

Metin Turan, Ertan Yildirim, Nurgul Kitir, Ceren Unek, Emrah Nikerel, Bahar Sogutmaz Ozdemir, Adem Güneş, and Mokhtari N.E.P

Abstract

Changes in climate, natural or man induced, urbanization and several other factors result in abiotic stress, for example, high winds, extreme temperatures, drought, flood, etc. Such factors in turn affect many plants including vegetables. Vegetables, being plants grown for their vegetative parts, are, however, more sensitive to abiotic stress, when compared to grass family. The abiotic stress limits soil/climate for vegetable plantation and consequently results in decreased vegetable yields. Plant growth-promoting bacteria (PGPB) are beneficial soil bacteria capable of stimulating physical, chemical and biological changes in plants. In particular, for vegetables, there are numerous applications of these beneficial microorganisms to alleviate the adverse effects of abiotic stress. This review focuses on alternative mechanisms employed by PGPB to enhance vegetable production under various abiotic stresses, including drought, salinity, extreme temperature, nutrient and heavy metal stresses.

M. Turan (✉) • N. Kitir • C. Unek • E. Nikerel • B.S. Ozdemir
Engineering Faculty, Genetics and Bioengineering Department, Yeditepe University, Istanbul, Turkey
e-mail: m_turan25@hotmail.com

E. Yildirim
Agricultural Faculty, Horticulture and Viticulture Department, Atataturk University, Erzurum, Turkey

A. Güneş
Agricultural Faculty, Soil Science and Plant Nutrition Department, Erciyes University, Kayseri, Turkey

M. N.E.P
Organic Farming Department, Islahiye Vocational School, Gaziantep University, Gaziantep, Turkey

© Springer International Publishing AG 2017
A. Zaidi, M.S. Khan (eds.), *Microbial Strategies for Vegetable Production*,
DOI 10.1007/978-3-319-54401-4_7

7.1   Introduction

Nutritional status, physical and biological properties of soil, constantly changing climate and other abiotic stresses are the primary causes for reduced agricultural productivity (Gopalakrishnan et al. 2015). Especially, abiotic stresses are the main reason for crop yield losses and food price increase in the world with growing population. The growth of plants in the field may be hampered by a large number of environmental abiotic stresses. These stresses include high and low temperature, drought, toxic metals, environmental organic contaminants and salinity (Glick et al. 2007). In addition to these stresses, climate change limits the geographical distribution and agricultural productivity of crops causing dramatic losses especially to vegetable species in several parts of the world (Olesen and Bindi 2002). Efforts to develop stress-tolerant vegetables via conventional breeding or transgenic approaches are challenging in itself since multiple genes and metabolic processes are involved in stress tolerance (Ashraf and Akram 2009). Apart from scientific and technical limitations, most of these techniques are time-consuming, cost intensive and not well accepted. Therefore, alternative approaches that would be affordable, eco-friendly and well accepted by the public need to be considered. A different approach to induce stress tolerance is the use of beneficial bacteria. Among variously distributed microbiota, the use of beneficial bacteria such as plant growth-promoting bacteria (PGPB or PGPR, henceforth denoted with the first one) has recently emerged as a potential new solution to protect crops against damages caused by abiotic stresses (Palaniyandi et al. 2014; Fatnassi et al. 2015; Wang et al. 2016).

Plant growth-promoting bacteria are beneficial soil bacteria capable of stimulating physical, chemical and biological changes in vegetables (Adesemoye et al. 2008), resulting both directly and indirectly in enhanced plant tolerance to abiotic stresses (Glick et al. 2007). Direct stimulation may include providing phytohormones to plants (Cassán et al. 2014), iron that has been sequestered by bacterial siderophores (Wandersman and Delepelaire 2004), soluble phosphate (Oteino et al. 2015) and fixing-free nitrogen (Santi et al. 2013; Yildirim et al. 2015), while indirect stimulation of plant growth includes preventing phytopathogens (biocontrol) and, thus, promotes plant growth and development (Glick and Bashan 1997). Particularly, production of ethylene in response to abiotic stresses leads to inhibition of root growth of the vegetables (Abeles et al. 2012). PGPR facing abiotic stress conditions regulate precipitated ethylene; examples are reported in several studies (Chen et al. 2013; Siddikee et al. 2012). Many of the studies have reported that PGPB strains improve the N_2 fixation (Nadeem et al. 2014; Gupta et al. 2014) and survive under stressed soil conditions, and when compared with agrochemicals, they have been found to be safe, inexpensive and rhizosphere competent.

Vegetables are plants grown for their vegetative parts, like leaves, fruits or stems. Vegetables consist of several plant families, grouped according to the plant organs: leafy vegetables (e.g. lettuce, spinach), stem vegetables (e.g. celery, asparagus), root vegetables (e.g. potatoes, carrots), legumes and pulses (e.g. beans, peas, lentils), crucifer family (*Brassicas*, e.g. cabbage, cauliflower, Brussel sprouts),

Allium family (bulb vegetables, e.g. onion, garlic), fruiting vegetables (botanical fruits, e.g. pumpkin, cucumber, tomato, zucchini), mushrooms and fungi. Vegetables contain vitamins, carbohydrates, salts and proteins, important for human nutrition. Worldwide, China, with 55% share, is the leading vegetable producer followed by India with 10.6%. Each vegetable species requires a specific growth condition where temperature, rainfall, humidity, chilling, density and length of sunlight, wind, etc. are important factors for vegetable growth. Among the horticultural plants, vegetables are especially more sensitive than the others for the extreme conditions (Schwarz et al. 2010). Temperature, moisture, soil physical characteristics, various cultural practices, disease and another stress factors can affect stand establishment in vegetable crop production (Grassbaugh and Bennett 1998).

7.2 Impact of Plant Growth-Promoting Bacteria on Vegetables Under Stressed Environment

7.2.1 Role of PGPB in Drought Stress

Drought is considered to be the most severe abiotic stress that limits growth and development of plants in arid and semiarid regions and attracts further attention considering climate-induced changes (Maybank et al. 1995). Furthermore, drought affects nearly all parts of the world (Wilhite 2000) and more than half of the earth is suffering from drought for long times (Kogan 1997). Plants respond to drought stress both at cellular and molecular levels. A well-studied response to drought is the increased level of ethylene (Singh et al. 2015). Drought accelerates ethylene production in plants leading to reduced or anomalous growth and premature senescence (Mattoo and Suttle 1991; et al. 2016; Hueso et al. 2011). PGPB promote plant growth and development under drought stress via lowering ethylene levels by hydrolyzing 1-aminocyclopropane-1-carboxylic acid (ACC), the immediate precursor of ethylene in plants (Zahir et al. 2008). Mayak et al. (2004a) reported that ACC deaminase containing *Achromobacter piechaudii* ARV8 substantially increased the fresh and dry weights of both tomato and pepper seedlings exposed to transient water stress. In most of the reported cases with increased ethylene levels due to drought, PGPB containing ACC deaminase significantly decreased ACC level in stressed plants, limiting ethylene synthesis, and hence relieved the damage to the plant. Saleem et al. (2007) reported such effect in tomato seedlings exposed to water stress. Interestingly, the review pointed that inoculation of tomato plants with the various bacteria resulted in continued plant growth both during water stress and when watering was resumed.

Similar effect of PGPB is expected in eliminating the growth-hampering effects of drought on the growth of peas (Akhtar and Azam 2014). Following this, Dodd et al. (2009) investigated the physiological responses of pea (*Pisum sativum* L.) to inoculation with ACC deaminase bacterium *Variovorax paradoxus* 5C-2 under drought stress. Within the same line, Figueiredo et al. (2008) reported an increased plant growth, N content and nodulation of common bean (*Phaseolus vulgaris* L.) even under drought due to co-inoculation of *Rhizobium tropici* and *Paenibacillus polymyxa*, leading to changes in hormone balance and stomatal conductance. During water scarcity, the bacteria did not influence the water content of plants; however, they significantly improved the recovery of plants when watering was resumed. Overall, positive effects of PGPB on plant growth parameters, e.g. chlorophyll content, trichome density, stomatal density and levels of secondary metabolites, are expected and/or reported. Indeed for peppermint, Cappellari et al. (2015) reported the application of PGPB, with the idea to illustrate the poorly known effects of PGPR on aromatic plant species. The authors measured several growth parameters and levels of secondary metabolites upon inoculation with strains of *Bacillus* and *Pseudomonas*. Also, the inoculation of PGPB has been reported to eliminate the effects of water stress on growth, yield and ripening of pea grown both under pot and field trials (Arshad et al. 2008).

7.2.2   Role of PGPB in Salinity Stress

Soil and water salinization is an important abiotic threat all over the world for agricultural production, currently affecting approximately 50% of fields, and the situation is only expected to get worse in the near future with the global climate changes. Moreover, lands facing salinity stress will also face drought due to increasing frequency of dry periods, leading to combined abiotic stresses. Salinity from soil or water has been causing significant decrease in the productiveness of many types of plants both in Turkey and in the world. Globally, 5% of the total 1.5 billion hectare land with agricultural production is affected by salinity (Tester and Davenport 2003). Salinity in the soil prevents the uptake of water (osmotic effect). Its effects on the roots, together with toxic effects of Na and Cl ions (increasing Na uptake while decreasing Ca and K uptake), are hampering plant development and productivity (Greenway and Munns 1980; Neel et al. 2002). Furthermore, it is reported that vegetables are more sensitive to salt than forages and grains (Waller and Yitayew 2016). Different methods have been adopted to obtain crops with increased tolerance to salinity stress. These methods include conventional breeding and selection, introgression with more resistant (wild) types and domesticating halophytes for the plant side and agrobiotechnological methods to handle the effect of flooding on crop production. Also, the use of PGPB to minimize salt stress on several plants has been reported (Mayak et al. 2004a; Rojas-Tapias et al. 2012; Yue et al. 2007).

All vegetables with the exception of beet and spinach are classified as either sensitive or semi-sensitive to salinity (Grattan and Grieve 1998). Similar to drought, the effects of salt stress can be reduced by the use of microorganisms that accelerate plant development (Mayak et al. 2004b). In coastal semiarid zones, the spreading of the halophyte *Salicornia bigelovii* was supported by using *Klebsiella pneumoniae* and *A. halopraeferens* as auxiliary biofertilizer combination (Rueda-Puente et al. 2004). Similarly, several PGPB such as *Rhizobium*, *Azospirillum* and mycorrhizal fungi species have been employed to inoculate crop seeds, such as lettuce, and seedlings to alleviate the salt and water stress (Barassi et al. 2006). The application of PGPB either to the seeds or environment reduces the adverse effects of salinity for eggplant (Bochow et al. 2001), tomato and pepper (Mayak et al. 2004b), beans (Yildirim and Taylor 2005), artichoke (Saleh et al. 2005), lettuce (Han and Lee 2005; Barassi et al. 2006; Sahin et al. 2015), squash (Yildirim et al. 2006), cabbage (Yildirim et al. 2015), chickpea (Elkoca et al. 2015), strawberry (İpek et al. 2014; Erdogan et al. 2016) and *Vicia pannonica* (Esringü et al. 2016). Research shows that positive effect of these bacteria results from the efficiency of water use by plants and stimulation of root development as a result of plant hormones such as auxin supplied to plants and released by these bacteria. A study carried on tomato shows that salinity negatively affects plant development, but the applications of *Streptomyces* sp. strain PGPA39 increased ACC deaminase activity and IAA production and phosphate solubilization in plants (Palaniyandi et al. 2014).

Similar to drought, an important mechanism to reduce the effect of salt stress by the use of PGPB is the reduction of ethylene synthesis, due to ACC deaminase enzyme (Mayak et al. 2004b; Sergeeva et al. 2006; Hontzeas et al. 2006). ACC deaminase is particularly useful in regulating ethylene concentration in roots (Glick 1995; Glick et al. 1999). A model has been suggested by Glick et al. (1998) to elicit the mechanism of how ACC deaminase helps to alleviate the stress conditions. The model suggests that phytohormone indole-3-acetic acid (IAA), which is one of the enzymes that is also synthesized by PGPB, can be consumed by plants. IAA not only has a contribution to the plant growth by stimulating the cell proliferation and elongation but also increases the activity of the enzyme ACC synthetase. The ACC synthetase has a role in production of ACC, an ethylene precursor. Significant amounts of the ACC synthesized by the plant are expelled into the soil. The soil microorganisms use this ACC as a nitrogen source via hydrolysis of ACC by ACC deaminase. In order to maintain the equilibrium of ACC levels inside the plant and the outside of the plant, plants secrete the ACC further into the soil. The ACC is a precursor for plant stress hormone ethylene; thus, when the ACC amount is lowered in plant's system, the ethylene production is constrained.

Salt stress was reduced in lettuce by application of *Azospirillum* (Barassi et al. 2006). It is speculated that this can be due to prevention of Na uptake and increase in the accumulation of osmolytes such as proline and glutamate (Barassi et al. 2006). Again in lettuce, it is reported that application of *Pseudomonas mendocina* in salty conditions increases plant's nutrition uptake and also ACC deaminase activity (Kohler et al. 2009). It is observed that application of PGPB like strains Mk1 of *Pseudomonas syringae*, Mk20 of *Pseudomonas fluorescens* and Mk25 of *Pseudomonas fluorescens* biotype G to mung beans subject to salt stress resulted in nodule formation and increase in ACC deaminase activity and effectiveness of water consumption (Ahmad et al. 2011, 2012). It is determined that *Rhizobium* and *Pseudomonas* strains, which play a major role in water consumption effectiveness in corns, increased proline and relative water content (RWC) levels in the leaves and facilitated uptake of K ions. Rueda-Puente et al. (2010) reported that after application of *Klebsiella pneumoniae* and *Azospirillum halopraeferens* to pepper grown under salt-stressed conditions, plant's total weight, root length and fresh and dry weight were increased. The effect of salinity stress with different salt concentration on pepper plants and alleviation of the stress with PGPB were investigated. Non-inoculated pepper plants died after 5 weeks by the time grown in the presence of high salt (120 mM NaCl); however, 80% of pepper plants inoculated with *P. fluorescens* 2112 survived under salinity stress (Lim et al. 2012). Specifically, *Azospirillum brasilense* has been reported to work well on alleviating the salinity stress on both seed germination and plant growth (Barassi et al. 2006). del Amor and Cuadra-Crespo (2012) in other investigation assessed the impact of *Azospirillum brasilense* and *Pantoea dispersa* on sweet pepper (*Capsicum annuum* L.) grown under saline stress. Plants were exposed to 0, 40, 80 and 120 mM NaCl in solution, and the effect on plant growth, leaf gas exchange, NO_3^-, Cl^-, K^+ and Na^+ accumulation and chlorophyll fluorescence and content was investigated. These results demonstrated that the benefit of these bacterial inoculants in ameliorating the deleterious effect of NaCl in a salt-sensitive vegetable as sweet pepper was apparent.

7.2.3   Role of PGPB in High Temperature

Increased CO_2 levels and other greenhouse gases are considered as major cause of global warming. It is also claimed that in the future, this effect will further be pronounced and will hamper the agricultural production (Kijne 2006). Plants react differently to the high temperature (both in soil and/or weather) during day and night. There are also substantial variations among vegetables regarding their response to varying temperature levels. Expectedly, vegetables of cool climates are more sensitive to hot weathers than vegetables of warm climates. Moreover, in vegetable production, level and duration of hot weather and plant's developmental stage play important roles in determining the level of damage (Hall 2000). Increased temperature negatively affects germination, shoot growth and nutrition uptake by vegetables and damages membrane stability. For instance, high temperature causes thermal dormancy of lettuce. During vegetative plant development period, it causes a decrease in the photosynthesis rate, carbon dioxide assimilation and metabolic activity (Al-Khatib and Paulsen 1999; Sam et al. 2001). Damages in membrane stability cause necrotic spots on the leaves similar to symptoms of drought stress and finally resulting in the death of the plant (Hall 2000). During plant generative development period, in turn, high temperature causes an important reduction in productivity by negatively affecting pollen germination, pollination, blossoming and formation of seed and fruit set (Hall 1992, 1993).

Plant growth-promoting bacteria have been reported to reduce the negative effect of heat on plants. For example, Martin and Stutz (2004) reported that isolates of *Glomus* increased dry substance quantity, positively affected the development and productivity of pepper plants and increased phosphorus uptake. Furthermore, in a similar study, *Microbacterium* M12M, *Bacillus* sp. B10M, *Pseudomonas* sp. P29M and *Pseudomonas* sp. P12M bacteria species decreased the negative effect of heat on soybean growth and productivity and increased the nutrient uptake of plants from the soil (Egamberdiyeva et al. 2004). Similar to other environmental stresses (drought and salinity), accelerated ethylene production under high temperatures has widely been reported both in plant tissues and microbial species in the rhizosphere. In order to cope with this, plants with ACC deaminase expression lower ethylene level (Timmusk and Wagner 1999). Bensalim et al. (1998) revealed that a *Burkholderia phytofirmans* strain PsJN improved the growth of potato plants by maintaining it even under high heat stress.

7.2.4   Role of PGPB in Low Temperature

Vegetables are greatly affected by the heat in the surrounding environment at all stages from seed germination to the final product. And hence, they need optimum temperature level. This optimum temperature (range) requirement varies between species and among plant varieties. Low temperature may negatively affect plant growth and consequently the productivity. It slows down the germination and shoots growth, limits the intake of nutrition and water, increases damages from the soil-borne diseases, affects negatively blooming, seed formation and ripening of fruits and finally may cause the death of the plant (Pierce 1987). Like other plants, warm climate vegetable species are even more sensitive to low temperature than the vegetables growing at temperate regions (Decoteau 2000).

Mechanistically, low temperature affects all metabolic activity of the cell due to the simple fact that enzymes work slower or inefficiently at low temperatures. Following this, low temperature may affect production of organic acids, sugars, phenolic compounds, phospholipids, protein and ATP in plants and in turn damages cell membranes (Lyons 1973). Particularly soluble sugars, as these are typical osmolytes in plants, play various roles in low-temperature tolerance. These play regulating roles similar to proline. Its accumulation is a significant metabolic adjustment by which plants exhibit low-temperature tolerance throughout cold acclimation (Janská et al. 2010; Turan et al 2012a and 2013).

Expectedly, application of PGPB relaxes the effect of low-temperature stress on vegetables such as cabbage by increasing the accumulation of osmoprotectants (e.g. proline) and hormones and the activities of antioxidants and the expression of genes that are associated with low-temperature stress tolerance (Wang et al. 2016; Bashan and Holguin 1997). These bacteria increase plant's nutrient uptake and hormone production and accumulation of starch, proline and phenolic compounds (Barka et al. 2006; Fernandez et al. 2012). Kang et al. (2015) reported that in pepper subjected to low-temperature stress and treated with PGPB *Serratia nematodiphila* PEJ1011, the level of gibberellin and abscisic acid was increased, while the level of jasmonic acid and salicylic acid did decrease. While it is observed that 15.56% of tomatoes were able to survive at 4 °C, while following *Bacillus cereus* AR156, *B. subtilis* SM21 and *Serratia* sp. XY21 application, the survival rate of tomato increased to 92.59% (Wang et al. 2016). Sun et al. (1995) reported that a possible mechanism of PGPB under low-temperature stress could be due to the release of antifreeze proteins which promote the growth of roots. It was also indicated that these bacteria, at low temperature, could increase the root length, fresh and dry weights and level of chlorophyll in the leaves, and these increases are being related to the reduction of ethylene level enzymatically (Glick et al. 1997).

7.2.5 Role of PGPB in Nutrient Stress

Another significant abiotic stress that affects the plant growth is the nutrient stress, which is indicated by decreased uptake of specific nutrients. Minerals, also called as macro- and micronutrients, are elements that directly/indirectly affect the physiological and biochemical processes of plants including vegetables. The deficiency of nutrients may slow down and even stop the vegetative and reproductive growth of plants (Gerloff 1987; Balakrishnan 1999) leading eventually to the death of the plant (Bennett 1993). However, the applications of PGPB as biofertilizer, in contrast, have been found to increase the uptake of nutrients by plants and facilitate plant growth (Calvo et al. 2015; Çakmakçi 2016). These inoculants, when applied, enhance plant growth and yield or protect plants from pests and diseases (Ramjegathesh et al. 2013). In this regard, several microbial inoculants have been used as biofertilizers, which supply nutrients like nitrogen, phosphorus, potassium, sulphur, iron, etc. to plants. The genera most commonly used as biocontrol agents are *Pseudomonas* (Tewari and Arora 2015), *Bacillus* (Alavo et al. 2015), *Burkholderia* (Pinedo et al. 2015), *Agrobacterium* (Bazzi et al. 2015), *Streptomyces* (Bhai et al. 2016; Viaene et al. 2016), etc. These organisms suppress plant disease by production of antibiotics (Prasannakumar et al. 2015) and siderophores (Patel et al. 2016; Adnan et al. 2016), by induction of systemic resistance (Zebelo et al. 2016; Annapurna et al. 2013) or any other mechanism.

Bacterial activity in the soil plays a major role in the functioning of the ecosystem since bacteria takes part in the biochemical cycle of many nutrient elements for plants. In an early work, Lin et al. (1983) reported the positive effect of PGPB in increasing the uptake of nutrients from the environment. It is found out that PGPB promote root formation and plant growth by increasing internal IAA level in the plant. Moreover, it also increases uptake of nitrogen, phosphorus and potassium under both stress and normal conditions (Singh and Singh 1993; Grichko and Glick 2001; Mayak et al. 2004b; Turan 2012b). A side, often underestimated feature is the stimulation of the ATPase proton pump to facilitate nutrient uptake. It is determined that application of *Enterobacter cloacae* CAL3 to tomatoes, peppers and beans under vermiculite environment without any nutrient medium for a period of 6–8 weeks caused an increase both in dry and fresh weights (Mayak et al. 2001). Yildirim et al. (2006) found out that bacteria belonging to *Bacillus* and *Paenibacillus* strains and fungi from *Trichoderma* strain were able to increase K and Ca uptake in squash grown under water stress. Similarly, Mayak et al. (2004b) also determined that application of PGPB increased phosphorous uptake by tomatoes grown under salt stress. Martin and Stutz (2004) found out that bacteria increased P uptake in peppers grown in high and low temperatures. Several studies report that some free-living bacteria in the rhizosphere fix free nitrogen and direct it to the use of plants

(Glick 1995; Glick et al. 1998; Sharma et al. 2003; Pii et al. 2015). Today, problems that can arise in soil as a result of excessive fertilizer application can be diminished by using PGPB in agricultural land that has been already damaged by unconscious and overdosing of fertilizers. Moreover, it could even be possible to obtain positive results by combining PGPB with the use of fertilizers at doses lower than normally prescribed for a plant. Hernández and Chailloux (2004) reported that in a greenhouse study, application of two different PGPB species supplemented with half of the normal fertilizer level resulted in increased productivity when compared with using only fertilizer.

7.2.6 Role of PGPB in Heavy Metal Stress

One of the significant environmental issues in soil is heavy metal contamination, and it has many negative effects on agricultural operations and human health. The increase in mining activities, new factories and growing industrialization contaminate large areas with heavy metals. It is reported that heavy metals become part of the food chain as they are accumulated by plants (Rubio et al. 1994). Some, but not all, metals are fundamental micronutrients required by plants for growth and development. Even if these are required, when present in excess, they may act as toxicants and suppress the growth (Ernst 1998). Those metals, which have negative effects on growth and productivity, if excessively accumulated, are cadmium, chromium, zinc, copper, lead and nickel (Prasad and Strzalka 2000; Brune and Dietz 1995). Such metals have toxic, as well as inhibitory, effects since they replace minerals that are necessary for plants such as iron by inhibiting the uptake of these. Additionally, there are public concerns about the accumulation of heavy metals present in soil, their transfer to the plants and eventual contribution of these heavy metals to the food chain (Kiziloglu et al. 2008).

Heavy metals cause reduction of chlorophyll, and, as a result, photosynthesis rate is decreased. Besides, high metal levels in the soil have also been shown to cause increased ethylene production, inhibiting, in turn, the plant development by minimizing CO_2 fixation and limiting sugar translocation (Buchanan et al. 2000). Arshad et al. (2008) reported successful application of PGPB, containing ACC deaminase activity, in phytoremediation of heavy metal polluted soil environment. Studies indicated that some plant growth-promoting bacteria decreased the negative effect of copper in beans (Fatnassi et al. 2015), zinc in potatoes (Gururani et al. 2013), cadmium in peas (Safronova et al. 2006) and nickel, lead and zinc in tomatoes, canola and Indian mustard (Burd et al. 2000). They had a protective effect and caused an increase in uptake of P, Ca and Fe. Similarly, it was also reported that ACC deaminase enzyme along with PGPB reversed and regulated the increase in ethylene levels caused by heavy metals (Grichko et al. 2000; Belimov et al. 2001; Nie et al. 2002). Vegetable growth and nutritional properties are decreased by heavy metal stress when compared to vegetables grown under normal conditions. This being said, vegetables inoculated with PGPB retained the biomass similar to healthy plants, even under metal stress. These results indicate that it is a multifunction of PGPB that can promote the growth and development of vegetables by alleviating the heavy metal stress. For example, in a study, it was focused on *Alcaligenes* sp. RZS3 and *P. aeruginosa* RZS3 producing siderophores as washing agents to clean up heavy metals from contaminated soils. It was reported that against several stress factors, different symbiotic and non-symbiotic bacteria (e.g. for symbiotics like *Rhizobium, Bradyrhizobium* and *Mesorhizobium* and for non-symbiotic like *Pseudomonas, Bacillus, Klebsiella, Azotobacter, Azospirillum* and *Azomonas*) were used without decreasing plant growth under these stresses (Munns et al. 2002).

Conclusion

In today's world, significant problems in agricultural production are due to abiotic stresses induced by environmental factors either in surrounding air or soil. In view of these environmental effects, especially in developed countries where vegetable production is common and much of this productivity is based on the extensive use of inexpensive chemicals and fertilizers, there is little immediate agricultural incentive. Besides these, in many of the less-developed countries of the world where vegetable growing is not as high, relatively low-cost labour and high chemical costs provide a situation where the use of plant growth-promoting bacteria provides an attractive commercial possibility. With the above-mentioned benefits, PGPB can help tolerate several abiotic stress conditions (drought, salt, heavy metals, etc.) in an affordable manner, preventing, at the same time, the excessive use of chemical fertilizers.

References

Abeles FB, Morgan PW, Saltveit ME Jr (2012) Ethylene in plant biology. Academic Press, San Diego, CA

Adesemoye AO, Obini M, Ugoji EO (2008) Comparison of plant growth-promotion with *Pseudomonas aeruginosa* and *Bacillus subtilis* in three vegetables. Braz J Microbiol 39:423–426

Adnan M, Shah Z, Saleem N, Basir A, Ullah H, Ibrahim M, Shah SRA (2016) Isolation and evaluation of summer legumes rhizobia as PGPR. Pure Appl Biol 5:127

Ahmad M, Zahir ZA, Asghar HN, Arshad M (2012) The combined application of rhizobial strains and plant growth promoting rhizobacteria improves growth and productivity of mung bean (*Vigna radiata* L.) under salt-stressed conditions. Ann Microbiol 62:1321–1330

Ahmad M, Zahir ZA, Asghar HN, Asghar M (2011) Inducing salt tolerance in mung bean through coinoculation with rhizobia and plant-growth-promoting rhizobacteria containing 1-aminocyc lopropane-1-carboxylate deaminase. Can J Microbiol 57:578–589

Akhtar MS, Azam T (2014) Effects of PGPR and antagonistic fungi on the growth, enzyme activity and fusarium root-rot of pea. Arch Phytopathol Plant Protect 47:138–148

Alavo TBC, Boukari S, Fayalo DG, Bochow H (2015) Cotton fertilization using PGPR *Bacillus amyloliquefaciens* FZB42 and compost: impact on insect density and cotton yield in North Benin, West Africa. Cogent Food Agric 1(1):1063829

Al-Khatib K, Paulsen GM (1999) High-temperature effects on photosynthetic processes in temperate and tropical cereals. Crop Sci 39:119–125

Annapurna K, Kumar A, Kumar LV, Govindasamy V, Bose P, Ramadoss D (2013) PGPR-induced systemic resistance (ISR) in plant disease management. In: Maheshwari DK (ed) Bacteria in agrobiology: disease management. Springer, Berlin Heidelberg, pp 405–425

Arshad M, Shaharoona B, Mahmood T (2008) Inoculation with plant growth promoting rhizobacteria containing ACC-deaminase partially eliminates the effects of water stress on growth, yield and ripening of *Pisum sativum* L. Pedosphere 18:611–620

Ashraf M, Akram NA (2009) Improving salinity tolerance of plants through conventional breeding and genetic engineering: an analytical comparison. Biotechnol Adv 27:744–752

Balakrishnan K (1999) Studies on nutrients deficiency symptoms in chilli (*Capsicim annum* L.) Indian J Plant Physiol 4:229–231

Barassi CA, Ayrault G, Creus CM, Sueldo RJ, Sobrero MT (2006) Seed inoculation with *Azospirillum* mitigates NaCl effects on lettuce. Sci Hortic 109:8–14

Barka EA, Nowak J, Clément C (2006) Enhancement of chilling resistance of inoculated grapevine plantlets with a plant growth-promoting rhizobacterium, Burkholderia phytofirmans strain PsJN. Appl Environ Microbiol 72:7246–7252

Bashan Y, Holguin G (1997) *Azospirillum*-plant relationships: environmental and physiological advances (1990–1996). Can J Microbiol 43:103–121

Bazzi C, Alexandrova M, Stefani E, Anaclerio F, Burr TJ (2015) Biological control of *Agrobacterium vitis* using non-tumorigenic agrobacteria. VITIS-J Grapevine Res 38:31

Belimov AA, Safronova VI, Sergeyeva TA, Egorova TN, Matveyeva VA, Tsyganov VE, Dietz KJ (2001) Characterization of plant growth promoting rhizobacteria isolated from polluted soils and containing 1-aminocyclopropane-1-carboxylate deaminase. Can J Microbiol 47:642–652

Bennett WF (1993) Nutrient deficiencies and toxicities in crop plants. American Phytopathological Society, Saint Paul, MN

Bensalim S, Nowak J, Asiedu SK (1998) A plant growth promoting rhizobacterium and temperature effects on performance of 18 clones of potato. Am J Potato Res 75(3):145–152

Bhai RS, Lijina A, Prameela TP, Krishna PB, Thampi A (2016) Biocontrol and growth promotive potential of *Streptomyces* spp. in Black Pepper (*Piper nigrum* L.) J Biol Control. doi:10.18641/jbc/0/0/93819

Bochow H, El-Sayed SF, Junge H, Stavropoulou A, Schmiedeknecht G (2001) Use of *Bacillus subtilis* as biocontrol agent. IV. Salt-stress tolerance induction by *Bacillus subtilis* FZB24 seed treatment in tropical vegetable field crops, and its mode of action. J Plant Dis Protect 108(1):21–30

Brune A, Dietz KJ (1995) A comparative analysis of element composition of roots and leaves of barley seedlings grown in the presence of toxic cadmium, molybdenum, nickel, and zinc concentrations 1. J Plant Nutr 18(4):853–868

Buchanan B, Gruissem W, Jones R (2000) Biochemistry and molecular biology of plants. American Society of Plant Physiologists, Courier Companies Inc, Waldorf, MD

Burd GI, Dixon DG, Glick BR (2000) Plant growth-promoting bacteria that decrease heavy metal toxicity in plants. Can J Microbiol 46(3):237–245

Çakmakçi R. (2016). Screening of multi-traits rhizobacteria for improving the growth, enzyme activities and nutrient uptake of tea (*Camellia sinensis*). Commun Soil Sci Plant Anal 47 (just-accepted)

Cappellari LDR, Santoro MV, Nievas F, Giordano W, Banchio E (2013) Increase of secondary metabolite content in marigold by inoculation with plant growth-promoting rhizobacteria. Appl Soil Ecol 70:16–22

Calvo P, Watts D, Torbert H, Kloepper J (2015). Application of microbial inoculants promote plant growth, increased nutrient uptake and improve root morphology of corn plants. In: American society of agronomy meetings, November, USA

Cassán F, Vanderleyden J, Spaepen S (2014) Physiological and agronomical aspects of phytohormone production by model plant-growth-promoting rhizobacteria (PGPR) belonging to the genus *Azospirillum*. J Plant Growth Regul 33(2):440–459

Chen L, Dodd IC, Theobald JC, Belimov AA, Davies WJ (2013) The rhizobacterium Variovorax paradoxus 5C-2, containing ACC deaminase, promotes growth and development of

Arabidopsis thaliana via an ethylene-dependent pathway. J Exp Bot 64(6):1565–1573. doi:10.1093/jxb/ert031

Decoteau DR (2000) Vegetable crops. Prentice-Hall, Upper Saddle River, NJ

del Amor FM, Cuadra-Crespo P (2012) Plant growth-promoting bacteria as a tool to improve salinity tolerance in sweet pepper. Funct Plant Biol 39(1):82–90

Dodd IC, Jiang F, Teijeiro RG, Belimoc A, Hartung W (2009) The rhizosphere bacterium *Variovorax paradoxus* 5C-2 containing ACC deaminase does not increase systemic ABA signaling in maize (*Zea mays* L.) Plant Signal Behav 4:519–521

Egamberdiyeva D, Qarshieva D, Davranov K (2004) Growth and yield of soybean varieties inoculated with *Bradyrhizobium* spp in N-deficient calcareous soils. Biol Fertil Soils 40(2):144–146

Elkoca E, Kocli T, Gunes A, Turan M (2015). The symbiotic performance and plant nutrient uptake of certain nationally registered chickpea (*Cicer arietinum* L.) cultivars of Turkey. J Plant Nutr 38(9): 1427–1443

Erdogan Ü, Çakmak Çi R, Varmazyari A, Turan M, Erdogan Y, Kitir N (2016) Role of inoculation with multi-trait rhizobacteria on strawberries under water deficit stress. Žemdirbystė/Agriculture 103(1):67–76

Ernst WHO (1998) Effects of heavy metals in plants at the cellular and organismic level: ecotoxicology. Wiley, New York, pp 587–620

Esringü A, Kaynar D, Turan M, Ercisli S (2016) Ameliorative effect of humic acid and plant growth-promoting rhizobacteria (PGPR) on hungarian vetch plants under salinity stress. Commun Soil Sci Plant Anal 47(5):602–618

Fatnassi IC, Chiboub M, Saadani O, Jebara M, Jebara SH (2015) Impact of dual inoculation with *Rhizobium* and PGPR on growth and antioxidant status of *Vicia faba* L. under copper stress. C R Biol 338(4):241–254

Fernandez O, Theocharis A, Bordiec S, Feil R, Jacquens L, Clément C, Fontaine F, Barka EA (2012) *Burkholderia phytofirmans* PsJN acclimates grapevine to cold by modulating carbohydrate metabolism. Mol Plant-Microbe Interact 25(4):496–504

Figueiredo MV, Burity HA, Martínez CR, Chanway CP (2008) Alleviation of drought stress in the common bean (*Phaseolus vulgaris* L.) by co-inoculation with *Paenibacillus polymyxa* and *Rhizobium tropici*. Appl Soil Ecol 40(1):182–188

Gerloff GC (1987) Intact-plant screening for tolerance of nutrient-deficiency stress. Plant Soil 99(1):3–16

Glick BR (1995) The enhancement of plant growth by free-living bacteria. Can J Microbiol 41(2):109–117

Glick BR, Bashan Y (1997) Genetic manipulation of plant growth-promoting bacteria to enhance biocontrol of phytopathogens. Biotechnol Adv 15(2):353–378

Glick BR, Cheng Z, Czarny J, Duan J (2007) Promotion of plant growth by ACC deaminase-producing soil bacteria. Eur J Plant Pathol 119(3):329–339

Glick BR, Liu C, Ghosh S, Dumbroff EB (1997) Early development of canola seedlings in the presence of the plant growth-promoting *Rhizobacterium* and *Pseudomonas putida* GR12-2. Soil Biol Biochem 29(8):1233–1239

Glick BR, Patten CL, Holguin G, Penrose DM (1999) Biochemical and genetic mechanisms used by plant growth promoting bacteria. World Scientific, London, UK

Glick BR, Penrose DM, Li J (1998) A model for the lowering of plant ethylene concentrations by plant growth-promoting bacteria. J Theor Biol 190(1):63–68

Gopalakrishnan S, Sathya A, Vijayabharathi R, Varshney RK, Gowda CL, Krishnamurthy L (2015) Plant growth promoting rhizobia: challenges and opportunities. 3 Biotech 5(4):355–377

Grassbaugh E, Bennett A (1998) Factors affecting vegetable stand establishment. Sci Agric 55:116–120

Grattan SR, Grieve CM (1998) Salinity–mineral nutrient relations in horticultural crops. Sci Hortic 78(1):127–157

Greenway H, Munns R (1980) Mechanisms of salt tolerance in nonhalophytes. Annu Rev Plant Physiol 31(1):149–190

Grichko VP, Filby B, Glick BR (2000) Increased ability of transgenic plants expressing the bacterial enzyme ACC deaminase to accumulate Cd, Co, Cu, Ni, Pb, and Zn. J Biotechnol 81(1):45–53

Grichko VP, Glick BR (2001) Amelioration of flooding stress by ACC deaminase-containingplant growth-promoting bacteria. Plant Physiol Biochem 39(1):11–17

Gupta S, Meena MK, Datta S (2014) Isolation, characterization of plant growth promoting bacteria from the plant Chlorophytum borivilianum and in-vitro screening for activity of nitrogen fixation, phosphate solubilization and IAA production. Int J Curr Microbiol Appl Sci 3(7):1082–1090

Gururani MA, Upadhyaya CP, Baskar V, Venkatesh J, Nookaraju A, Park SW (2013) Plant growth-promoting rhizobacteria enhance abiotic stress tolerance in *Solanum tuberosum* through inducing changes in the expression of ROS-scavenging enzymes and improved photosynthetic performance. J Plant Growth Regul 32(2):245–258

Hall AE (1992) Breeding for heat tolerance. In: Janick J (ed) Plant breeding reviews, vol 10. John Wiley & Sons, Inc., Oxford, UK. doi:10.1002/9780470650011.ch5

Hall AE (1993). Physiology and breeding for heat tolerance in cowpea, and comparisons with other crops. Adaptation of food crops to temperature and water stress. Publ No. 93-410, USA, pp 271–284

Hall AE (2000) Crop responses to environment. CRC Press, LLC, Boca Raton, FL

Han HS, Lee KD (2005) Plant growth promoting rhizobacteria effect on antioxidant status, photosynthesis, mineral uptake and growth of lettuce under soil salinity. Res J Agric Biol Sci 1(3):210–215

Hernández MI, Chailloux M (2004) Las micorrizas arbusculares y las bacterias rizosfericas como alternativa a la nutricion mineral del tomate. Cultivos Tropicales 25(2):5–13

Hontzeas N, Hontzeas CE, Glick BR (2006) Reaction mechanisms of the bacterial enzyme 1-ami nocyclopropane-1-carboxylate deaminase. Biotechnol Adv 24(4):420–426

Hueso S, Hernández T, García C (2011) Resistance and resilience of the soil microbial biomass to severe drought in semiarid soils: the importance of organic amendments. Appl Soil Ecol 50:27–36

Ipek M, Pirlak L, Esitken A, Figen Dönmez M, Turan M, Sahin F (2014) Plant Growth-Promoting Rhizobacteria (PGPR) increase yield, growth and nutrition of strawberry under high-calcareous soil conditions. J Plant Nutr 37(7):990–1001

Janská A, Maršík P, Zelenková S, Ovesna J (2010) Cold stress and acclimation–what is important for metabolic adjustment? Plant Biol 12(3):395–405

Kang SM, Khan AL, Waqas M, You YH, Hamayun M, Joo GJ, Lee IJ (2015) Gibberellin-producing Serratia nematodiphila PEJ1011 ameliorates low temperature stress in *Capsicum annuum* L. Eur J Soil Biol 68:85–93

Kijne JW (2006) Abiotic stress and water scarcity: identifying and resolving conflicts from plant level to global level. Field Crop Res 97(1):3–18

Kiziloglu FM, Turan M, Sahin U, Kuslu Y, Dursun A (2008) Effects of untreated and treated wastewater irrigation on some chemical properties of cauliflower (*Brassica oleracea* L. var. botrytis) and red cabbage (*Brassica oleracea* L. var. rubra) grown on calcareous soil in Turkey. Agric Water Manage 95(6):716–724

Kogan FN (1997) Global drought watch from space. Bull Am Meteorol Soc 78(4):621–636

Kohler J, Hernández JA, Caravaca F, Roldán A (2009) Induction of antioxidant enzymes is involved in the greater effectiveness of a PGPR versus AM fungi with respect to increasing the tolerance of lettuce to severe salt stress. Environ Exp Bot 65(2):245–252

Lim JH, An CH, Kim YH, Jung BK, Kim SD (2012) Isolation of auxin- and 1-aminocyclopropane-1-carboxylic acid deaminase-producing bacterium and its effect on pepper growth under saline stress. J Kor Soc Appl Biol Chem 55(5):607–612

Lin W, Okon Y, Hardy RW (1983) Enhanced mineral uptake by *Zea mays* and *Sorghum bicolor* roots inoculated with *Azospirillum brasilense*. Appl Environ Microbiol 45(6):1775–1779

Lyons JM (1973) Chilling injury in plants. Annu Rev Plant Physiol 24(1):445–466

Martin CA, Stutz JC (2004) Interactive effects of temperature and arbuscular mycorrhizal fungi on growth, P uptake and root respiration of *Capsicum annuum* L. Mycorrhiza 14(4):241–244

Mattoo AK, Suttle JC (1991) The plant hormone ethylene. CRC Press, Boca Raton, FL

Mayak S, Tirosh T, Glick BR (2001) Stimulation of the growth of tomato, pepper and mung bean plants by the plant growth-promoting bacterium *Enterobacter cloacae* CAL3. Biol Agric Hortic 19(3):261–274

Mayak S, Tirosh T, Glick BR (2004a) Plant growth-promoting bacteria that confer resistance to water stress in tomatoes and peppers. Plant Sci 166(2):525–530

Mayak S, Tirosh T, Glick BR (2004b) Plant growth-promoting bacteria confer resistance in tomato plants to salt stress. Plant Physiol Biochem 42(6):565–572

Maybank J, Bonsai B, Jones K, Lawford R, O'brien EG, Ripley EA, Wheaton E (1995) Drought as a natural disaster. Atmosphere-Ocean 33(2):195–222

Munns R, Husain S, Rivelli AR, James RA, Condon AT, Lindsay MP, Hare RA (2002). Avenues for increasing salt tolerance of crops, and the role of physiologically based selection traits. In: Horst WJ, Bürkert A, Claassen N, Flessa H, Frommer WB, Goldbach H, Merbach W, Olfs H.-W, Römheld V, Sattelmacher B, Schmidhalter U, Schenk MK, Wirén, NV, eds. Progress in plant nutrition: plenary lectures of the XIV international plant nutrition colloquium. Springer, Netherlands, pp 93–105

Nadeem SM, Ahmad M, Zahir ZA, Javaid A, Ashraf M (2014) The role of mycorrhizae and plant growth promoting rhizobacteria (PGPR) in improving crop productivity under stressful environments. Biotechnol Adv 32(2):429–448

Neel JPS, Alloush GA, Belesky DP, Clapham WM (2002) Influence of rhizosphere ionic strength on mineral composition, dry matter yield and nutritive value of forage chicory. J Agron Crop Sci 188(6):398–407

Nie L, Shah S, Rashid A, Burd GI, Dixon DG, Glick BR (2002) Phytoremediation of arsenate contaminated soil by transgenic canola and the plant growth-promoting bacterium *Enterobacter cloacae* CAL2. Plant Physiol Biochem 40(4):355–361

Olesen JE, Bindi M (2002) Consequences of climate change for European agricultural productivity, land use and policy. Eur J Agron 16(4):239–262

Oteino N, Lally RD, Kiwanuka S, Lloyd A, Ryan D, Germaine KJ, Dowling DN (2015) Plant growth promotion induced by phosphate solubilizing endophytic *Pseudomonas* isolates. Front Microbiol 6:745

Palaniyandi SA, Damodharan K, Yang SH, Suh JW (2014) *Streptomyces* sp. strain PGPA39 alleviates salt stress and promotes growth of 'Micro Tom' tomato plants. J Appl Microbiol 117(3):766–773

Patel PR, Shaikh SS, Sayyed RZ (2016) Dynamism of PGPR in bioremediation and plant growth promotion in heavy metal contaminated soil. Indian J Exp Biol 54:286–290

Pierce LC (1987) Vegetables: characteristics, production and marketing, John Willey and Sons Inc., New York, NY, 433 p

Pii Y, Penn A, Terzano R, Crecchio C, Mimmo T, Cesco S (2015) Plant-microorganism-soil interactions influence the Fe availability in the rhizosphere of cucumber plants. Plant Physiol Biochem 87:45–52

Pinedo I, Ledger T, Greve M, Poupin MJ (2015) *Burkholderia phytofirmans* PsJN induces long-term metabolic and transcriptional changes involved in *Arabidopsis thaliana* salt tolerance. Front Plant Sci 6:466

Prasad MNV, Strzalka K (2000) Physiology and biochemistry of metal toxicity and tolerance in plants. Kluwer Academic Publishers, Boston, pp 153–160

Prasannakumar SP, Gowtham HG, Hariprasad P, Shivaprasad K, Niranjana SR (2015) Delftia tsuruhatensis WGR–UOM–BT1, a novel rhizobacterium with PGPR properties from *Rauwolfia serpentina* (L.) Benth. ex Kurz also suppresses fungal phytopathogens by producing a new antibiotic—AMTM. Lett Appl Microbiol 61(5):460–468

Ramjegathesh R, Samiyappan R, Raguchander T, Prabakar K, Saravanakumar D (2013) Plant–PGPR interactions for pest and disease resistance in sustainable agriculture. In: Maheshwari DK (ed) Bacteria in agrobiology: disease management. Springer, Berlin Heidelberg, pp 293–320

Rojas-Tapias D, Moreno-Galván A, Pardo-Díaz S, Obando M, Rivera D, Bonilla R (2012) Effect of inoculation with plant growth-promoting bacteria (PGPB) on amelioration of saline stress in maize (*Zea mays*). Appl Soil Ecol 61:264–272

Rubio MI, Escrig I, Martinez-Cortina C, Lopez-Benet FJ, Sanz A (1994) Cadmium and nickel accumulation in rice plants. Effects on mineral nutrition and possible interactions of abscisic and gibberellic acids. Plant Growth Regul 14(2):151–157

Rueda-Puente EO, Castellanos T, Troyo-Diéguez E, de León-Alvarez JD (2004) Effect of *Klebsiella pneumoniae* and *Azospirillum halopraeferens* on the growth and development of two *Salicornia bigelovii* genotypes. Anim Prod Sci 44(1):65–74

Rueda-Puente EO, Murillo-Amador B, Castellanos-Cervantes T, García-Hernández JL, Tarazòn-Herrera MA, Medina SM, Barrera LEG (2010) Effects of plant growth promoting bacteria and mycorrhizal on *Capsicum annuum* L. var. aviculare ([Dierbach] D'Arcy and Eshbaugh) germination under stressing abiotic conditions. Plant Physiol Biochem 48(8):724–730

Safronova VI, Stepanok VV, Engqvist GL, Alekseyev YV, Belimov AA (2006) Root-associated bacteria containing 1-aminocyclopropane-1-carboxylate deaminase improve growth and nutrient uptake by pea genotypes cultivated in cadmium supplemented soil. Biol Fertil Soils 42(3):267–272

Sahin U, Ekinci M, Kiziloglu FM, Yildirim E, Turan M, Kotan R, Ors S (2015) Ameliorative effects of plant growth promoting bacteria on water-yield relationships, growth, and nutrient uptake of lettuce plants under different irrigation levels. HortScience 50(9):1379–1386

Saleem M, Arshad M, Hussain S, Bhatti AS (2007) Perspective of plant growth promoting rhizobacteria (PGPR) containing ACC deaminase in stress agriculture. J Ind Microbiol Biotechnol 34(10):635–648

Saleh SA, Heuberger H, Schnitzler WH (2005) Alleviation of salinity effect on artichoke productivity by *Bacillus subtilis* FZB24, supplemental Ca and micronutrients. J Appl Bot Food Qual 79(1):24–32

Sam O, Núñez M, Ruiz-Sánchez MC, Dell'Amico J, Falcón V, De La Rosa MC, Seoane J (2001) Effect of a brassinosteroid analogue and high temperature stress on leaf ultrastructure of *Lycopersicon esculentum*. Biol Plant 44(2):213–218

Santi C, Bogusz D, Franche C (2013) Biological nitrogen fixation in non-legume plants. Ann Bot 111(5):743–767

Schwarz D, Rouphael Y, Colla G, Venema JH (2010) Grafting as a tool to improve tolerance of vegetables to abiotic stresses: thermal stress, water stress and organic pollutants. Sci Hortic 127:162–171

Sergeeva E, Shah S, Glick BR (2006) Growth of transgenic canola (*Brassica napus* cv. Westar) expressing a bacterial 1-aminocyclopropane-1-carboxylate (ACC) deaminase gene on high concentrations of salt. World J Microbiol Biotechnol 22(3):277–282

Sharma A, Johri BN, Sharma AK, Glick BR (2003) Plant growth-promoting bacterium *Pseudomonas* sp. strain GRP 3 influences iron acquisition in mung bean (*Vigna radiata* L. Wilzeck). Soil Biol Biochem 35(7):887–894

Siddikee MA, Chauhan PS, Sa T (2012) Regulation of ethylene biosynthesis under salt stress in red pepper (*Capsicum annuum* L.) by 1-aminocyclopropane-1-carboxylic acid (ACC) deaminase-producing halotolerant bacteria. J Plant Growth Regul 31(2):265–272

Singh HP, Singh TA (1993) The interaction of rock phosphate, *Bradyrhizobium*, vesicular-arbuscular mycorrhizae and phosphate-solubilizing microbes on soybean grown in a sub-Himalayan mollisol. Mycorrhiza 4(1):37–43

Singh RP, Shelke GM, Kumar A, Jha PN (2015) Biochemistry and genetics of ACC deaminase: a weapon to "stress ethylene" produced in plants. Front Microbiol 6:937

Sun X, Griffith M, Pasternak JJ, Glick BR (1995) Low temperature growth, freezing survival, and production of antifreeze protein by the plant growth promoting rhizobacterium *Pseudomonas putida* GR12-2. Can J Microbiol 41(9):776–784

Tester M, Davenport R (2003) Na+ tolerance and Na+ transport in higher plants. Ann Bot 91(5):503–527

Tewari S, Arora NK (2015) Plant growth promoting fluorescent *Pseudomonas* enhancing growth of sunflower crop. Int J Sci Technol Soc 1(1):51–53

Timmusk S, Wagner EGH (1999) The plant-growth-promoting rhizobacterium *Paenibacillus polymyxa* induces changes in *Arabidopsis thaliana* gene expression: a possible connection between biotic and abiotic stress responses. Mol Plant-Microbe Interact 12(11):951–959

Turan M, Güllüce M, Çakmak R, Şahin F (2013) Effect of plant growth-promoting rhizobacteria strain on freezing injury and antioxidant enzyme activity of wheat and barley. J Plant Nutr 36(5):731–748

Turan M, Gulluce M, Şahin F (2012a) Effects of plant-growth-promoting rhizobacteria on yield, growth, and some physiological characteristics of wheat and barley plants. Commun Soil Sci Plant Anal 43(12):1658–1673

Turan M, Gulluce M, von Wirén N, Sahin F (2012b) Yield promotion and phosphorus solubilization by plant growth-promoting rhizobacteria in extensive wheat production in Turkey. J Plant Nutr Soil Sci 175(6):818–826

Viaene T, Langendries S, Beirinckx S, Maes M, Goormachtig S (2016) Streptomyces as a plant's best friend? FEMS Microbiol Ecol 92(8):fiw119. doi:10.1093/femsec/fiw119

Waller P, Yitayew M (2016) Water and salinity stress. In: Waller P, Yitayew M, eds. Irrigation and drainage engineering. Springer International Publishing, Switzerland, pp 51–65

Wandersman C, Delepelaire P (2004) Bacterial iron sources: from siderophores to hemophores. Annu Rev Microbiol 58:611–647

Wang C, Wang C, Gao YL, Wang YP, Guo JH (2016) A consortium of three plant growth-promoting rhizobacterium strains acclimates *Lycopersicon esculentum* and confers a better tolerance to chilling stress. J Plant Growth Regul 35(1):54–64

Wilhite DA (ed) (2000) Published in drought: a global assessment, vol I. Routledge, London, pp 3–18. Copyright © 2000 Donald A. Wilhite for the selection and editorial matter; individual chapters, the contributors

Yildirim E, Taylor AG, Spittler TD (2006) Ameliorative effects of biological treatments on growth of squash plants under salt stress. Sci Hortic 111(1):1–6

Yildirim E, Turan M, Ekinci M, Dursun A, Gunes A, Donmez M (2015) Growth and mineral content of cabbage seedlings in response to nitrogen fixing rhizobacteria treatment. Rom Biotechnol Lett 20(6):10929–10935

Yue H, Mo W, Li C, Zheng Y, Li H (2007) The salt stress relief and growth promotion effect of Rs-5 on cotton. Plant Soil 297(1-2):139–145

Zahir ZA, Munir A, Asghar HN, Shaharoona B, Arshad M (2008) Effectiveness of rhizobacteria containing ACC deaminase for growth promotion of peas (*Pisum sativum*) under drought conditions. J Microbiol Biotechnol 18(5):958–963

Zebelo S, Song Y, Kloepper JW, Fadamiro H (2016) Rhizobacteria activates (+)-δ-cadinene synthase genes and induces systemic resistance in cotton against beet armyworm (*Spodoptera exigua*). Plant Cell Environ 39(4):935–943

Metal Toxicity to Certain Vegetables and Bioremediation of Metal-Polluted Soils

8

Saima Saif, Mohd. Saghir Khan, Almas Zaidi, Asfa Rizvi, and Mohammad Shahid

Abstract

The production of quality vegetables is a crucial issue worldwide due to consistently deteriorating soil health. Plants including vegetables absorb a number of metals from soil, some of which have no biological function, but some are toxic at low concentrations, while others are required at low concentration but are toxic at higher concentrations. As vegetables constitute a major source of nutrition and are an important dietary constituent, the heavy metal uptake and bioaccumulation in vegetables is important since it disrupts production and quality of vegetables and consequently affects human health via food chain. Considering the serious threat of metals to vegetables, an attempt in this chapter is made to highlight the effects of certain metals on vegetables grown in different agroclimatic regions of the world. Also, the bioremediation strategies adopted to clean up the metal-contaminated soil is discussed. The results of different studies conducted across the globe on metal toxicity and bioremediation strategies presented in this chapter are likely to help vegetable growers to produce fresh and contaminant-free vegetables.

8.1 Introduction

Vegetables play an important role in humans' diet by providing and assisting the body with a variety of important constituents such as minerals, vitamins, complex carbohydrate, high dietary fibre, low levels of fat and high amount of water. Due to

S. Saif (✉) • M.S. Khan • A. Zaidi • A. Rizvi • M. Shahid
Faculty of Agricultural Sciences, Department of Agricultural Microbiology, Aligarh Muslim University, Aligarh 202002, Uttar Pradesh, India
e-mail: saima.saif3@gmail.com

© Springer International Publishing AG 2017
A. Zaidi, M.S. Khan (eds.), *Microbial Strategies for Vegetable Production*,
DOI 10.1007/978-3-319-54401-4_8

these, the consumption of vegetables is encouraged in human dietary systems. Consequently, there is an increasing demand of fresh and healthy vegetables which, however, may be contaminated by pathogens, heavy metals and/or toxins (Mello 2003). Mostly, growers engaged in vegetable production in horticulture practices worldwide often use poor-quality irrigation water due to unavailability of good-quality water (Drechsel and Keraita 2014). Apart from poor-quality waters, soil, human handling, organic fertilizers and wastewater are the major factors in contaminating the fresh vegetables. Among these factors, organic fertilizers and wastewater are considered the major source of vegetable contamination (Grant 2011). Depending on the source of contamination, the industrial wastewater contributes significant amounts of metals, metalloids and volatile or semi-volatile compounds, while domestic wastewater is most harmful due to its pathogenic load (Fiona et al. 2003). International Water Management Institute (IWMI 2006) has reported that at least 3.5 million ha is irrigated globally with untreated, partly treated, diluted or treated wastewater. Such poor-quality water containing toxic materials after uptake by plants may cause severe toxicity to vegetables. The consumption of such contaminated vegetables in turn affects the human health. And, hence, due to increasing concern of food safety, proper practices and methods of production have to be developed, and the hazards and the risks associated with toxic elements like heavy metals have to be fully understood before they pose any serious and consequential threat to consumer health. The heavy metals cannot be degraded by any biological, physical or chemical processes (Naz et al. 2015) and, hence, persist in the environment. However, numerous traditional physicochemical processes are available for remediations of polluted sites which are expensive and quite often inefficient as they do not permanently eradicate the pollutants. Also, the byproducts generated in the process become hazardous to human health (Singh et al. 2011). On the other hand, biological methods are more acceptable as they do not pose such problems, are easy to operate and do not produce secondary pollution. The biological approach generally called as bioremediation is considered a safe and inexpensive technique since they are based on the use of living organisms, microorganisms and plants (Karigar and Rao 2011) For instance, microorganisms adopt several mechanisms such as biotransformation (Xiong et al. 2010) and have varied ability of interacting with heavy metals. Another heavy metal removal strategy involves the use of plants, called as phytoremediation, wherein plants partially or completely remediate selected contaminants from soil, sludge, sediments, wastewater and groundwater. It is a cost-effective, efficient and eco-friendly in situ remediation technology driven by solar energy. There are however, certain drawbacks associated with this technology such as the pollutants or their metabolites accumulate within plant tissues, which in turn shorten plant life and releases contaminants into the atmosphere via volatilization.

8.2 Heavy Metals: A Brief Account

A heavy metal is defined as a member of a loosely defined subset of elements that exhibits metallic properties and mainly includes the transition metals, some metalloids, lanthanides and actinides. However, based on density, atomic number or atomic weight and chemical properties or toxicity, heavy metals have been defined variously (John 2002). For instance, any metallic chemical element that has a relatively higher density and is toxic or poisonous even at low concentration is defined as heavy metal (Alloway 1990). On the other hand, the elements such as cadmium, copper, nickel, mercury and lead which are commonly associated with pollution and exhibiting significant toxicity are considered heavy metals by Fiona et al. (2003). However, based on their importance as a nutrient, metals have been classified as (1) essential (e.g. Zn, Cu, Fe, Mn and Se), (2) probably essential (e.g. V and Co) and (3) potentially toxic (As, Cd, Pb, Hg and Ni) (Ebdon 2001). Besides this, all metals, in general, have toxic effects when there is excessive exposure (Woimant and Trocello 2014).

Heavy metals are a serious concern throughout the globe due to their toxic, mutagenic and teratogenic effects even at very low concentrations (Oluwole et al. 2013). While growing in metal-polluted soils, the plant can absorb metals through roots, or they can be deposited on foliar surfaces (Jassir et al. 2005). Heavy metal enters the human body mainly through inhalation of dust, direct ingestion of soil, consumption of food plants grown in metal-contaminated soil and drinking contaminated water. Due to non-destructive nature, heavy metals consequently accumulate in human vital organs and cause varying degrees of illnesses (Lenntech 2006). Elimination of heavy metals deposited on the surface, however, can often be accomplished simply by washing prior to consumption, whereas bio-accumulated metals are difficult to remove and are, therefore, of major concern (Michio 2005).

8.3   Sources of Vegetable Contamination by Heavy Metals

Because of soil contamination, heavy metal stress is becoming a major challenge to crop plants, particularly to vegetable crops. The heavy metals are derived from city/industrial effluent (Cai et al. 2012; Wang et al. 2013), mining and smelting (Zhao et al. 2012), fertilizers and pesticides (Nacke et al. 2013; Yu et al. 2013), electronic waste recycling/dismantling activities (Liu et al. 2013) and auto mobile depositions (Turer and Maynard 2003). Additionally, wastewater/sewage water can be another major source of heavy metals in areas where raw sewage water is used for irrigation (Li et al. 2013; Wang et al. 2013). Vegetable growing areas which are often situated in or near sources of atmospheric deposits have an elevated risk of potential contamination. Ingestion of vegetables that have been produced with contaminated water poses a serious risk to human health including various chronic diseases, particularly after prolonged dietary intakes (Sharma et al. 2009). Different vegetable species, however, tend to accumulate different metals based on environmental conditions, metal species and plant available forms of heavy metals (Lokeshwari and Chandrappa 2006). Uptake through roots depends on many factors such as soil pH, plant growth stages, the soluble content of heavy metals in soil, as well as type of crops, fertilizers and soil (Sharma et al. 2006). Most common heavy metals often

Table 8.1 Guidelines for safe limits of heavy metals

Sample/source	Standards	Cd	Cu	Pb	Zn	Mn	Ni	Cr
Soil (μg/g)	Indian Standard (Awashthi 1999)	03–06	135–270	250–500	300–600	–	75–150	–
	European Union Standards (EU 2002)	3.0	140	300	300	–	75	150
Water (μg/ml)	Indian Standard (Awashthi 2000)	0.01	0.05	0.1	5.0	0.1	–	0.05
	FAO (1985)	0.01	0.2	5.0	2.0	0.2	0.2	0.1
Plant (μg/g)	Indian Standard (Awashthi 1999)	1.5	30	2.5	50	–	1.5	20
	WHO/FAO (2007)	0.2	40	5	60	–	–	–
	European Union Standards (EU 2006)	0.2	–	0.3	–	–	–	–

Adapted from Singh et al. (2010)

found in vegetables include Cd, Cu, As, Cr, Pb, Zn, Co and Ni. When present in trace quantities, some of them act as micronutrients. Comparing the accumulated concentrations of metals with permissible limits of the Indian Standard (Awashthi 1999) and safe limits given by WHO/FAO (WHO/FAO 2007), several studies have found that metal concentrations were higher in vegetables grown in metal-polluted soil as compared to the safe limits given by commission regulation (EU 2006) (Table 8.1).

Other than safety concerns, excessive heavy metals significantly deteriorate the fertility of soil and consequently affect the growth and quality of crops (Muchuweti et al. 2006). Several studies have indicated that vegetables, particularly leafy vegetables grown in heavy metal-contaminated soils, have higher concentrations of heavy metals as compared to those grown in non-polluted soil. The symptoms of phytotoxicity of heavy metals, however, vary from metal to metal (Table 8.2). Routine monitoring of heavy metal concentrations in soils and also in crops is, therefore, essential to know the pollution levels and to devise strategies to minimize contamination, in order to reduce the risks to human health.

8.4 Bioaccumulation of Heavy Metals: A Serious Concern

Contamination and subsequent accumulation of heavy metals in leafy (Table 8.3) and non-leafy (Table 8.4) vegetables from different sources have been widely reported. However, the concentration of heavy metals in vegetables varies from below the detection limit to above the safe limit depending upon the source of contamination. Among heavy metals, Cd, a relatively rare element (WHO 1992), is

Table 8.2 Some examples of phytotoxicity symptoms of heavy metals in plants

Metal	Symptoms	Reference
Cadmium	Brown margin in leaves, chlorosis, necrosis, curled leaves, stunted roots, reddish veins and reduction in growth, purple coloration	Singh (2006)
Lead	Dark green leaves, stunted foliage, increased amounts of shoots	
Zinc	Chlorosis, stunted growth, reduction of root elongation	
Copper	Chlorosis, yellow coloration, purple coloration of the lower side of the midrib, less branched roots, inhibition of root growth, brown, stunted, coralloid roots, inhibition of plant growth	
Iron	Dark green foliage, stunted top and root growth, thickening of roots, brown spots on leaves starting from the tip of lower leaves, dark brown and purple leaves	
Chromium	Reduction in root growth, leaf chlorosis, necrosis, inhibition of seed germination and depressed biomass, disturb water balance	Ghani (2011), Pederno et al. (1997)
Arsenic	Wilting leaves, violet coloration (due to increased anthocyanin levels), root discoloration, inhibition of root growth, cell plasmolysis and plant death	Kabata-Pendias and Pendias (2001), Quaghebeur and Rengel (2003), Liu et al. (2005), Barrachina et al. (1995)

used in electroplating and galvanization processes, in batteries, in the production of pigments, as chemical reagent and in miscellaneous industrial processes such as smelting (ATSDR 1989). Cadmium compounds have varying degrees of solubility ranging from highly soluble to nearly insoluble which affects their absorption and toxicity (ATSDR 1989). Cadmium among metals is the most toxic heavy metal because it bioaccumulates, has a long half-life (about 30 years) and may cause health disorders even at low doses (Lenntech 2006). The increase in Cd uptake by plant tissues occurs due to the use of contaminated water for irrigation, fertilizers, sewage and composts. The absorption of Cd by plants, however, depends on genotypes and physical and chemical properties of plants (Jing and Logan 1992). Several workers have reported that the concentration of Cd was high and not suitable for human consumption in vegetables such as lettuce, spinach, radish, etc. (Prabu 2009), brinjal (Jamali et al. 2007), carrot and potato (Ding et al. 2014) and cucumber, tomato, green pepper, parsley, onion, bean, eggplant, pepper mint, pumpkin and okra (Demirezen and Aksoy 2006). In a study, Jassir et al. (2005) reported that the levels of Cd in the garden rocket vegetable species were high in both washed and unwashed samples which could possibly be due to the relatively easy uptake of Cd by food crops, especially by leafy vegetables. Also it may be due to the foliar absorption of atmospheric deposits on plant leaves (Midio and Satake 2003). In a similar investigation, significant variation in Cd concentration within different

Table 8.3 Heavy metal concentrations (mg/kg) in edible portion of leafy vegetables

Vegetables	Source of heavy metals	Heavy metals										References
		Cu	Zn	Cd	Pb	Ni	Cr	Mn	As	Hg		
Amaranthus blitum (*Amaranthus*)	Agricultural activities	42.82	–	0.16	1.91	–	1.85	–	0.67	0.27	Lui et al. (2006)
	Wastewater irrigation	1.981	54.65	2.918	2.361	–	–	–	1.780	–	Zhou et al. (2016)
	Contaminated riverside	15.60	78.45	0.15	2.54	2.46	2.28	–	0.19	–	Islam and Hoque (2014)
	Tannery effluents	–	–	0.08	2.06	–	9.08		0.14	0.04	Islam et al. (2014)
Beta vulgaris var. all green (palak)	Wastewater irrigation	28.58	41.51	4.36	15.74	7.57	27.83	117.94	–	–	Sharma et al. (2007)
	Sewage sludge	25.30	79.0	23.70	1.90	5.65	2.90	56.0	–	–	Singh and Agrawal (2007)
Apium graveolens (celery)	Sewage sludge	0.91	–	0.020	0.42	–	–	–	–	–	Dogheim et al. (2004)
	Agricultural activities	109.89	–	0.10	1.76	–	0.08	–	0.49	0.31	Lui et al. (2006)
	Atmospheric deposition	8.21	49.70	0.23	2.98	8.50	1.21	75.82	–	–	Stalikas et al. (1997)
Coriandrum sativum (coriander)	Sewage sludge	1.62	–	0.020	0.35	–	–	–	–	–	Dogheim et al. (2004)
	Atmospheric deposition	25.42	41.05	0.495	0.143	–	–	–	–	–	Jassir et al. (2005)
	Wastewater irrigation	22.24	186.40	–		–	83.06	65.64	–	–	Sinha et al. (2006)

Lactuca sativa (lettuce)	Industrial effluents, vehicular pollution	59.93	39.50	0.34	9.70	6.30	–	–	–	Demirezen and Aksoy (2006)
	Urban and industrial activities	0.92	–	0.01	0.07	–	–	–	–	Dogheim et al. (2004)
	Compost amendment	13.5	1171.0	–	4.90	–	1246.6	–	–	Intawongse and Dean (2006)
	Atmospheric deposition	–	–	0.07	0.58	–	–	–	–	Radwan and Salama (2006)
Mentha piperita (mint)	Sewage sludge	2.15	–	0.010	0.59	–	–	–	–	Dogheim et al. (2004)
	Atmospheric deposition	–	–	1.89	43.00	–	–	–	–	Kachenko and Singh (2006)
	Wastewater irrigation	17.34	192.00	–	–	–	–	–	–	Sinha et al. (2006)
Nasturtium officinale (watercress)	Atmospheric deposition	17.19	46.45	0.495	0.106	–	–	–	–	Jassir et al. (2005)
	Urban and industrial activities	1.96	105.20	1.22	14.37	43.62	18.77	–	–	Mohamed et al. (2003)
	Sewage sludge	1.11	–	0.080	0.29	–	–	–	–	Dogheim et al. (2004)

(continued)

Table 8.3 (continued)

Vegetables	Source of heavy metals	Heavy metals									References
		Cu	Zn	Cd	Pb	Ni	Cr	Mn	As	Hg	
Petroselinum crispum (parsley, lettuce)	Atmospheric deposition	24.89	43.54	0.062	0.099	–	–	–	–	–	Jassir et al. (2005)
	Sewage sludge	1.82	–	0.010	0.43	–	–	–	–	–	Dogheim et al. (2004)
	Industrial effluents, vehicular pollution	53.12	259.20	0.84	9.90	3.47	–	–	–	–	Demirezen and Aksoy (2006)
	Urban and industrial activities	3.34	21.00	–	3.29	0.60	–	13.09	–	–	Mohamed et al. (2003)
	Atmospheric deposition	–	–	0.067	0.34	–	–	–	–	–	Kachenko and Singh (2006)
		–	–	–	3.29	–	–	–	–	–	
Spinacia oleracea (spinach)	Industrial effluents, vehicular pollution	50.0	282	9.2	9.00	–	–	–	–	–	Singh and Kumar (2006)
	Atmospheric deposition	4.48	20.9	0.11	0.34	–	–	–	–	–	Radwan and Salama (2006)
	Compost amendment	32.3	632	–	5.20	–	–	6631	–	–	Intawongse and Dean (2006)
	Urban and industrial activities	1.18	–	0.03	0.56	–	–	–	–	–	Dogheim et al. (2004)
	Wastewater irrigation	23.0	85.08	–	–	–	8.55	155.6	–	–	Sinha et al. (2005)
	Agricultural activities	37.62	–	0.076	1.0	–	0.66	–	0.81	0.21	Lui et al. (2006)

Modified from Agrawal et al. (2007)

Table 8.4 Heavy metal concentrations (mg/kg) in edible portion of non-leafy vegetables

Vegetables	Source of heavy metals	Cu	Zn	Cd	Pb	Ni	Cr	Mn	As	Hg	References
Abelmoschus esculentus L (lady's finger)	Wastewater irrigation	11.30	116.26	–	–	–	6.00	29.22	–	–	Sinha et al. (2005)
	Industrial effluents, vehicular pollution	37.54	15.56	0.58	10.70	2.70	–	–	–	–	Demirezen and Aksoy (2006)
	Wastewater irrigation	5.10	132.70	6.60	28.0	10.60	12.80	–	–	–	Sharma et al. (2007)
Allium cepa (onion)	Industrial effluents, vehicular pollution	53.83	21.34	0.97	8.70	4.60	–	–	–	–	Demirezen and Aksoy (2006)
	Urban and industrial activities	2.81	17.60	0.76	10.29	18.37	–	3.26	–	–	Mohamed et al. (2003)
	Atmospheric deposition	1.49	11.40	0.02	0.14	–	–	–	–	–	Radwan and Salama (2006)
Brassica oleracea (cauliflower)	Industrial effluents, vehicular pollution	1.70	21.5	–	–	–	–	–	–	–	Singh and Kumar (2006)
	Wastewater irrigation	12.08	173.21	–	–	–	40.30	28.40	–	–	Sinha et al. (2005)

(continued)

Table 8.4 (continued)

Vegetables	Source of heavy metals	Cu	Zn	Cd	Pb	Ni	Cr	Mn	As	Hg	References
Daucus carota (carrot)	Atmospheric deposition	1.51	8.03	0.18	0.01	–	–	–	–	–	Radwan and Salama (2006)
	Compost amendment	37.60	149.50	27.40	8.50	–	–	758.90	–	–	Intawongse and Dean (2006)
	Agricultural activities	27.12	–	0.085	0.92	–	0.38	–	0.15	0.24	Lui et al. (2006)
	Urban and industrial activities	0.98	9.60	0.81	7.94	17.54	–	6.14	–	–	Mohamed et al. (2003)
	Wastewater irrigation	–	59.58	6.32	–	–	–	–	–	–	Sharma and Agrawal (2006)
Lycopersicon esculentum (tomato)	Agricultural activities	201.8	–	0.11	5.23	–	0.34	–	0.46	0.13	Lui et al. (2006)
	Atmospheric deposition	1.83	7.69	0.26	0.01	–	–	–	–	–	Radwan and Salama (2006)
	Wastewater irrigation	8.70	42.45	7.20	29.00	–	–	–	–	–	Sharma et al. (2006)
	Industrial effluents, vehicular pollution	32.60	3.56	0.41	9.70	3.10	–	–	–	–	Demirezen and Aksoy (2006)
	Urban and industrial activities	4.47	14.40	0.77	2.59	14.64	–	7.39	–	–	Mohamed et al. (2003)
Raphanus sativus (radish)	Agricultural activities	8.65	–	0.083	0.47	–	0.38	–	0.22	0.21	Lui et al. (2006)
	Compost amendment	26.90	500.30	68.20	11.80	–	–	271.0	–	–	Intawongse and Dean (2006)
	Tannery effluent irrigation	–	–	0.55	1.05	–	3.44	–	0.06	0.05	Islam et al. (2014)

Species	Source										References
Solanum melongena (brinjal)	Urban and industrial activities	2.93	50.70	0.69	4.57	11.87	–	21.66	–	–	Mohamed et al. (2003)
	Industrial effluents, vehicular pollution	37.38	9.35	0.43	7.20	4.6	–	–	–	–	Demirezen and Aksoy (2006)
	Atmospheric deposition	1.41	11.50	0.02	0.21	–	–	–	–	–	Radwan and Salama (2006)
	Agricultural activities	41.37	–	0.16	1.30	–	1.15	–	0.98	0.26	Lui et al. (2006)
	Wastewater irrigation	20.89	53.92	0.47	27.76	–	–	–	–	–	Parashar and Prasad (2013)
	Contaminated riverside	17.04	18.68	0.24	0.07	4.52	1.02	–	0.04	–	Islam and Hoque (2014)
Solanum tuberosum (potato)	Atmospheric deposition	0.83	7.16	0.02	0.01	–	–	–	–	–	Radwan and Salama (2006)
	Urban and industrial activities	0.88	4.50	0.84	2.81	10.74	–	5.67	–	–	Mohamed et al. (2003)
Cucurbita maxima (pumpkin)	Contaminated riverside	11.44	46.10	0.01	0.25	5.82	1.45	–	0.02	–	Islam and Hoque (2014)
	Sewage irrigation	–	–	0.365	1.79		12.4	53.5	–	–	Ashfaq et al. (2015)
Cucumis sativus (cucumber)	Sewage irrigation	20.38	54.84	0.5	26.27	–		–	–	–	Parashar and Prasad (2013)

Modified from Agrawal et al. (2007)

tomato genotypes was found (Hussain et al. 2015). The heavy metals that accumulated within tissues of various metal-tolerant tomato genotypes followed the order, shoot > fruit > leaf > root, while in susceptible genotypes, the order was fruit, shoot, leaf and root. The genotypic variation of a crop species makes it possible to select either Cd-accumulating cultivars to remediate contaminated soils or Cd-excluding cultivars to avoid Cd excessive uptake.

Chromium is yet another important metal which may enter through air, drinking water or eating food containing Cr or even through skin contact (Dinis and Fiúza 2011). However, for human and animals, it is considered as an essential metal for carbohydrate and lipid metabolism, and the recommended dietary intake of Cr is 50–200 µg/day for adults. However, exceeding normal concentrations (50–200 ug/day) lead to accumulation and toxicity that can result in hepatitis, gastritis, ulcers and lung cancer (Garcia et al. 2001). Several studies have demonstrated that some vegetables like cabbage and lettuce accumulate higher levels of Cr and could contribute to dietary problems (Biego et al. 1998; Castro 1998). Chromium has a low mobility and moves very slowly from roots to above-ground parts of plants (Skeffington et al. 1996). And, hence, the concentration of Cr is low in the upper organs of plants. Several studies have shown that the Cr concentration was higher than the maximum permitted metal concentration in lettuce, cabbage (Itanna 2002), spinach (Banerjee et al. 2011), luffa, brinjal, ladyfinger, cucumber and gourd (Kumar et al. 2016). In contrast, the Cr contents in different vegetables grown in the lands irrigated with wastewater were found within the safe limits (Sharma et al. 2006).

Mankind is exposed to the highest levels of lead (Pb) that occurs naturally as a sulphide compound and is a soft bluish-white, silvery grey metal that melts at 327.5° C (Budavari 2001). There are different sources of environmental pollution with Pb as Pb alkyl additives in petrol and manufacturing processes. This can bring Pb into the human food chain through uptake of food (about 65%), water (up to 20%) and from air and dust (about 15%) (IPCS 1992). Like other heavy metals, Pb can bioaccumulate over time and remain in the body for long periods. It therefore becomes important to detect such metals even at very low concentrations. In Sudan, for example, Dafelseed (2007) determined the level of Pb in selected fresh vegetables and reported 0.35, 0.86, 0.60 and 0.48 mg/kg in carrot, sweet pepper, garden rock and tomato, respectively. On the contrary, the FAO/WHO (2001) has reported that the maximum permissible level of Pb in vegetables is 0.3 mg/kg.

Human carcinogen, arsenic, is a well-known toxic element, widely distributed in the environment in both inorganic and organic forms (Hughes et al. 2011). It is well-recognized that inorganic arsenic is probably the most dangerous form of arsenic in food, being As (III) more toxic than As (V) (Pizarro et al. 2016). There are many routes by which As can enter the human body (1) via food chain (ingestion by water and food sources) and (2) occupational exposure (Rahman et al. 2009). There have been several reports of arsenic speciation in vegetables growing in natural or contaminated soils (Pell et al. 2013). Broccoli, lettuce, potato, carrots, etc., for example, can accumulate arsenic when such crops are grown in soil irrigated with As (V) containing water. In most of the vegetables, arsenic is taken up by plant roots via macronutrient transporters (Zhao et al. 2010; Wu et al. 2011). In a study, it was observed that the calculated accessible doses of As expressed as microgram arsenic per year are about 470 for carrots, 550 for beets and 180 for quinoa considering the maximum intake of 2.5 kg per year of quinoa and of 6 kg per year of carrots and beets. Therefore, quinoa seems to be the vegetable with the lower toxicological risk (Pizarro et al. 2016). When taken up by plants, significant changes have been observed in the growth, yield and accumulation characteristics of okra spiked with 20 and 50 mg/kg of As(III), As(V) and dimethylarsinic acid (DMA) (Chandra et al. 2016). The arsenic concentration in the aerial part followed the order As(V) > As(III) > DMA while it was As(III) > As(V) > DMA in the roots. Thus, the plant has the capacity to tolerate As stress and can be considered as a resistive variety. Similarly, arsenic accumulation has also been reported in different vegetables beyond the permissible limit. For instance, Santra et al. 2013 found that tuberous vegetables accumulated higher amount of arsenic than leafy vegetables which was followed by fruity vegetable. This is supported by the fact that the accumulation of As in plants occurs primarily through the root system and, hence, the highest As concentrations have been reported in plant roots and tubers (Marin et al. 1993). In this study, the highest arsenic accumulation was observed in potato, brinjal, arum, amaranth, radish, lady's finger and cauliflower, whereas lower level of arsenic accumulation was observed in beans, green chilli, tomato, bitter guard, lemon and turmeric. Rehman et al. (2016) in contrast reported that the As concentration in edible portions of vegetable ranged from 0.03 to 1.38 mg/kg. Similarly, the trend of As bioaccumulation in vegetables irrigated with arsenic contaminated water was spinach > tomato > radish > carrot, and this distribution of As in vegetable tissues was species dependent; As was mainly found in the roots of tomato and spinach, but accumulated in the leaves and skin of root crops as reported by Bhatti et al. (2013).

8.4.1   Vegetable Toxicity by Multiple Metals

The interactions between different metals in soil may lead to increased uptake of one or the other metal by plants which in effect may cause toxicity to animals and humans via food chain. In a study carried out in Slovakia, Musilova et al. (2016) reported the accumulation of Cd, Pb and Zn in potato tubers in a concentration-dependent manner. The correlation between heavy metal content in soil and its content in potato tubers followed the order cv. Laura-Spissky Stvrtok (Cd), cv. Red Anna-Odorin (Pb) and Marabel and Red Anna-Odorin, cv. Marabel-Belusa and cv. Volumia-Imel (Zn). Also, heavy metals have been found several folds higher than the safe limit in other vegetables like *Colocasia*, *Amaranthus*, cauliflower, etc. (Saha et al. 2015). Of the various metals detected, the concentrations of Pb, Cd and Ni were above the permissible limit in all vegetables, while *Colocasia* and *Amaranthus* accumulated highest metal contents. The highest mean transfer coefficients (TCs) values recorded for Zn, Cu, Pb, Cd and Ni were 0.59 (*cauliflower*), 0.67 (*Colocasia*), 0.93 (*Amaranthus*), 1.02 (*Colocasia*) and 1.09 (*Amaranthus*), respectively. The results further revealed that the maximum single element pollution index (SEPI) value was found for Cd which ranged from 2.93 to 6.03 with a mean of 5.32. In yet another investigation, Tiwari et al.(2011) assessed the edible parts of five vegetables such as spinach, radish, tomato, chilli and cabbage growing in field receiving mixed industrial effluent and reported a high level of toxic metals (As, Cd, Cr, Pb and Ni). It was concluded from this study that the cultivation of such vegetables is not safe under heavy metal-stressed soil. Similarly, parsley, followed by spinach, contained the highest concentration of heavy metals besides onion that contained high levels of toxic heavy metals. The content of Cu in parsley and spinach and Pb in onion exceeded the Codex limits (Osaili et al. 2016). In the western region of Saudi Arabia, the human health problems were found associated with the consumption of metal-contaminated okra (Balkhair and Ashraf 2016). The levels of Ni, Pb, Cd and Cr in the edible parts were 90, 28, 83 and 63%, respectively, above the safe limit. The uptake and accumulation of heavy metals by the edible portions followed the order Cr > Zn > Ni > Cd > Mn > Pb > Cu > Fe. Moreover, the health risk index (HRI) was >1 which indicated a significant health risk, and hence, okra among vegetables was not safe for human consumption.

Antonious et al. (2012) reported that regardless of soil amendments, the overall bioaccumulation factor (BAF) of seven heavy metals in cabbage leaves and broccoli heads were poor. For leafy vegetables collected in 15 ha of squatted land belonging to the international airport of Cotonou, total concentrations of metal (loids) measured in consumed parts of *Lactuca sativa* L. and *Brassica oleracea* were 52.6–78.9, 0.02–0.3, 0.08–0.22, 12.7–20.3, 1.8–7.9 and 44.1–107.8 mg/kg for Pb, Cd, As, Sb, Cu and Zn, respectively (Uzu et al. 2014). Ferri et al. (2015) reported that 60 and 10% of spinach samples exceeded maximum Pb and Cd European standards and recommended that washing before consuming vegetables can reduce toxicity risk to humans. Moreover, crude or untreated wastewater, the treated wastewater and groundwater used for irrigating vegetables also contribute significantly to bioaccumulation of heavy metals. As an example, Ghosh et al. (2012) in a study found that radish, turnip and spinach, irrigated with sewage water, were grouped as hyperaccumulator of heavy metals, whereas brinjal and cauliflower accumulated less heavy metal though the metal concentrations did not exceed the permissible limit and, hence, were considered safe for consumption.

Industrial activities also add heavy metals to the environment that pose risk to both human and plant health. Also, the atmospheric deposition through the particulate matter released from transport creates heavy metal pollution among vegetables grown along roadside or during marketing. The health risk assessment methods of United States Environmental Protection Agency (USEPA) are used to establish the potential health hazards of heavy metals in soils growing vegetables in different regions of the world. In a study conducted in three economically developed areas of Zhejiang Province, China suffering from increasing heavy metal damages from various pollution sources including agriculture, traffic, mining and Chinese typical local private family-sized industry, 268 vegetable samples which included celery, cabbages, carrots, asparagus lettuces, cowpeas, tomatoes and cayenne pepper and their corresponding soils were collected. Metal concentrations were measured in soil, settled atmospheric particulate matter (PM) and vegetables at two different sites near a waste incinerator and a highway. A risk assessment was performed using both total- and bio-accessible metal concentrations in vegetables. At both sites, total Cr, Cd and Pb concentrations in vegetables were found above or just under the maximum limit levels for foodstuffs according to Chinese and European Commission regulations. High metal bio-accessibility in the vegetables (60–79%, with maximum value for Cd) was observed (Xiong et al. 2016). In another study, leaves of mature cabbage and spinach were exposed to manufactured mono-metallic oxide particles (CdO, Sb_2O_3 and ZnO) or to complex process. Particulate matter was mainly enriched with lead, and it was found that high quantities of Cd, Sb, Zn and Pb were taken up by the plant leaves. The levels of metals depend on both the plant species and nature of the PM. A maximum of 2% of the leaf surfaces were covered up to 12% of stomatal openings. Metal (loid) bio-accessibility was significantly higher for vegetables due to chemical speciation changes (Xiong et al. 2014).

8.4.2   Effect of Metals on Physiological Processes of Vegetables Grown in Metal-Stressed Soil

Heavy metals have strong influence on nutritional quality of vegetables when grown in metal-contaminated soil. Therefore, the consumption of such vegetables may lead to nutritional deficiency in developing countries which are already facing malnutrition problems. In a study, effects of four different levels of Cd, Pb and mixture of Cd and Pb on different nutrients of three vegetables, potato, tomato and lettuce, grown in pots containing soil contaminated with Cd, Pb and Cd-Pb mixture were evaluated. The edible portions of each plant were analysed for Cd, Pb and different macro- and micronutrients including protein, vitamin C, N, P, K, Fe, Mn, Ca and Mg. Results revealed a significant variation in elemental concentrations in all the three vegetables. The projected daily dietary intake values of selected metals were significant for Fe, Mn, Ca and Mg but it was not significant for protein, vitamin C, N and P. The elemental contribution to recommended dietary allowance (RDA) was significant for Mn. Similarly, RDA for Fe and Mg was higher while Ca, N, P, K, protein and vitamin C showed the minimal contribution for different age groups. This study suggests that there can be substantial negative effects on nutritional composition when such vegetables cultivated in soil poisoned with Cd and Pb are consumed (Khan et al. 2015a). However, dosage of Cd higher than critical level (≥ 25 mg/kg soil in treatments) drastically alters plant growth (stunted), reduced yield as well as dietary contents (sugar and vitamin C) of these important vegetables especially its antioxidant content, and the hazardous effect was more visible at higher bioaccumulation of heavy metals during vegetative growth stage (Mani et al. 2012). Heavy metals also adversely affect the mineral uptake and metabolic processes in plants when present in excess. In a recent study, the accumulation of Cr in various plant tissues and its relation to the antioxidant activity and root exudation was evaluated (Uddin et al. 2015). The results revealed that 1 mM of Cr enhanced the weight of shoots and roots of *Solanum nigrum*, whereas weight of shoots and roots of *Parthenium hysterophorus* decreased when compared with lower levels of

Cr (0.5 mM) or control plants. In both plants, the concentrations of Cr and Cl were increased while Ca, Mg and K contents in root, shoot and root exudates were declined with increasing levels of Cr. The higher levels of Cr augmented SOD and POD activities and proline content in foliage of *S. nigrum*, while Cr at lower levels had stimulatory effects in *P. hysterophorus*. Citric acid concentration in root exudates increased with increasing rates of Cr by 35% and 44% in *S. nigrum*, while it was 20 and 76% in *P. hysterophorus*. Generally, *P. hysterophorus* exuded maximum amounts of organic acids. Moreover, the increasing amounts of Cd showed a differential impact on the content and translocation of micro- and macronutrients in tomato (Bertoli et al. 2012). Among different organs, the aerial part had 2.25 g/kg, 2.80 g/kg, 18.93 mg/kg and 14.15 mg/kg of K, Ca, Mn and Zn, respectively, compared to the control.

Apart from these effects, heavy metals in some studies have also been found to adversely affect the water and iron content in some vegetables. For example, 100 and 400 µM Cr had an obvious effect on iron metabolism and water relations of spinach (Gopal et al. 2009). Visual symptoms and increased accumulation of Cr were observed in roots than in leaves, when spinach was exposed to Cr. Moreover, the concentration of chlorophylls and the activities of heme enzymes, catalase and peroxidase decreased following exposure to excess Cr suggesting the intervention of Cr in iron metabolism of plants. These changes coupled with reduction in Fe concentration in Cr-exposed plants further indicate that by declining Fe absorption, Cr disrupts the chlorophyll-forming process and heme biosynthesis. Additionally, the transpiration rate along with proline accumulation was found to decrease in the leaves of Cr-treated plants which indicated water stress. In contrast, heavy metal has also been found to improve growth of celery more than lettuce and spinach, when irrigated with sludge containing heavy metals (Haghighi 2011). The stimulatory effect of sludge on growth rate of all three vegetables occurred via photosynthesis. It was, therefore, concluded from this study that the increasing element uptake induces photosynthesis and concurrently enhances the growth of leafy vegetable. In yet similar experiment, the impact of mixing native soil with municipal sewage sludge (MSS) or yard waste (YW) mixed with MSS (YW + MSS) was assayed to determine (1) yield and quality of sweet potato; (2) concentration of Cd, Cr, Mo, Cu, Zn, Pb and Ni in different organs of sweet potato (edible roots, leaves, stem and feeder roots); and (3) concentrations of ascorbic acid, total phenols, free sugars and β-carotene in edible roots at harvest (Antonious et al. 2011). The results revealed that even though the total concentrations of Pb, Ni and Cr were greater in plants grown with MSS and YW, applied together, compared to control plants, the mixture of MSS and YW increased yield, ascorbic acid, soluble sugars and phenols in edible roots of sweet potato by 53, 28, 27 and 48%, respectively, compared to plants grown in native soil. β-Carotene was greater (157.5 µg/g fresh weight) in the roots of plants grown in MSS compared to roots of plants grown in MSS + YW treatments (99.9 µg/g fresh weight). In a follow-up study, the concentrations of capsaicin, dihydro-capsaicin, β-carotene, ascorbic acid, phenols and soluble sugars in the fruits of *Capsicum annuum* L. (cv. Xcatic) grown under four soil management practices including YW, SS, chicken manure (CM) and no-much (NM) bare soil were determined. The total marketable pepper yield was increased by 34% and 15%, when it was grown in SS- and CM-treated soil, respectively, compared to NM bare soil. However, the number of culls (fruits that fail to meet the requirements of foregoing grades) was lower in YW-treated soils compared to SS- and CM-treated soils (Antonious 2012).

Elevated levels of heavy metals also affects the plants at the cellular and at the whole-plant level (Burzynski and Klobus 2004; Shaw et al. 2004). For instance, Cd and Cu have been reported to modify plasma membrane H^+-ATPase activity (Janicka-Russak et al. 2012). Also, an increased level of heat-shock proteins (hsp) in the tissues was observed as an adaptive process to survive under adverse conditions, and increased PM H^+-ATPase activity could further enhance the repair processes in heavy metal-stressed plants. In other investigations, metal ions have been found to inhibit root elongation, photosynthesis and enzyme activity and cause oxidative damage to membranes (Hernandez and Cooke 1997; Shaw et al. 2004; Sheoran et al. 1990). In a similar study, the inhibitory impact of metals on physiological, biochemical and morphological characteristics of spinach grown at 20 and 40% sewage sludge-amended soil is reported (Singh and Agrawal 2007). At 40% sewage sludge application, a substantial decrease in length of root and shoot and leaf area of spinach was observed. Among the biochemical parameters, photosynthetic rate was reduced by 23.6 and 28.8% in palak at 20 and 40% sewage sludge amendment, respectively. As compared to untreated soil, foliar thiol content decreased at 20 and 40% sewage sludge amendment. There was an increase in lipid peroxidation at different concentrations of sewage sludge used, and this is attributed to the formation of reactive oxygen species (ROS) and free radicals induced by Cd, Ni and Pb leading to disorganization of membrane structures of cells. In addition to these, chlorophyll content, fluorescence ratio (Fv/Fm) and protein content were also decreased, but peroxidase activity increased with increasing sewage sludge amendment ratio. These destructive effects in turn make plants more susceptible to additional stresses such as drought which reduces water uptake capacity and water use efficiency of the smaller root system and possibly blocks aquaporins (Yang et al. 2004; Ionenko et al. 2006 and Ryser and Emerson 2007). Heavy metal toxicity and drought stresses are likely to occur simultaneously, as metal-contaminated soils tend to have poor water-holding capacity (Derome and Nieminen 1998), and evaporation rates are high due to sparse vegetation cover (Johnson et al. 1994).

8.5   Bioremediation Strategies Adopted for Heavy Metal Removal

8.5.1   Phytoremediation

Remediation of metal-contaminated soils is indeed a major challenge before the scientists working in different countries. The conventional technologies employed to remove heavy metals from soils often involve stringent physicochemical agents (Neilson et al. 2003), which can destruct soil fertility and also negatively affect the agroecosystem.

Despite these, numerous methods including chlorination, use of chelating agents and acid treatments have been proposed to remove metals from contaminated sites. However, such methods are considered ineffective due to operational difficulties, high cost and low metal leaching efficiency. Due to these problems, there is an urgent need to find some viable alternative. In this regard, bioremediation which is the process of cleaning up of hazardous wastes involving the use of microorganisms or plants is considered the safest method of clearing polluted soil (Dixit et al. 2015). Among various bioremediation strategies, phytoremediation, also called as botanoremediation, green remediation and agro remediation, has been found inexpensive and more practicable method for minimizing/clearing metals from soil and water (Lasat 2000). Also, during phytoremediation practices, no hazardous product is generated. Broadly, this remediation system is plant based which is a solar-driven biological system (Santiago and Bolan 2010). Plants involved in phytoremediation have been categorized as:

1. Excluders: plants that survive through restriction mechanisms and are sensitive to metals over a wide range of soil. Members of the grass family, for example, sudan grass, bromegrass, fescue, etc., belong to this group.
2. Indicators: plants that show poor control over metal uptake and transport processes and correspondingly respond to metal concentrations in soils. This group includes the grain and cereal crops like corn, soybean, wheat, oats, etc.
3. Accumulators: plants which do not prevent metals from entering the roots, but they have evolved specific mechanisms for detoxifying high metal levels that accumulated in the cells (Baker 1981). Tobacco, mustard and Compositae families (e.g. lettuce, spinach, etc.) fall within this category.

Apart from these three categories, extreme accumulators, often called hyperaccumulators, form a fourth category which includes plants with exceptional metal-accumulating capacity. This property of accumulating excessive metal concentration, allows plants to survive and even thrive in heavily contaminated soils (or near ore deposits). To date, about 400 plant species (Table 8.5) have been identified as metal hyperaccumulators, representing <0.2% of all angiosperms (Brooks 1998).

Table 8.5 Numbers of known hyperaccumulating plants and their families

Heavy metals	Total number of plants	Families
Cadmium	1	Brassicaceae
Cobalt	26	Lamiaceae, Scrophulariaceae
Copper	24	Lamiaceae, Cyperaceae, Poaceae, Scrophulariaceae
Manganese	11	Apocynaceae, Cunoniaceae, Proteaceae
Nickel	290	Brassicaceae, Cunoniaceae, Euphorbiaceae, Flacourtiaceae, Violaceae
Selenium	19	Fabaceae
Thallium	1	Brassicaceae
Zinc	16	Brassicaceae, Violaceae

Adapted from Brooks (1998)

Phytoremediation involves many steps and techniques to clean up the contaminants from the polluted sites (Santiago and Bolan 2010). Some of the most commonly practised phytoremediation strategies are:

1. *Phytoextraction*: contaminants are taken up by roots and translocated within the plants and are removed by harvesting the plants. In this process, toxic metals from contaminated soils, sediment and sludge are absorbed, concentrated and precipitated into the above-ground biomass such as shoots, leaves, etc. (Singh et al. 2012).
2. *Phytodegradation*: involves the breakdown of organics, taken up by the plant to simpler molecules that are incorporated into the plant tissues (Dermentzis 2009).
3. *Rhizofiltration*: is primarily used to remediate extracted groundwater, surface water and wastewater with low contaminant concentrations. Rhizofiltration can be used for Pb, Cd, Cu, Ni, Zn and Cr, which are primarily retained within the roots.
4. *Phytostabilization*: primarily used for the remediation of soil, sediment and sludges. It involves the use of plant roots to limit contaminant mobility and bioavailability in the soil. Phytostabilization can occur through the sorption, precipitation, complexation or metal valence reduction. It is useful for the removal of Pb and As, Cd, Cr, Cu and Zn (Flora et al. 2008).
5. *Phytovolatilization*: involves the use of plants to take up contaminants from the soil, transforming them into volatile forms and releasing them into the atmosphere. Phytovolatilization occurs as growing trees and other plants take up water and the organic and inorganic contaminants. Some of these contaminants can pass through the plants to the leaves and volatilize into the atmosphere at comparatively low concentrations. Phytovolatilization has been primarily used for the removal of mercury (Durnibe et al. 2007).
6. *Phytostimulation*: using plants to stimulate bacteria and fungi to mineralize pollutant using exudates and root sloughing. Some plants can release as much as 10–20% of their photosynthates in the form of root sloughing and exudates (Pilon-Smits 2005).

Considering the importance of phytoremediation technology in metal clean up from the contaminated soils, several vegetables have also been explored for their phytoremediation ability in order to detoxify or reduce the heavy metal contamination in vegetable growing fields. For example, the growth response, metal tolerance and phytoaccumulation properties of water spinach and okra were assessed under different contaminated spiked metals by Ng et al. (2016) using control, 50 mg Pb/kg soil, 50 mg Zn/kg soil and 50 mg Cu/kg soil. Of the two vegetables, okra accumulated highest concentrations of Pb (80.20 mg/kg) in its root followed by Zn in roots (35.70 mg/kg) and shoots (34.80 mg/kg) of water spinach, respectively. Moreover, the accumulation of Pb, Zn and Cu in both water spinach and okra differed considerably. Though the accumulation of Pb in the shoots of water spinach and okra exceeded the maximum permissible limits of the National Malaysian Food Act 1983 and Food Regulations 1985 and the International Codex

Alimentarius Commission, both crops were found as good Pb and Zn phytoreme-diators. Generally, leafy vegetables have a higher tendency for uptake and accu-mulation of heavy metals; these can be used as indicator and also for removal of toxic heavy metals from polluted agricultural field. In yet other study, *Ipomoea aquatica* Forsk., an aquatic macrophyte, was assessed for its ability to accumulate Pb by exposing it to graded concentrations of this metal. Accumulation of Pb was the highest in root followed by stem and leaf. Furthermore, Pb at 20 mg/l induced colour changes in the basal portion of stem which had significantly higher Pb con-centration than in the unaffected apical part. This resulted in sequestration of excess metal in affected stem tissue, which could take up Pb by the process of caulofiltration or shoot filtration and served as a secondary reservoir of Pb in addi-tion to the root. The ability of the plant to store Pb in its root and lower part of stem coupled with its ability to propagate by fragmentation through production of adventitious roots and lateral branches from nodes raises the possibility of utiliz-ing *I. aquatica* for Pb phytoremediation (Chanu and Gupta 2016). Even among different varieties of vegetables, difference in the bioaccumulation property that can be exploited for remediation of polluted soils is reported. The high accumula-tor genotypes may be useful for phytoremediation, while the low accumulator varieties might be appropriate selections for growing on metal-contaminated soils to prevent potential human exposure to heavy metals and health hazards through the food chain. To categorize the pepper accessions of *Capsicum chinense* Jacq, collected from eight different countries, grown in a silty-loam soil under field conditions as low or high heavy metal accumulators, Antonious et al. (2010) col-lected mature fruits and analyse Cd, Cr, Ni, Pb, Zn, Cu and Mo. Fruits accumu-lated significant concentrations of Cd (0.47 μg/g dry fruit), Pb (2.12 μg/g dry fruit) and Ni (17.2 μg/g).

8.5.2 Microbe-Assisted Remediation

Numerous microbial communities belonging to different genera have evolved cer-tain mechanisms to tolerate and detoxify metals from contaminated environment (Mosa et al. 2016). Interestingly, the high surface to volume ratio of microorgan-isms and their ability to circumvent metal toxicity makes such organisms a viable and inexpensive alternative to chemical methods of metal remediation (Kapoor et al. 1999; Magyarosy et al. 2002). Biological mechanisms involved in microbial survival under metal-stressed environment include complexation, biosorption to cell wall and pigments, extracellular precipitation and crystallization, transforma-tion of metals, decreased transport or impermeability, efflux, intracellular compart-mentation and sequestration (Kang et al. 2016). One or many of these strategies may be adopted by microbiota to overcome metal problems. For example, synthesis of metallothioneins or γ-glutamyl peptides is a mechanism of Cu resistance in *Saccharomyces cerevisiae*, but Cu binding or precipitation around the cell wall and intracellular transport are also components of the total cellular response (Gadd and White 1989).

Considering the importance of microbes in metal detoxification/removal, identification of metal-tolerant microorganisms has become important to remediate the metal-polluted soils so that larger area can be used for vegetable cultivation. In this regard, the effect of two strains of *Trichoderma* (*T. harzianum* strain T22 and *T. atroviride* strain P1) on the growth of lettuce plants irrigated with As-contaminated water was assessed (Caporale et al. 2014). The results revealed the accumulation of this element mainly into the root system which subsequently reduced both biomass development and net photosynthesis rate (while altering the plant P status). However, both species of plant growth-promoting fungi (PGPF) *Trichoderma* alleviated, at least in part, the phytotoxicity of and eventually decreased As accumulation in tissues and concurrently enhanced plant growth, P status and net photosynthesis rate (Caporale et al. 2014). In a similar experiment, heavy metal-resistant strain J62 of *Burkholderia* sp. has been reported to promote the growth of tomato and maize (Jiang et al. 2008). In a follow-up study, the biological properties such as dry weights of fruit, roots, stem, leaf and whole tomato plants were increased by single or combined remediation of ryegrass and arbuscular mycorrhiza, while MDA contents and antioxidant enzyme activities of foliage and roots were declined in two varieties of tomato when exposed to Cd (20 mg/kg). Cadmium accumulation in tomato followed the order leaf > stem > fruit > root. However, the Cd concentrations in leaf, stem, root and fruit of both varieties were decreased by single or combined application of ryegrass and AM-fungi (Jiang et al. 2014).

Adequate nutrients are required for proper growth and development of a plant (Anil et al. 2003). Also, it is essentially required for maintaining normal metabolic reactions of plants. In contrast, the metal-contaminated soil is generally deficient in plant nutrients, and the plants that remain under constant stress fail to take up sufficient amounts of nutrients from soil. To overcome these problems, several metal-tolerant microbes possessing one or many plant growth-promoting activities such as ability to solubilize insoluble phosphate (Kim et al. 2013), phytohormone production (Franco-Hernández et al. 2010) or by some indirect mechanisms such as biocontrol activity (Khan and Bano 2016) involving siderophore production (Rajkumar et al. 2010) have been used. Besides these, microbes also aide in the phytoremediation process (Ullah et al. 2015), and as a result, the plants grow better in metal-stressed soils. As an example, two Cd- and Ni-resistant plant growth-promoting bacteria, *Pseudomonas* sp. ASSP 5 and ASSP 29, were isolated from fly ash-contaminated sites, and their plant growth promotion ability was tested by inoculating *Lycopersicon esculentum* plants grown in fly ash-amended soil (Kumar and Patra 2013). In most cases, strain ASSP 29 of *Pseudomonas* sp. produced more pronounced effect on biological (plant height and wet and dry weights) and chemical (protein and chlorophyll content in leaves) characteristics of plants and accumulation of metals in root and shoot of plants. Both the strains ASSP 5 and ASSP 29 showed a remarkable ability to protect the plants against the inhibitory effect of Ni and Cd besides promoting the growth of plants through production of IAA and siderophore and solubilization of P. Similarly, Dourado et al. (2014) evaluated Cd–*Burkholderia*–tomato interaction studies by inoculating a Cd-tolerant *Burkholderia* strain SCMS54 that exhibited a higher metabolic diversity and plasticity. Inoculated tomato plants in the presence of Cd grew well compared to non-inoculated plants indicating that the strain SCMS54 abated the toxicity of Cd and consequently enhanced tomato production grown under Cd stress. Based on this study, it was suggested that the bacterial strain isolated from Cd-contaminated soil could be used for tomato cultivation in soils even contaminated with Cd.

An endophytic bacterium *Serratia* sp. RSC-14 isolated from the roots of *S. nigrum* RSC-14 was used as an inoculant against *S. nigrum* plants grown in metal-stressed soils. In this study, the toxic effect of Cd-induced stress was relieved, and there were some significant improvements in root/shoot growth, biomass production and chlorophyll content, while MDA and electrolyte contents were found to decrease considerably. Besides the ability to tolerate Cd concentration up to 4 mM, the strain RSC-14 exhibited P solubilizing activity and secreted plant growth-promoting phytohormones such as IAA (54 µg/ml). The regulation of metal-induced oxidative stress enzymes such as catalase, peroxidase and polyphenol peroxidase had ameliorative effects on host growth. Activities of these enzymes were significantly reduced in RSC-14-inoculated plants as compared to control plants under Cd treatments. The current findings thus supported the hypothesis that *Serratia* sp. RSC-14 endowed with improved phytoextraction abilities could be used as metal-tolerant microbial inoculants to enhance the overall performance of *S. nigrum* plants when grown intentionally or inadvertently in Cd-contaminated soil (Khan et al. 2015b). Similarly, Luo et al. (2011) isolated endophytic bacterium *Serratia* sp. LRE07 from Cd-hyperaccumulator *S. nigrum* plants which, besides expressing the ability to promote plant growth, had high metal removal efficiencies also. Cadmium tolerant endophytic fungal community associated with *S. nigrum* has also been isolated and characterized for host plant growth modulation under Cd contamination (Khan et al. 2016). Owing to the levels of Cd tolerance detected, in order to simulate a tripartite plant–microbe–metal interaction, *S. nigrum* plants were inoculated with *Glomerella truncata* PDL-1 and *Phomopsis fukushii* PDL-10 under Cd spiking of 0, 5, 15, and 25 mg/kg. The results indicated that PDL-10-inoculated plants had significantly higher Cd content in shoots and in roots than those observed in the PDL-1-inoculated plants. Additionally, irrespective of Cd stress, PDL-1 and PDL-10 inoculation significantly improved plant growth attributes. He et al.(2009) reported that two Cd-resistant strains *Pseudomonas* sp. RJ10 and *Bacillus* sp. RJ16 increased plant growth through Cd and lead (Pb) solubilization and by secreting IAA, sidero-phore and 1-aminocyclopropane-1-carboxylate deaminase (ACC deaminase) besides enhancing Cd and Pb uptake ability of a Cd-hyperaccumulator tomato. Significant increase in Cd and Pb contents of above-ground tissues varied from 92 to 113% and from 73 to 79% respectively in inoculated plants grown in heavy metal-contaminated soil compared to the uninoculated control. These results show that the bacteria could be exploited for bacteria enhanced-phytoextraction of Cd- and Pb-polluted soils. Also, the effects of metal-resistant microorganisms and metal chelators on the ability of plants to accumulate heavy metals have been investigated. Though the application of Cd- or Pb-resistant microorganisms improved the ability of *S. nigrum* to accumulate heavy metals and increased plant yield, but the effects of microorganisms on phytoextraction were smaller than the effects of citric acid (CA).

When plants were grown in the presence of Cd contamination, the co-application of CA and metal-resistant strains enhanced biomass by 30–50% and increased Cd accumulation by 25–35%. In the presence of CA and the metal-resistant microorganisms, the plants were able to acquire 15–25% more Cd and 10–15% more Pb than control plants. It was therefore suggested from this study that the synergistic combination of plants, microorganisms and chelators can enhance phytoremediation efficiency in the presence of multiple metal contaminants (Gao et al. 2012).

Conclusion

Heavy metals are one of the major toxic pollutants whose removal from contaminated areas is urgently required in order to reduce their impacts on various food chains and ultimately human health. Among food commodities, vegetables are one of the main components of human dietary system because they provide essential micro and macronutrients, proteins, antioxidants and vitamins to the human body. All vegetables are often grown in suburban areas experiencing high concentrations of heavy metals both through aerial deposition and contamination accumulators through soil and irrigation water. Among vegetables, leafy vegetables have more potential to accumulate heavy metals from a contaminated environment due to their higher capacity of absorption both from contaminated soil and aerial deposits. The advantage of high biomass production and easy disposal also makes vegetables useful to remediate heavy metals from a contaminated environment, but the excessive intake and consequent accumulation in human beings through long-term consumption of contaminated food may result in negative effect on human health. Remediation and safe consumption of vegetables are, therefore, the two opposite concerns of heavy metal impact on the environment. Stringent enforcement of standards should therefore be followed for maximum allowable intake of heavy metals to avoid risk to human health. Heavy metals besides contaminating food also reduce the nutritional value of vegetables affecting other biochemical and physiological processes reducing the yield and quality of the crops. Thus regular monitoring of heavy metal contamination in the vegetables grown at wastewater irrigated area is necessary, and consumption of contaminated vegetables should be avoided in order to reduce the health risk caused by taking the contaminate vegetables. Furthermore, a safe and inexpensive metal-removing strategy like the use of plants and microbes both in isolation and association should be promoted to grow fresh and contaminant-free vegetables.

References

Agency for Toxic Substances and Disease Registry (ATSDR) (1989) Toxicological profile for cadmium. Agency for Toxic Substances and Disease Registry, Atlanta, Georgia. Publication No. ATSDR/TP-88/08 PB89-194476

Agrawal SB, Singh A, Sharma RK, Agrawal M (2007) Bioaccumulation of heavy metals in vegetables: a threat to human health. Terres Aqua Environ Toxicol 1:13–23

Alloway BJ (1990) Heavy metal in soils. J Environ Sci 3:97–99

Anil KG, Yunus M, Pandey PK (2003) Bioremediation: ecotechnology for the present century. Int Soc Environ Botanists 9(2):9–19

Antonious GF, Dennis SO, Unrine JM et al (2011) Ascorbic acid, β-carotene, sugars, phenols, and heavy metals in sweet potatoes grown in soil fertilized with municipal sewage sludge. J Environ Sci Health, Part B 46:112–121

Antonious GF, Kochhar TS, Coolong T (2012) Yield, quality, and concentration of seven heavy metals in cabbage and broccoli grown in sewage sludge and chicken manure amended soil. J Environ Sci Health Part A: Toxic/Hazard Subs Environ Eng 476:1955–1965

Antonious GF, Snyder JC, Berke T et al (2010) Screening *Capsicum chinense* fruits for heavy metals bioaccumulation. J Environ Sci Health B 45:562–567

Ashfaq A, Khan ZI, Bibi Z et al (2015) Heavy metals uptake by *Cucurbita maxima* grown in soil contaminated with sewage water and its human health implications in peri-urban areas of Sargodha city. Pak J Zool 47:1051–1058

Awashthi SK (2000) Prevention of food adulteration, Act no. 37 of 1954. Central and state rules as amended for 1999. Ashoka Law House, New Delhi

Baker AJM (1981) Accumulators and excluders—strategies in the response of plants to heavy metals. J Plant Nutr 3:643–654

Balkhair KS, Ashraf MA (2016) Field accumulation risks of heavy metals in soil and vegetable crop irrigated with sewage water in western region of Saudi Arabia. Saudi J Biol Sci 23:32–44

Banerjee D, Kuila P, Ganguli A et al (2011) Heavy metal contamination in vegetables collected from market sites of Kolkata. Indian J Elec Environ Agric Food Chem 10:2160–2165

Barrachina AC, Carbonell FB, Beneyto JM (1995) Arsenic uptake, distribution, and accumulation in tomato plants: effect of arsenite on plant growth and yield. J Plant Nutr 18:1237–1250

Bertoli AC, Cannata MG, Carvalho R et al (2012) *Lycopersicon esculentum* submitted to Cd-stressful conditions in nutrition solution: nutrient contents and translocation. Ecotoxicol Environ Saf 86:176–181

Bhatti SM, Anderson CWN, Stewart RB et al (2013) Risk assessment of vegetables irrigated with arsenic-contaminated water. Environ Sci Process Impacts 15:1866–1875

Biego GH, Joyeux M, Hartemann P et al (1998) Daily intake of essential minerals and metallic micro pollutants from foods in France. J Anal Sci 20:85–88

Brooks RR (1998) Geobotany and hyperaccumulators. In: Brooks RR (ed) Plants that hyperaccumulate heavy metals. CAB International, Wallingford, UK, pp 55–94

Budavari SE (2001) The Merck index, 11th edn. Merck and Co, Inc, Rahway, pp 432–435

Burzynski M, Klobus G (2004) Changes of photosynthetic parameters in cucumber leaves under Cu, Cd and Pb stress. Photosynthetica 42:505–510

Cai L, Xu Z, Ren M, Guo Q et al (2012) Source identification of eight hazardous heavy metals in agricultural soils of Huizhou, Guangdong Province, China. Ecotoxicol Environ Saf 78:2–8

Caporale AG, Sommella A, Lorito M et al (2014) *Trichoderma* spp. alleviate phytotoxicity in lettuce plants (*Lactuca sativa* L.) irrigated with arsenic-contaminated water. J Plant Physiol 171:1378–1384

Castro A (1998) Chromium in a series of Portuguese plants used in the herbal treatment of diabetes. J Biol Trace Elem Res 62:101–106

Chandra S, Saha R, Pal P (2016) Arsenic uptake and accumulation in okra (*Abelmoschus esculentus*) as affected by different arsenical speciation. Bull Environ Contam Toxicol 96:395–400

Chanu LB, Gupta A (2016) Phytoremediation of lead using *Ipomoea aquatica* Forsk. in hydroponic solution. Chemosphere 156:407–411

Dafelseed M (2007) Heavy metals and pesticides residues in selected fresh vegetables. B.Sc., University of Khartoum, pp 45–49

Demirezen D, Aksoy A (2006) Heavy metal levels in vegetables in Turkey are within safe limits for Cu, Zn, Ni and exceeded for Cd and Pb. J Food Qual 29:252–265

Dermentzis K (2009) Copper removal from industrial wastewaters by means of electrostatic shielding driven electrodeionization. J Eng Sci Tech Rev 2:131–136

Derome J, Nieminen T (1998) Metal and macronutrient fluxes in heavy-metal polluted Scots pine ecosystems in SW Finland. Environ Poll 103:219–228

Ding C, Zhang T, Wang X, Zhou F, Yang Y, Yin Y (2014) Effects of soil type and genotype on cadmium accumulation by rootstalk crops: implications for phytomanagement. Int J Phytoremediation 16:1018–1030

Dinis MDL, Fiúza A (2011) Exposure assessment to heavy metals in the environment: measures to eliminate or reduce the exposure to critical receptors. In: Simeonov LI et al (eds) Environmental heavy metal pollution and effects on child mental development. Springer, Netherlands, pp 27–50

Dixit R, Malaviya D, Pandiyan K et al (2015) Bioremediation of heavy metals from soil and aquatic environment: an overview of principles and criteria of fundamental processes. Sustainability 7:2189–2212

Dogheim SM, El-Ashraf MM, Alla SG et al (2004) Pesticides and heavy metals levels in Egyptian leafy vegetables and some aromatic medicinal plants. J Food Addit Contam 21: 323–330

Dourado MN, Souza LA, Martins PF et al (2014) *Burkholderia* sp. SCMS54 triggers a global stress defense in tomato enhancing cadmium tolerance. Water Air Soil Pollut 225:1–16

Drechsel P, Keraita B (eds) (2014) Irrigated urban vegetable production in Ghana: characteristics, benefits and risk mitigation, 2nd edn. International Water Management Institute (IWMI), Colombo, Sri Lanka

Ebdon L (2001) Trace element speciation for environment, food and health. Royal Society of Chemistry, Cambridge

European Union (EU) (2002) Heavy metals in wastes. European Commission on Environment. http://ec.europa.eu/environment/waste/studies/elv/heavymetals_report.pdf

EU (2006) Commission regulation (EC) No. 1881/2006 of 19 December 2006 setting maximum levels for certain contaminants in foodstuffs. Off J Eur Union L364:5–24

FAO (1985) Water quality for agriculture. Paper no. 29 (Rev. 1) UNESCO, Publication, Rome

FAO/WHO (2001) Food additives and contaminants. Joint Codex Alimentarius Commission, FAO/WHO Food Standards Programme 34:745–750

Ferri R, Hashim D, Smith DR et al (2015) Metal contamination of home garden soils and cultivated vegetables in the province of Brescia, Italy: implications for human exposure. Sci Total Environ 518:507–517

Fiona M, Ravi A, Rana PB (2003) Executive summary of technical report. Heavy metal contamination of vegetables in Delhi, Imperial College, London, pp 248–256

Flora SJS, Mittal M, Mehta A (2008) Heavy metal induced oxidative stress and it's possible reversal by chelation therapy. Division of Pharmacology and Toxicology, Defence Research and Development Establishment, Gwalior, India. Indian J Med Res 128:501–502

Franco-Hernández, MO, Montes-Villafàn, S, Ramírez-Melo M et al (2010) Comparative analysis of two phytohormone and siderophores rhizobacteria producers isolated from heavy metal contaminated soil and their effect on *Lens esculenta* growth and tolerance to heavy metals. In: Mendez-Vilas A (ed) Current research technology and education topics in applied microbiology and microbial biotechnology, Formatex 2010, Spain, pp 74–80

Gadd GM, White C (1989) Havy metal and redionuclide accumulation and toxicity in fungi and yeast. In: Poole RK, Gadd GM (eds) Metal-microb interactions. IRL Press, Oxford, pp 19–38

Gao Y, Miao C, Wang Y et al (2012) Metal-resistant microorganisms and metal chelators synergistically enhance the phytoremediation efficiency of *Solanum nigrum* L. in Cd- and Pb-contaminated soil. Environ Technol 33:1383–1389

Garcia E, Cabrera C, Lorenzo ML et al (2001) Daily dietary intake of chromium in southern Spain measured with duplicate diet sampling. J Nutr 86:391–396

Ghani A (2011) Effect of chromium toxicity on growth, chlorophyll and some mineral nutrients of *Brassica juncea* L. Egypt Acad J Biol Sci 2:9–15

Ghosh AK, Bhatt MA, Agrawal HP (2012) Effect of long-term application of treated sewage water on heavy metal accumulation in vegetables grown in northern India. Environ Monit Assess 184:1025–1036

Gopal R, Rizvi AH, Nautiyal N (2009) Chromium alters iron nutrition and water relations of spinach. J Plant Nutr 32:1551–1559

Grant CA (2011) Influence of phosphate fertilizer on cadmium in agricultural soils and crops. Pedologist 54:143–155

Haghighi M (2011) Sewage sludge application in soil improved leafy vegetable growth. J Biol Environ Sci 5(15)

He LY, Chen ZJ, Ren GD et al (2009) Increased cadmium and lead uptake of a cadmium hyperaccumulator tomato by cadmium-resistant bacteria. Ecotoxicol Environ Saf 72:1343–1348

Hernandez LE, Cooke DT (1997) Modification of the root plasma membrane lipid composition of cadmium-treated *Pisum sativum*. J Exp Bot 48:1375–1381

Hughes MF, Beck BD, Chen Y et al (2011) Arsenic exposure and toxicology: a historical perspective. Toxicol Sci 123:305–332

Hussain MM, Saeed A, Khan AA et al (2015) Differential responses of one hundred tomato genotypes grown under cadmium stress. Genet Mol Res 14:13162–13171

Intawongse M, Dean JR (2006) Uptake of heavy metals by vegetable plants grown on contaminated soil and their bioavailability in the human gastrointestinal tract. Food Addit Contam 23:36–48

Ionenko IF, Anisimov AV, Karimova FG (2006) Water transport in maize roots under the influence of mercuric chloride and water stress: a role of water channels. Biol Plant 50:74–80

IPCS (1992) International Programme on Chemical Safety (IPCS). Environmental health criteria 134, 135: cadmium

Islam MS, Hoque MF (2014) Concentrations of heavy metals in vegetables around the industrial area of Dhaka city, Bangladesh and health risk assessment. Int Food Res J 21:2121–2126

Islam GBR, Khan FE, Hoque MM et al (2014) Consumption of unsafe food in the adjacent area of Hazaribag tannery campus and Buriganga River embankments of Bangladesh: heavy metal contamination. Environ Monit Assess 186:7233–7244

Itanna F (2002) Metals in leafy vegetables grown in Addis Ababa and toxicological implications. Ethiop J Health Dev 16:295–302

IWMI (2006) Recycling realities: managing health risks to make wastewater an asset. Water Policy Brief 17:432–434

Jamali MK, Kazi TG, Arain MB et al (2007) Heavy metal contents of vegetables grown in soil, irrigated with mixtures of wastewater and sewage 53 sludge in Pakistan, using ultrasonic-assisted pseudo-digestion. J Agron Crop Sci 193:218–228

Janicka-Russak M, Kabała K, Burzyński M (2012) Different effect of cadmium and copper on H+-ATPase activity in plasma membrane vesicles from *Cucumis sativus* roots. J Exp Bot 63:4133–4142

Jassir MS, Shaker A, Khaliq MA (2005) Deposition of heavy metals on green leafy vegetables sold on roadsides of Riyadh City, Saudi Arabia. J Environ Contam Toxicol 75:1020–1027

Jiang CY, Sheng XF, Qian M et al (2008) Isolation and characterization of a heavy metal-resistant *Burkholderia* sp. from heavy metal-contaminated paddy field soil and its potential in promoting plant growth and heavy metal accumulation in metal polluted soil. Chemosphere 72:157–164

Jiang L, Yang Y, Xu WH et al (2014) Effects of ryegrass and arbuscular mycorrhiza on activities of antioxidant enzymes, accumulation and chemical forms of cadmium in different varieties of tomato. Huan Jing Ke Xue 35:2349–2357

Jing J, Logan T (1992) Measurement of levels of heavy metal contamination in vegetables grown and sold in selected areas in Lebanon. J Environ Qual 21:1–8

John HD (2002) Heavy metals: a meaningless term (IUPAC Technical Report). Pure Appl Chem 74:793–807

Johnson MS, Cooke JA, Stevenson JKW (1994) Revegetation of metalliferous wastes and land after metal mining. In: Hester RE, Harrison RM (eds) Mining and its environmental impact, issues in environmental science and technology. Royal Society of Chemistry, London, pp 31–48

Kabata-Pendias A, Pendias H (2001) Trace elements in soils, 3rd edn. CRC Press, Boca Raton and London

Kachenko AG, Singh B (2006) Heavy metals contamination in vegetables grown in urban and metal smelter contaminated sites in Australia. Water Air Soil Pollut 169:101–123

Kang CH, Kwon YJ, So JS (2016) Bioremediation of heavy metals by using bacterial mixtures. Ecol Eng 89:64–69

Kapoor A, Viraraghavan T, Roy Cullimore D (1999) Removal of heavy metals using the fungus *Aspergillus niger*. Bioresour Technol 70:95–104

Karigar C, Rao SS (2011) Role of microbial enzymes in the bioremediation of pollutants: a review. Enz Res 2011:11

Khan N, Bano A (2016) Role of plant growth promoting rhizobacteria and Ag-nano particle in the bioremediation of heavy metals and maize growth under municipal wastewater irrigation. Int J Phytoremediation 18:211–221

Khan A, Khan S, Khan MA et al (2015a) The uptake and bioaccumulation of heavy metals by food plants, their effects on plants nutrients, and associated health risk: a review. Environ Sci Pollut Res Int 22:13772–13799

Khan AR, Ullah I, Khan AL et al (2015b) Improvement in phytoremediation potential of *Solanum nigrum* under cadmium contamination through endophytic-assisted *Serratia* sp. RSC-14 inoculation. Environ Sci Poll Res 22:14032–14042

Khan AR, Waqas M, Ullah I et al (2016) Culturable endophytic fungal diversity in the cadmium hyperaccumulator *Solanum nigrum* L. and their role in enhancing phytoremediation. Environ Exp Bot. doi:10.1016/j.envexpbot.2016.03.005

Kim JO, Lee YW, Chung J (2013) The role of organic acids in the mobilization of heavy metals from soil. KSCE J Civil Eng 17:1596–1602

Kumar KV, Patra DD (2013) Influence of nickel and cadmium resistant PGPB on metal accumulation and growth responses of *Lycopersicon esculentum* plants grown in fly ash amended soil. Water Air Soil Pollut 224:1–10

Kumar M, Rahman MM, Ramanathan AL et al (2016) Arsenic and other elements in drinking water and dietary components from the middle Gangetic plain of Bihar, India: health risk index. Sci Total Environ 539:125–134

Lasat MM (2000) Phytoextraction of metals from contaminated soil: a review of plant, soil, metal interaction and assessment of pertinent agronomic issues. J Hazard Subs Res 2:1–25

Lenntech (2006) Water treatment and air purification. http://www.lenntech.com/heavymetals.htm. Accessed 23 Jan 2006

Li F, Zeng XY, Wu CH, Duan ZP et al (2013) Ecological risks assessment and pollution source identification of trace elements in contaminated sediments from the pearl river delta, China. Biol Trace Elem Res 155:301–313

Liu M, Huang B, Bi X, Ren Z et al (2013) Heavy metals and organic compounds contamination in soil from an e-waste region in South China. Environ Sci Process Impacts 15:919–929

Liu X, Zhang S, Shan X et al (2005) Toxicity of arsenate and arsenite on germination seedling growth and amylolytic activity of wheat. Chemosphere 61:293–301

Lokeshwari H, Chandrappa GT (2006) Impact of heavy metal contamination of Bellandur lake on soil and cultivated vegetation. Curr Sci 91:622–627

Lui WX, Li HH, Li SR et al (2006) Heavy metal accumulation of edible vegetables cultivated in agricultural soil in the suburb of Zhengzhou City, People's Republic of China. Bull Environ Contam Toxicol 76:163–170

Luo S, Wan Y, Xiao X et al (2011) Isolation and characterization of endophytic bacterium LRE07 from cadmium hyperaccumulator *Solanum nigrum* L. and its potential for remediation. Appl Microbiol Biotechnol 89:1637–1644

Magyarosy A, Laidlaw RD, Kilaas R et al (2002) Nickel accumulation and nickel oxalate precipitation by *Aspergillus niger*. Appl Microbiol Biotechnol 59:382–388

Mani D, Sharma B, Kumar C, Balak S (2012) Cadmium and lead bioaccumulation during growth stages alters sugar and vitamin C content in dietary vegetables. Proc Natl Acad Sci India Sect B: Biol Sci 82:477–488

Marin AR, Pezeshki SR, Masschelen PH et al (1993) Effect of dimethylarsenic acid (DMAA) on growth, tissue arsenic, and photosynthesis of rice plants. J Plant Nutr 16:865–880

Mello JP (2003) Food safety: contaminants and toxins. CABI Publishing, Wallingford, Oxon, UK; Cambridge, MA, p 480

Michio X (2005) Bioaccumulation of organochlorines in crows from an Indian open waste dumping site: evidence for direct transfer of dioxin-like congeners from the contaminated soil. J Environ Sci Technol 39:4421–4430

Midio Y, Satake M (2003) Chemicals and toxic metals in the environment. Discovery Publishing House, New Delhi, pp 45–68

Mohamed AE, Rashed MN, Mofty A (2003) Assessment of essential and toxic elements in some kinds of vegetables. Ecotoxicol Environ Saf 55:251–260

Mosa KA, Saadoun I, Kumar K et al (2016) Potential biotechnological strategies for the cleanup of heavy metals and metalloids. Front Plant Sci 7:303. PMCID: PMC4791364

Muchuweti M, Birkett JW, Chinyanga E et al (2006) Heavy metal content of vegetables irrigated with mixtures of wastewater and sewage sludge in Zimbabwe: implications for human health. Agric Ecosyst Environ 112:41–48

Musilova J, Bystricka J, Lachman J et al (2016) Potatoes—a crop resistant against input of heavy metals from the metallicaly contaminated soil. Int J Phytoremediation 18:547–552

Nacke H, Gonçalves A Jr, Schwantes D et al (2013) Availability of heavy metals (Cd, Pb, and Cr) in agriculture from commercial fertilizers. Arch Environ Contam Toxicol 64:537–544

Naz T, Khan MD, Ahmed I et al (2015) Biosorption of heavy metals by *Pseudomonas* species isolated from sugar industry. Toxicol Ind Health 32:1619–1627

Neilson JW, Artiola JF, Maier M (2003) Characterization of lead removal from contaminated soils by non-toxic soil washing agents. J Environ Qual 32:899–908

Ng CC, Rahman MM, Boyc AM, Abbas MR (2016) Heavy metals phyto assessment in commonly grown vegetables: water spinach (*I. aquatica*) and okra (*A. esculentus*). Springer plus 5:469

Oluwole SO, Olubunmi Makinde SC, Yusuf KA et al (2013) Determination of heavy metals contaminants in leafy vegetables cultivated by the road side. Int J Eng Res Dev 7:01–05

Osaili TM, Al Jamali AF, Makhadmeh IM et al (2016) Heavy metals in vegetables sold in the local market in Jordan. Food Addit Contam Part B 9:223–229

Parashar P, Prasad FM (2013) Study of heavy metal accumulation in sewage irrigated vegetables in different regions of Agra District, India. J Soil Sci 3:1

Pederno NJI, Gomez R, Moral G et al (1997) Heavy metals and plant nutrition and development. Rec Res Dev Phytochem 1:173–179

Pell A, Márquez A, López-Sánchez JF et al (2013) Occurrence of arsenic species in algae and freshwater plants of an extreme arid region in northern Chile, the Loa River Basin. Chemosphere 90:556–564

Pilon-Smits E (2005) Phytoremediation. Annu Rev Plant Biol 56:15–39

Pizarro I, Gomez M, Roman D et al (2016) Bioavailability, bioaccesibility of heavy metal elements and speciation of as in contaminated areas of Chile. J Environ Anal Chem 3:175

Prabu PC (2009) Impact of heavy metal contamination of Akaki river of Ethiopia on soil and metal toxicity on cultivated vegetable crops. Elec J Environ Agric Food Chem 8:818–827

Quaghebeur M, Rengel Z (2003) The distribution of arsenate and arsenite in shoots and roots of *Holcus lanatus* is influenced by arsenic tolerance and arsenate and phosphate supply. Plant Physiol 132:1600–1609

Radwan MA, Salama AK (2006) Market basket survey for some heavy metals in Egyptian fruits and vegetables. J Food Chem Toxicol 44:1273–1278

Rahman MM, Ng JC, Naidu R (2009) Chronic exposure of arsenic via drinking water and its adverse health impacts on humans. Environ Geochem Health 31:189–200

Rajkumar M, Ae N, Prasad MNV et al (2010) Potential of siderophore-producing bacteria for improving heavy metal phytoextraction. Trends Biotechnol 28:142–149

Rehman ZU, Khan S, Qin K et al (2016) Quantification of inorganic arsenic exposure and cancer risk via consumption of vegetables in southern selected districts of Pakistan. Sci Total Environ 550:321–329

Ryser P, Emerson P (2007) Growth, root and leaf structure, and biomass allocation in *Leucanthemum vulgare* Lam. (Asteraceae) as influenced by heavy-metal-containing slag. Plant Soil 301:315–324

Saha S, Hazra GC, Saha B et al (2015) Assessment of heavy metals contamination in different crops grown in long-term sewage-irrigated areas of Kolkata, West Bengal, India. Environ Monit Assess 187:1–12

Santiago M, Bolan NS (2010) Phytoremediation of arsenic contaminated soil and water. In: Proceedings of 19th world congress of soil science. Soil Solutions for a Changing World, Brisbane, Australia

Santra SC, Samal AC, Bhattacharya P et al (2013) Arsenic in food chain and community health risk: a study in Gangetic West Bengal. Procedia Environ Sci 18:2–13

Sharma RK, Agrawal M (2006) Single and combined effects of cadmium and zinc on carrots: uptake and bioaccumulation. J Plant Nutr 29:1791–1804

Sharma RK, Agrawal M, Marshall F (2006) Heavy metals contamination in vegetables grown in wastewater irrigated areas of Varanasi, India. Bull Environ Contam Toxicol 77:312–318

Sharma RK, Agrawal M, Marshall F (2007) Heavy metal contamination of soil and vegetables in suburban areas of Varanasi, India. Ecotoxicol Environ Saf 66:258–226

Sharma RK, Agrawal M, Marshall FM (2009) Heavy metals in vegetables collected from production and market sites of a tropical urban area of India. Food Chem Toxicol 47:583–591

Shaw BP, Sahu SK, Mishra RK (2004) Heavy metal induced oxidative damage in terrestrial plants. In: Prasad MNV (ed) Heavy metal stress in plants. Springer, Berlin Heidelberg, pp 84–126

Sheoran IS, Singal HR, Singh R (1990) Effect of cadmium and nickel on photosynthesis and the enzymes of the photosynthetic carbon reduction cycle in pigeonpea (*Cajanus cajan* L.) Photosynth Res 23:345–351

Singh VP (2006) Metal toxicity and tolerance in plants and animals. Sarup and Sons, New Delhi, India, p 238

Singh RP, Agrawal M (2007) Effects of sewage sludge amendment on heavy metal accumulation and consequent responses of *Beta vulgaris* plants. Chemosphere 67:2229–2240

Singh S, Kumar M (2006) Heavy metal load of soil, water and vegetables in peri-urban Delhi. Environ Monit Assess 120:79–91

Singh R, Gautam N, Mishra A et al (2011) Heavy metals and living systems: an overview. Indian J Pharm 43:246–253

Singh A, Sharma RK, Agrawal M et al (2010) Risk assessment of heavy metal toxicity through contaminated vegetables from waste water irrigated area of Varanasi, India. Trop Ecol 51:375–387

Singh D, Tiwari A, Gupta R (2012) Phytoremediation of lead from wastewater using aquatic plants. J Agric Tech 8:1–11

Sinha S, Gupta AK, Bhatt K et al (2006) Distribution of metals in the edible plants grown at Jajmau, Kanpur (India) receiving treated tannery wastewater: relation with physico-chemical properties of the soil. Environ Monit Assess 115:1–22

Sinha S, Pandey K, Gupta AK et al (2005) Accumulation of metals in vegetables and crops grown in the area irrigated with river water. Bull Environ Contam Toxicol 74:210–218

Skeffington RA, Shewry R, Peterson PJ (1996) Chromium uptake and transport in barley seedlings (*Hordeum vulgare* L.) J Int Environ 132:209–214

Stalikas CD, Mantalovas AC, Pilidis GA (1997) Multielement concentrations in vegetable species grown in two typical agricultural areas of Greece. Sci Total Environ 206:17–24

Tiwari KK, Singh NK, Patel MP et al (2011) Metal contamination of soil and translocation in vegetables growing under industrial wastewater irrigated agricultural field of Vadodara, Gujarat, India. Ecotoxicol Environ Saf 74:1670–1677

Turer DG, Maynard BJ (2003) Heavy metal contamination in highway soils. Comparison of Corpus Christi, Texas and Cincinnati, Ohio shows organic matter is key to mobility. Clean Technol Environ Policy 4:235–245

UdDin I, Bano A, Masood S (2015) Chromium toxicity tolerance of *Solanum nigrum* L. and *Parthenium hysterophorus* L. plants with reference to ion pattern, antioxidation activity and root exudation. Ecotoxicol Environ Saf 113:271–278

Ullah A, Heng S, Munis MFH et al (2015) Phytoremediation of heavy metals assisted by plant growth promoting (PGP) bacteria: a review. Environ Exp Bot 117:28–40

Uzu G, Schreck E, Xiong T et al (2014) Urban market gardening in Africa: foliar uptake of metal (loid) s and their bioaccessibility in vegetables; implications in terms of health risks. Water Air Soil Pollut 225:1–13

Wang Y, Qiu Q, Xin G, Yang Z et al (2013) Heavy metal contamination in a vulnerable mangrove swamp in South China. Environ Monit Assess 185:5775–5787

WHO (1992) Environmental health criteria 134: cadmium. Geneva, p 156

WHO/FAO (2007) Joint FAO/WHO Food Standard Programme Codex Alimentarius Commission 13th session. Report of the thirty eight session of the Codex Committee on Food Hygiene, Houston, USA, ALINORM 07/30/13

Woimant F, Trocello JM (2014) Disorders of heavy metals. Handb Clin Neurol 120:851–864

Wu Z, Ren H, McGrath SP et al (2011) Investigating the contribution of the phosphate transport pathway to arsenic accumulation in rice. Plant Physiol 157:498–508

Xiong T, Dumat C, Pierart A et al (2016) Measurement of metal bioaccessibility in vegetables to improve human exposure assessments: field study of soil–plant–atmosphere transfers in urban areas, South China. Environ Geochem Health 38:1–19

Xiong TT, Leveque T, Austruy A et al (2014) Foliar uptake and metal (loid) bioaccessibility in vegetables exposed to particulate matter. Environ Geochem Health 36:897–909

Xiong J, Wu L, Tu S et al (2010) Microbial communities and functional genes associated with soil arsenic contamination and the rhizosphere of the arsenic-hyperaccumulating plant *Pteris vittata* L. Appl Environ Microbiol 76:7277–7284

Yang HM, Zhang XY, Wang GX (2004) Effects of heavy metals on stomatal movements in broad bean leaves. Russ J Plant Physiol 51:464–468

Yu HY, Li FB, Yu WM, Li YT et al (2013) Assessment of organochlorine pesticide contamination in relation to soil properties in the Pearl River Delta, China. Sci Total Environ 447:160–168

Zhao FJ, McGrath SP, Meharg AA (2010) Arsenic as a food chain contaminant: mechanisms of plant uptake and metabolism and mitigation strategies. Annu Rev Plant Biol 61:535–559

Zhao H, Xia B, Fan C, Zhao P et al (2012) Human health risk from soil heavy metal contamination under different land uses near Dabaoshan mine, southern China. Sci Total Environ 417:45–54

Zhou H, Yang WT, Zhou X et al (2016) Accumulation of heavy metals in vegetable species planted in contaminated soils and the health risk assessment. Int J Environ Res Public Health 13:289

Recent Advances in Management Strategies of Vegetable Diseases

9

Mohammad Shahid, Almas Zaidi, Mohd. Saghir Khan, Asfa Rizvi, Saima Saif, and Bilal Ahmed

Abstract

Vegetables are one of the most important components of human foods since they provide proteins, vitamins, carbohydrates and some other essential macro- and micronutrients required by the human body. Phytopathogenic diseases, however, cause huge losses to vegetables during cultivation, transportation and storage. To protect vegetable losses, various strategies including chemicals and biological practices are used worldwide. Pesticides among agrochemicals have however been found expensive and disruptive. Due to the negative health effects of chemical fungicides via food chain, the recent trend is shifting towards safer and more eco-friendly biological alternatives for the control of vegetable diseases. Of the various biological approaches, the use of antagonistic microorganisms is becoming more popular throughout the world due to low cost and environment safety. Numerous phytopathogenic diseases can now be controlled by microbial antagonists which employ several mechanisms such as antibiosis, direct parasitism, induced resistance, production of cell wall-lysing/cell wall-degrading enzymes, and competition for nutrients and space. The most commonly used biological control agents belong to the genera, *Bacillus*, *Pseudomonas*, *Flavobacterium*, *Enterobacter*, *Azotobacter*, *Azospirillum* and *Trichoderma*, and some of the commercial biocontrol products developed and registered for the use against phytopathogens are Aspire, BioSave, Shemer etc. Here, an attempt is made to highlight the mechanistic basis of vegetable disease suppression by some commonly applied microbiota. This information is likely to help vegetable growers to reduce dependence on chemicals and to produce fresh and healthy vegetables in different production systems.

M. Shahid (✉) • A. Zaidi • M.S. Khan • A. Rizvi • S. Saif • B. Ahmed
Faculty of Agricultural Sciences, Department of Agricultural Microbiology,
Aligarh Muslim University, Aligarh 202002, Uttar Pradesh, India
e-mail: shahidfaiz5@gmail.com

© Springer International Publishing AG 2017
A. Zaidi, M.S. Khan (eds.), *Microbial Strategies for Vegetable Production*,
DOI 10.1007/978-3-319-54401-4_9

197

9.1   Introduction

Vegetables are important components of human diet since they provide essential nutrients (Table 9.1) that are required to run most of the metabolic reactions of the human body (Pujeri et al. 2015). Among various vegetable producing countries, India is the second largest producer of vegetables after China and accounts for 13.4% (more than 97 million tons) of world production. More importantly, vegetables produce higher returns per unit area and time. In the year 2000, the vegetable production in India was 92.8 million tons, grown over an area of 6 million hectares, which is about 3% of the gross cropped area of the country (Pujeri et al. 2015). However, as the country's population is increasing at the rate of 1.8 per cent per year, the demand of vegetables will be 225 million tons by 2020 and 350 million tons by 2030 (Anonymous 2011a, b).

Table 9.1 Nutritional value of some commonly grown vegetables

Nutritional value (per 100 g)	Examples of common vegetables					
	Spinach	Tomato	Onion	Cucumber	Lettuce	Broccoli
Energy (kcal)	23	18	40	16	13	34
Carbohydrate (g)	3.6	3.9	9.34	3.63	2.23	6.64
Starch (g)	0.4	2.6	4.24	1.67	1.9	1.7
Dietary fibre (g)	2.2	1.2	1.7	0.5	1.1	2.6
Fat (g)	0.4	0.2	0.1	0.11	0.22	0.37
Protein (g)	2.9	0.9	1.1	0.65	1.35	2.82
Vitamins						
Thiamine (B1) (mg)	0.08	0.042	0.046	0.027	0.057	0.031
Riboflavin (B2) (mg)	0.078	0.449	0.027	0.033	0.062	0.361
Niacin (B3) (mg)	0.724	0.123	0.116	0.098	0.15	0.071
Pantothenic acid (B5) (mg)	0.296	0.037	0.123	0.0259	0.082	0.117
Vitamin B6 (mg)	0.195	0.594	0.12	0.04	0.184	0.639
Folate (B9) (µg)	194	80.0	19	7	73	175
Vitamin C (mg)	28	14	7.4	2.8	3.7	0.063
Vitamin E (mg)	2	0.57	–	–	0.18	89.2
Vitamin K (mg)	0.483	7.9	–	0.0164	0.1023	0.78
Minerals					–	–
Calcium (mg)	99		23	16	35	0.47
Iron (mg)	2.71	–	0.21	0.28	1.24	0.73
Magnesium (mg)	79	11	10	13	13	21
Manganese (mg)	0.897	0.114	0.129	0.079	0.179	0.21
Phosphorous (mg)	49	24	29	24	34	66
Potassium (mg)	558	237	146	174	238	316
Sodium (mg)	7.9		–	2	5	0.58
Zinc	0.53		0.17	0.2	0.27	33
Other elements					–	–
Water (g)	91.4	94.5	89.11	95.23	95.63	89.3

Vegetables are the inseparable components of world cuisine and are consumed in different forms and preparations. They are the major source of vitamins and nutrients and hence fulfil the requirements of our balanced diet. Vegetables (especially Brassica vegetables, e.g. broccoli, cabbage, cauliflower, brussels sprouts and kale) contain high levels of bioactive compounds which include phenolics, glucosinolates, vitamin C, vitamin E, carotenoids and selenium. These vegetables are among the most important vegetables consumed all over the world owing to their availability at local markets, cheapness and consumer preference (Herr and Buchler 2010). Phytochemicals of vegetables have been reported to prevent oxidative stress, induce detoxification enzymes, stimulate immune system, decrease the risk of cancers and inhibit malignant transformation and carcinogenic mutations (Herr and Buchler 2010; Kestwal et al. 2011). Some authors suggest that addition of glucosinolates, obtained from Brassica, to the diet may decrease both oxidative stress and also inflammation (Wu et al. 2004; Noyan Ashraf et al. 2006). Despite these health benefits, the production of vegetables in different countries is very low. Also, the yield loss due to plant diseases in vegetables grown in different agronomic regions is very high. In order to produce more and more vegetables and to protect quality and to prevent losses due to plant diseases, various agrochemicals such as synthetic fertilizers and pesticides are used indiscriminately across the globe (Ngowi et al. 2007). Such chemicals, however, when used beyond their recommended level cause changes in soil fertility and, hence, adversely affect the production of vegetables. Considering the importance of vegetables in human dietary system, the threat caused by phytopathogens to vegetable production across the globe and ill effects of agrochemicals used to control phytopathogens, scientists are searching for a safe and viable alternative for management of vegetable diseases. In this context, the use of microbes has provided some solution to the expensive chemicals (Kanjanamaneesathian 2015). Here, emphasis is given to better understand the mechanistic basis of disease suppression by soil microbiota and to identify most suitable organisms that could enhance vegetable production while reducing the dependence on agrochemicals used in the management of vegetable diseases.

9.2   Soilborne Phytopathogenic Diseases of Vegetables: A General Account

Soilborne plant pathogens including bacteria, fungi, viruses and nematodes are a major threat to many vegetables and other horticultural crops. These pathogens pose a serious challenge because intensive farming systems with narrow rotational crop practices frequently increase their population in the soil. Further, such nuisance organisms survive for many years and cannot be eradicated completely from the soil. Hence, the infection from multiple pathogens in soil results in a disease complex which can further damage subsequent crops grown in the same field. Of these, fungi are the most important and major group of pathogens which infect a wide range of host plants and cause destructive and economical losses of vegetables (Salau et al. 2015). The most common fungal pathogens are *Fusarium*, *Rhizoctonia*, *Verticillium*, etc. causing wilt and powdery scab of most of the vegetables and play a major role in yield loss of vegetable crops. Fungal diseases account preharvest losses in crop production up to 12% or even more in developing countries (Kim et al. 2003), while postharvest diseases account for 10–30% yield losses in developing countries (Tripathi et al. 2008; Fatima et al. 2009). Among vegetables, the world is facing 70–100% yield losses of cucumbers due to infection by *Fusarium* spp. In India alone, the loss is estimated at above 70% of the yield by the soilborne diseases (Egel and Martyn 2007). Vegetables are highly susceptible to pathogenic fungi due to their higher moisture content, low pH and nutrient compositions. The infection is activated further by the improper handling, packaging, storage and transportation (Mari and Guizzardi 1998; Sharma and Tripathi 2006). The presence, growth and colonization of fungi may adversely affect the quality and quantity of vegetables. Some of the notable fungi like *Aspergillus flavus*, *A. parasiticus*, *Penicillium* spp. and *Fusarium* spp. in addition to causing the severe diseases in various vegetables also make the food unfit for human consumption due to mycotoxins secretion (Brewer et al. 2013). Over 25% of the world's vegetables are contaminated with known mycotoxins, and more than 300 fungal metabolites are reported to be toxic to man and animals. The main toxic effects of mycotoxins are carcinogenicity, genotoxicity, teratogenicity, nephrotoxicity, hepatotoxicity and reproductive disorders causing damages such as toxic hepatitis, haemorrhage, oedema, immunosuppression and hepatic carcinoma (Makun 2013). Some of the common field and storage fungi in India include *Alternaria alternata*, *Cladosporium cladosporioides*, *Curvularia* spp., *Phoma* spp., *Fusarium* spp., *Aspergillus flavus*, *A. niger*, *A. parasiticus*, *A. tamarii*, *A. nidulans*, *A. candidus* and *Penicillium* spp. that cause severe diseases on a number of vegetables (Reddy et al. 2009). These fungi predominantly inhabit soil and have been found to infect more than 400 host plants including vegetables, pulses and cereals. For example, yield losses in tomato due to *A. solani* that varied between 0.75 and 0.77 t/ha for every 1% increase in disease severity for the 2 consecutive years in Gangetic Plains of West Bengal is reported (Poly and Srikanta 2012). The different species of *Alternaria* also cause black spot or leaf spot disease in *Brassica* vegetables and Solanaceae crops (Mamgain et al. 2013). Once infected, the damaged seeds usually have fungus both inside and on to the seed surface, and the yield losses due to this disease differed between 35 and 70% (Srivastava et al. 2011). *Fusarium* spp. is yet another important soilborne plant pathogens that severely reduces the vegetables' production. Several species of *Fusarium* causes symptoms such as cortical decay of roots, root rot, wilting, yellowing, rosette and premature death on infected plants. Late blight disease, caused by *Phytophthora infestans*, is one of the most serious threats to the tomato production worldwide. The foliage and stem of the tomato can be killed by *P. infestans* which spreads through airborne asexual sporangia during the growing season. The late blight also causes fruit rot either in the field or postharvest. A brief account of common diseases, their causative agents, symptoms and control measures of some vegetable crops is listed in Table 9.2.

Table 9.2 Examples of some common diseases of vegetables and their control measures

Vegetables	Diseases	Causal agent	Symptoms	Control: chemical/biological	References
Onion	Basal rot	*Fusarium oxysporum* f. sp. *cepae*	Yellowing and dieback from the tips of leaves, roots of the plants become pink in colour and rotting occurs thereafter	Fungicides Antracol, Carbendazim, copper oxychloride and Kingmil MZ	Behrani et al. (2015)
	Purple blotch	*Alternaria porri*	Symptoms occur on leaves and flower stalks as small, sunken, whitish flecks with purple-coloured centres	Fungicides Hexaconazole, tebuconazole, propiconazole, difenoconazole	Priya et al. (2015)
Potato	Fusarium wilt	*F. oxysporum* f. sp. *tuberose*	Wilting, chlorosis, necrosis, premature leaf drop, browning of vascular system, stunting, and damping-off	Use of biopesticides *T. harzianum* and *P. fluorescens*	Abeer and Rehab (2015)
	Potato dry rot	*Fusarium* sp.	Dark depressions on tuber, wrinkled skin in concentric rings as underlying dead tissue desiccates	Using biocontrol bacteria *B. subtilis*, *B. pumilus*, *Burkholderia cepacia*, *P. putida*, *B. amyloliquefaciens*, *B. atrophaeus*, *B. macerans* and *Flavobacterium balastinium*	Kotan et al. (2016)
	Late blight	*Phytophthora infestans*	Water-soaked lesions on leaves, spots turn black as the leaves start rotting	Fungicides metalaxyl + mancozeb (Ridomil Gold) and IproWelcarb + mancozeb (Melody Dew)	Subhani et al. (2015)
	Common scab	*Streptomyces scabiei*	Damaged tubers have rough, cracked skin, with scab-like spots, infected potato skins covered with rough black welts, initial infections result in superficial reddish-brown spots on the surface of tubers, As the tubers grow, lesions expand, becoming corky and necrotic	*Bacillus subtilis*, *Enterobacter cloacae*, (biopesticides) chloropicrin, Pic-Plus, manganese sulphate and mustard meal, fludioxonil and mancozeb	Al Mughrabi et al. (2016)

(continued)

Table 9.2 (continued)

Vegetables	Diseases	Causal agent	Symptoms	Control: chemical/biological	References
Tomato	Early blight	*Alternaria solani*	Symptoms typically appear soon after fruit set, starting on the lower leaves as tiny dark brown spots. The spots enlarge to over 1/2 inch in diameter and develop a greyish-white centre with a darker border	Chlorothalonil (Daconil 2787, Ortho multipurpose fungicide) and EBDC fungicides (such as mancozeb and maneb)	Kemmitt (2002)
	Late blight	*Phytophthora infestans*	Dark, olivaceous greasy spots develop on green fruit; a thin layer of white mycelium, water-soaked spots that enlarge rapidly into brown to black lesions that cover large areas of the petioles and stems	Taegro (biofungicide), mancozeb (fungicide)	Manukinda et al. (2016)
	Damping-off	*Pythium aphanidermatum*	Soft, mushy, brown and decomposed due to seed infection, seedling emerges from the soil but dies shortly afterwards. The affected portions (roots, hypocotyls and perhaps the crown of the plant) are pale brown, soft, water soaked, and thinner than non-affected tissue, infected stems collapse, stunting of plants	*Streptomyces* isolate DBTB 13, *Trichoderma viride*, *T. harzianum* and *P. fluorescens* in combination with different biofertilizers *Azotobacter* and *Azospirillum*	Dhanasekaran et al. (2005), Thakur and Tripathi (2015)

Table 9.2 (continued)

Vegetables	Diseases	Causal agent	Symptoms	Control: chemical/ biological	References
Lettuce	Downy mildew	*Bremia lactucae*	Outer leaves have pale green/yellow areas that later turn brown. Affected areas often have an angular margin where they are limited by a leaf vein. White, fluffy growth develops on the undersides of these areas. Spores spread with air currents from infected lettuce plants or crop debris	*Pseudomonas aeruginosa, Trichoderma* sp.	Shafique et al. (2015), Vinale et al. (2008)
	Septoria leaf spot	*Septoria lactucae*	Infected seedlings have yellow leaf markings with tiny black dots. Larger plants have brown spots with an angular outline on the older leaves or yellowish markings covered with tiny black dots	Kototine, Apron plus, benlate, captan and Dithane M-45 (fungicides)	Okoi et al. (2015)
	Bottom rot	*Rhizoctonia solani*	Outer leaves of field plants wilt and are associated with a rot at ground level with rusty markings on the midribs of the undersides of lower leaves. Affected tissue offers entry sites for bacterial soft rots. In warm, humid weather the rusty lesions expand quickly and the whole heart may rot and die	*Bacillus amyloliquefaciens* strain FZB42	Chowdhury et al. (2013)

(continued)

Table 9.2 (continued)

Vegetables	Diseases	Causal agent	Symptoms	Control: chemical/biological	References
Broccoli	Alternaria leaf spot	*Alternaria brassicae*	Small dark spots on leaves which turn brown to grey; lesions may be round or angular and may possess a purple-black margin; lesions may form concentric rings, become brittle and crack in the centre; dark brown elongated lesions may develop on stems and petioles	Fungicide Dithane M-45	Kumar et al. (2013)
	Clubroot	*Plasmodiophora brassicae*	Slow-growing, stunted plants; yellowish leaves which wilt during day and rejuvenate in part at night; swollen, distorted roots; extensive gall formation	Bavistin (systemic fungicide)	Sharma and Sohi (1981)
	Downy mildew	*Hyaloperonospora parasitica*	Small angular lesions on upper surface of leaves which enlarge into orange or yellow necrotic patches; white fluffy growth on undersides of leaves		
Cucumber	Powdery mildew	*Podosphaera xanthii, Golovinomyces cichoracearum*	Tiny white superficial spots appear on leaves and stem. Spots become powdery white and expand to cover all portions of the plant	*Ampelomyces quisqualis* Ces.	Quarles (2004)
	Blue mould rot	*Penicillium oxalicum*	Blue-grey to blue-green fungal growth on the surface giving off a cloud of spore, cankers expand to a few centimetres above and below the node. They have dry, pale-brown edges	Fungicides iprodione + carbendazim (Quintal), chlorothalonil (Kavach), copper hydroxide (Kocide), difenoconazole (Score), mancozeb (Dithane M-45)	Shankar et al. (2014)
	Belly rot	*Rhizoctonia solani*	Yellowish-brown superficial discoloration, dark brown water-soaked decay most often on the fruit side	Biocontrol agent *Trichoderma viride, T. harzianum,* Azoxystrobin + chlorothalonil, flusilazole (Nustar) (fungicides)	Shankar et al. (2014)

9.3 General Effects of Diseases on Vegetable Crop Production

Diseases cause lots of damage to vegetable crops that include (1) reduction in growth, (2) reduction in the yield or productivity, (3) reduction in the quality of crops, (4) malformation of plant organs or the whole plants, (5) the death of a whole plant, (6) increase in the cost of production, (7) making vegetables unattractive and unmarketable and finally (8) reducing the income of the growers/farmers.

9.4 Management Practices for Control of Vegetable Diseases

Vegetable diseases can be controlled by different methods such as (1) cultural practices, (2) chemical control measures and (3) biological control methods.

9.4.1 Chemical Control of Vegetable Phytopathogens

Agrochemicals play an important role in vegetables production by protecting them from different diseases (Ogundana and Dennis 1981). Chemicals used to control phytopathogens are, however, considered safe only when they are used within regulatory limits and are applied properly. Among agrochemicals, pesticides including fungicides are used frequently worldwide to protect crops including vegetables (Aktar et al. 2009) before and after harvest from insect pests in order to increase food security despite the fact that pesticides can have negative health effects on consumers via food chain (Damalas and Eleftherohorinos 2011). The use of pesticides in recent times has, however, further increased due to its rapid action, and they are easy to apply compared to other pest control measures (Gilden et al. 2010). For the production of vegetables in India, about 13–14% pesticides are used due to heavy pest infestation throughout the cropping season of horticultural crops (Agnihotri 1999).

Among pesticides, certain protective fungicides even though hazardous to the environment are still in use for the control of fungal diseases (Vaish and Sinha 2003). In this regard, the use of commercially available fungicides, for example, captan, carbendazim, Dithane M-45 and Antracol to control *Phomopsis vexans* causing phomopsis leaf blight of brinjal has been reported (Rohini et al. 2015). Of these four fungicides, carbendazim had the highest percentage of disease protection, and, hence, it was suggested that this fungicide may be used to control the phomopsis leaf blight of brinjal. Also, the efficacy of carbendazim in other study was found most effective against *P. vexans* in field resulting in lowest disease incidence (4.3%) and highest fruit yield of 222.83q/ha (Singh and Agrawal 1999). Similarly, Habib et al. (2007) evaluated four fungicides, viz. Topsin, benlate, Dithane M-45 and captan against seed-borne mycoflora of eggplant and found that these fungicides were inhibitory to pathogenic fungi *F. solani* and *A. alternata*. In yet other experiment, carboxin and mancozeb were found effective and reduced phomopsis fruit rot of brinjal in field (Thippeswamy et al. 2006). Similarly, effectiveness of six different fungicides against *P. vexans* causing phomopsis blight and fruit rot of brinjal is reported (Beura et al. 2008). Of the six fungicides, carbendazim at 0.1% resulted in better control of twig blight and fruit rot disease of brinjal grown under field trials. Islam and Meah (2011) used different fungicides such as Bavistin, Vitavax, botanicals like garlic and Allamanda and bioagent *Trichoderma harzianum* CP, *T. harzianum* and T22 and *Trichoderma* sp. EP as seed treatment of brinjal. All treatments were found effective in controlling seedling diseases of eggplant in the nursery bed. In contrast, Hossain et al. (2013) applied four fungicides and three micronutrients against *P. vexans* causing phomopsis blight and fruit rot disease of eggplant in vitro. They observed that Bavistin at 0.1% concentration very effectively arrested the spore germination and mycelial growth of *P. vexans*, while micronutrients even though had little effect against this disease but were significantly better than control. In a similar investigation, the efficacy of four systemic fungicides (Topsin-M, difenoconazole, aliette and nativo) and a bioagent *Bacillus subtilis* against *Fusarium oxysporum* f. sp. melongenae were tested under laboratory and greenhouse conditions by poison food technique (Sahar et al. 2013). Of these formulations, Topsin-M and *B. subtilis* were found very effective and substantially reduced the mycelial growth and disease incidence. The efficacy of two fungicides and bioagent against the mycelial growth of *P. vexans* causing brinjal leaf blight and fruit rot disease was, however, variable (Muneeshwar et al. 2012). In a follow-up study, a field trial was conducted during the year 2011 and 2012 for the management of phomopsis blight and fruit rot of brinjal caused by *P. vexans* (Pani et al. 2013). They used seed treatment with carboxin + thiram at 2 g/kg and foliar application of copper oxychloride at 0.3% in field. Both the treatments significantly reduced the incidence of seedling mortality, seedling blight and fruit rot infection and considerably increased the yield of brinjal.

Carbendazim and mancozeb in other study completely inhibited the growth and sporulation of many fungi causing postharvest fruit rot of chilli (Pan and Acharya 1995; Suryawanshi and Deokar 2001). Also, Bavistin and Dithane M-45 were found very effective against anthracnose and fruit rot of chilli (Das and Mohanty 1988). Rahman and Bhattiprolu (2005) used fungicides and mycorrhizal fungi for seed and soil treatment to control the fungal diseases like damping-off of tomato, brinjal and chilli in the nursery stage. Fungicides like carbendazim (Bavistin 50% wp), thiophanate methyl (Topsin-M 45–75% wp), mancozeb (Dithane-M 45–75% wp), propiconazole (Tilt 25% EC), copper oxychloride (Blitox 50% wp), copper hydroxide (Kocide 75% wp), iprodione 25% + carbendazim 25% (Quintal 50% wp), flusilazole (Nustar 40% EC), fenamidone 10% + mancozeb 50% (Secure 60% WDG), carbendazim 12% + mancozeb 63% (Sixer 75% wp) and aureofungin (Aureofungin 46.15% SP) were tested against *P. vexans* causing fruit rot of brinjal by poisoned food technique in PDA medium (Sabalpara et al. 2009). Carbendazim and thiophanate methyl at 250, 500 and 1000 ppm concentration showed high inhibition (93–100%) of the mycelial growth of the pathogen. Rahman et al. (1988) evaluated different fungicides against the purple blotch of onion; among them, mancozeb was reported as the best fungicide for the management of the disease. Among the nonsystemic fungicides, mancozeb at 0.3 per cent completely inhibited the growth of *Alternaria porri* (Chethana et al. 2011), while 0.2% of mancozeb was found effective against *A. porri* by Mishra and Gupta (2012). In a similar study, Madhavi et al. (2012) reported that out of the five fungicides, namely, mancozeb, carbendazim, Blitox, captafol and benlate tested, mancozeb was highly effective against *A. porri*, followed by Blitox and benlate. A gradual reduction in fungal growth was, however, found as the concentrations of the fungicides increased from 10 to 500 ppm. The efficacy of different fungicides like mancozeb and dicloran against the purple blotch disease has also been reported, and mancozeb was found as the best fungicide for the treatment of the disease (Rahman et al. 1988). In a follow-up study, the efficacy of few systemic fungicides such as chlorothalonil (Kavach 75%WP), zineb (Indofil Z-78 75% WP), mancozeb (Agastya M-45 75%WP) and propineb (Antracol 70% WP) and nonsystemic fungicides, for example, hexaconazole (Contaf 5 EC), tebuconazole (Folicur 25 EC), propiconazole (Tilt 25 EC) and difenconazole (Score 25 EC) was tested against *A. porri* causing purple blotch of onion under in vitro conditions. Among the nonsystemic fungicides evaluated in vitro against *A. porri*, mancozeb 75WP showed maximum inhibitory (89.51%) effect on the mycelial growth of the pathogen which was followed by propineb 70%WP (86.30%), chlorothalonil (84.94%) and zineb (78.15%). Among the systemic fungicides assayed, propiconazole 25 EC had largest inhibitory (92.59%) effects on the mycelial growth of the pathogen followed in order by hexaconazole 5 EC 70% (91.11%), difenconazole 25 EC (86.30%) and tebuconazole 25 EC (80.86%) (Priya et al. 2015).

9.4.2   Biomanagement of Vegetable Diseases

Among various phytopathogens, fungal pathogens in general contribute significantly to the yield losses in agriculture. However, in order to reduce such losses and to optimize crop production, fungicides have been used in agricultural practices around the world. But, the environmental threats of fungicides (Thabet et al. 2016) have forced the scientists to look for alternate disease control strategies. Beside their toxic impact on useful soil microflora (Solecki et al. 2005), emergence of resistance to fungicide among pathogens (Hobbelen et al. 2014), cost and lack of appropriate application/delivery technologies to resource-poor farmers further supports the need for identifying alternate strategies of insect pest controls. Considering these factors, the focus in recent times has been shifted towards the use of some biological resources which are considered environmentally safe, inexpensive and easily applicable (Satish et al. 2007; Taiga et al. 2008; Kotan et al. 2016). In this regard, the use of biological materials especially non-pathogenic soil microbiota to control plant pathogens often called "biological control" method has provided some solution to the expensive and environmentally toxic chemicals. Broadly, the term "biological control" involves the use of one or more organisms in isolation or combination to eliminate detrimental microbes. Some important groups of biocontrol agents used in the management of plant diseases are discussed in Table 9.3. Among soil

Table 9.3 Some important groups of biocontrol agents used in the management of plant diseases

Characteristics	Mechanism of action	Uses	Advantages	Limitation	Ways to improve efficacy
Bacillus spp.					
Gram-positive rod-shaped, endospore-forming bacteria, exploited as biopesticides, e.g. *B. subtilis*, *B. amyloliquefaciens*, *B. firmus* and *B. pumilus*	Competition, direct antibiosis and induced resistance, production of several microbial metabolites	Effective against a broad spectrum of plant pathogens, used either as foliar or root application	Long shelf life, low cost of production	Efficacy is low in case of high pathogen's inoculum	Verification of compatibility between strains and cultivars, application in strategies where another biocontrol agents are targeted to reducing soilborne inoculum
Pseudomonas spp.					
Gram-negative aerobic bacteria	Antibiosis through production of a wide spectrum of bioactive metabolites, competition	Commonly used to treat seeds or roots of plants before planting, can also be used to treat tubers and bulbs	Grow rapidly and can be mass produced at relatively low cost, good compatibility with several fungicides	Inability to produce resting spores	Identification of new formulations, combination between the specific strain and plant species
Streptomyces spp.					
Ubiquitous and naturally-occurring bacteria, e.g. *Streptomyces* spp. and *S. lydicus*	Colonize growing root tips and acts as a mycoparasite of fungal root pathogens, produces antifungal metabolites	Used as an antifungal agent	Once colonize the root tip, it can continuously protect the plant	Efficacy is not extremely high, and the rhizosphere competence is poor	Combine well with other biocontrol agents, more active in reducing the inoculums

Trichoderma spp.				
Asexually reproducing fungi with greater genetic diversity. e.g. *T. atroviride*, *T. asperellum*, *T. harzianum*, *T. viride*, *T. gamsii*, *T. polysporum*, etc.	Hyperparasitic activity against pathogens, induction of resistance, produce a great variety of lytic enzymes, direct antibiosis	Used against soilborne pathogens, wood disease and grey mould	Not toxic for mammals and environment, no production of toxic residues, no adverse effect on soil microflora and microfauna	Poor viability of conidia, insufficient concentration in the soil, short time for activity	Increase of the applied quantity per unit of soil, use of carriers or substrates which can guarantee a longer persistency
Pythium oligandrum				
Mycoparasite belonging to the Pythium family	Hyperparasitism, produces the protein oligandrin and other compounds that destroys phytopathogens	Applied as seed dressing, pre-plant soak, overhead spray, soil drench or through irrigation system application	Wide spectrum of activity	Requires a moist environment and a temperature range of 20–35 °C	Implementation of pre-planting treatments with the irrigation systems and the reduction of the cost per hectare
Phlebiopsis gigantea				
Common and widely distributed saprophytic wood-decay fungus	Competition	Used as a biological control of annosum root rot	Support the phase out of the use of chemical	Expensive	–

(continued)

Table 9.3 (continued)

Characteristics	Mechanism of action	Uses	Advantages	Limitation	Ways to improve efficacy
Coniothyrium minitans					
Coelomycete with a worldwide distribution.	Enters through the small pores or a lacerated surface of the target organism and/or lysing by hydrolytic enzymes	Used mainly in high value crops as lettuce, beans and other vegetables	Controls specifically *Sclerotinia* species without any interference with the soil microflora, easy to apply without any	Costly, longer duration needed to reduce the inoculum	–
Gliocladium catenulatum					
Naturally-occurring saprophytic fungus widespread in environment	Enzymatic degradation	Used to control damping-off, seed-root, stem-rots, and wilt diseases	Not toxic for mammals and environment, do not leave toxic residues, have no adverse effect on soil microflora and microfauna	Poor viability of conidia, insufficient concentration in the soil, short time for activity	Increase of the applied quantity per unit of soil, use of carriers or substrates which can guarantee a longer persistency

Adapted from Pertot et al. (2015)

microflora, plant growth-promoting rhizobacteria (PGPR) (Kloepper and Schroth 1978) are the most common biological control agents that have extensively been used to control disease (Beneduzi et al. 2012) and to concurrently enhance the yield of crop plants including vegetables (Zaidi et al. 2015). Among PGPR, species of *Actinoplanes* (El-Tarabily et al. 2009), *Alcaligenes* (Sayyed and Patel 2011), *Arthrobacter* (Joseph et al. 2007), *Azotobacter* (Bjelic et al. 2015), *Bacillus* (Kumar et al. 2011), *Cellulomonas* (Martinez et al. 2006), *Enterobacter* (Chen et al. 2016), *Erwinia* (Kang et al. 2012), *Flavobacterium* (Sang and Kim 2012), *Hafnia* (Gunes et al. 2015), *Actinobacteria* (Zamoum et al. 2015), *Pseudomonas* (Panpatte et al. 2016), *Pasteuria* (Kokalis 2015), *Rhizobium* (Ahmed et al. 2016), *Bradyrhizobium* (Deshwal et al. 2003), *Serratia* (Guo et al. 2004) and *Xanthomonas* (Massomo et al. 2004) have been identified as the most common biocontrol agents. Of these, *Bacillus* (Rahman et al. 2016) and *Pseudomonas* spp. (Mehrabi et al. 2016) are widely used as biocontrol agent for disease suppression. Among these, *Bacillus* spp. are Gram positive and can survive under unfavourable environmental conditions by producing endospores and are highly suitable for production of commercial formulations. Likewise, *Pseudomonas* spp. are Gram negative and have received particular attention as biocontrol agents because of their catabolic versatility, excellent root-colonizing abilities and production of broad range antifungal metabolites such as 2,4-diacetylphloroglucinol (DAPG) (Bottiglieri and Keel 2006), pyoluteorin (Yan et al. 2007), pyrrolnitrin (Hammer et al. 1997) and phenazines (Mavrodi et al. 2006).

9.4.2.1 Mechanism of Disease Suppression by Biocontrol Bacteria

Antibiosis

Antibiotics secreted by PGPR play an important role in disease suppression (Reddy 2014). The antibiotics commonly produced by different antagonistic bacteria include butyrolactones, DAPG (2,4-diacetylphloroglucinol) (Lanteigne et al. 2012), kanosamine, oligomycin A, oomycin A, phenazine-1-carboxylic acid, pyoluteorin, pyrrolnitrin, viscosinamide, xanthobaccin, and zwittermicin A (Whipps 2001). Moreover, other antibiotics such as phenazine-1-carboxamide, aerugine, rhamnolipids, cepaciamide A, pseudomonic acid, azomycin, antitumor antibiotics FR901463, caspofungins and antiviral antibiotic karalicin have also known to be antiviral, antimicrobial, insecticidal, antihelminthic, phytotoxic, antioxidant and cytotoxic effects and can also produce a plant growth-promoting effect (Ulloa-Ogaz et al. 2015). Many of these antibiotics possess a broad-spectrum activity (Raaijmakers et al. 2002).

Siderophores

The low molecular weight, iron-chelating compounds often called siderophores, produced by certain microbes (Gupta et al. 2015) and plants, are also involved in antibiosis and nutrient competition (Shanmugaiah et al. 2015). Production of siderophores (pyoverdine and pseudobactin) has been identified and considered as a new strategy adopted by PGPR to control the deleterious phytopathogens affecting many crops (Hassen et al. 2016) including vegetables (Iqbal et al. 2012). Siderophores primarily help the producing organism in iron acquisition under iron-limiting conditions. The siderophore produced by *B. subtilis* CAS15 has been observed to control Fusarium wilt (*F. oxysporum* Schl. f. sp. *capsici*) of pepper (Yu et al. 2011), while the siderophore secreted by *Bacillus amyloliquefaciens* (DSBA-11 and DSBA-12) were found to control bacterial wilt (*Ralstonia solanacearum*) of tomato (Singh et al. 2015).

Parasitism or Lysis

Parasitism of pathogenic fungi facilitated by the production of hydrolytic enzymes is yet another major mechanism involved in biological control of fungal diseases. Some PGPR strains excrete a high level of lytic enzymes, which has antifungal activity (Huang et al. 2005; Xiao et al. 2009). The enzymes for instance, chitinase and β-1,3-glucanase produced by *B. subtilis* strain EPCO 16 strongly inhibited *F. oxysporum* f. sp. *lycopersici* of tomato (Ramyabharathi et al. 2012). *B. subtilis* BSK17 is known to produce chitinase and β-1,3-glucanase to help in their competence and antagonistic activity (Dubey et al. 2014). Chitinase and β-1,3-glucanase have been reported as a major class of lytic enzyme that dissolve the major constituent of fungal cell wall-like chitin and laminarin (Kumar et al. 2012). Chitinases, among the hydrolytic enzymes, are of prime importance since chitin, a linear polymer of p-(1, 4)-A'-acetylglucosamine, is a major cell wall constituent in majority of the phytopathogenic fungi. Purified chitinases of *Trichoderma harzianum* (El-Katatny et al. 2000), *Serratia plymuthica* (Frankowski et al. 2001) and *Streptomyces* sp. (Gomes et al. 2001) were highly antifungal. Other important groups of hydrolytic enzymes like amylase, urease (Shrivastava et al. 2015), catalase (Twisha and Desai 2014), etc. are secreted by several strains of PGPR that possess antifungal activity. In addition to these, some antagonistic PGPR also secrete lipases, proteases (Pereg and McMillan 2015), glucanases (Figueroa-López et al. 2016), chitin-/chitosan-modifying enzyme chitosanases (Kumar et al. 2015) to ward off the pathogens by inhibiting their growth.

9.5 Examples of Some Vegetable Diseases Controlled by PGPR

9.5.1 Onion (*Allium cepa*)

Onion often called as "queen of kitchen" is one of the oldest known and an important vegetable crop grown all over the world including India. Globally, India ranks first in total cultivation area (1064 thousand hectares) and second in production after China with total production of 15,118 thousand million tons with a productivity of 14.2 million tons per hectare and contributing 19.9% of total world production. It is estimated that about 55 million tons of onions are produced annually all over the world where China and India contribute almost half of the world's onion production. China produces nearly 18.03 million tons, and India produces about 5.50 million tons (PHDEB 2006). In India, it is cultivated in many states like, Maharashtra, Karnataka, Gujarat,

Bihar, Madhya Pradesh, Andhra Pradesh, Rajasthan and Haryana (Anon. 2011). Onion is one of the most widely used vegetables both at mature and immature bulb stage for its flavouring and seasoning of food. Medicinal values of onion are numerous and is one of the ancient crops being utilized in medicine. Several factors have, however, been attributed for the low productivity of onion in many countries; chief among them is the phytopathogenic diseases which cause huge yield loses. Among diseases, fungal diseases such as leaf blight, purple blotch, basal rot, downy mildew, blight, storage rots, etc. are the most severe one. Of these, purple blotch is one of the most destructive diseases, commonly found in almost all onion-growing pockets of the world, which causes heavy losses to onions under field conditions. Though many researchers have worked on this pathogen and its management, the disease still remains a major challenge in onion cultivation. Generally, these diseases are mostly controlled by the use of synthetic fungicides (Gade et al. 2007). However, Naguleswaran et al. (2014) in a field experiment found that bulb treatment together with foliar application of *Trichoderma viride* improved the yield and other yield-related parameters such as basal diameter, circumference of bulb and mean number of bulb per bunch. It is also reported that some pathogens, for example, *Penicillium* species having antagonistic effect has been used as biological control agent against onion fungal pathogen *A. niger* (Khokhar et al. 2013). Also, *Penicillium roqueforti* and *P. viridicatum* greatly inhibited the growth of *A. niger* by 66% and 60%, respectively.

In a similar experiment, Malathi (2013) and Sudhasha et al. (2008) also reported that seed treatment with *T. harzianum* significantly reduced the basal rot incidence on onion under pot and field conditions. Recently, some biocontrol agents were isolated from the rhizosphere soil of onion, cultivated in different places of Tamil Nadu, India, and in vitro efficacy of such biocontrol agents was evaluated against basal rot of onion. Among the isolates of *Trichoderma* sp., *T. harzianum* (TH 3) resulted in the greatest (83%) inhibition while *Pseudomonas* sp. (Pf 12) significantly (75%) reduced the mycelial growth of *F. oxysporum* f. sp. *cepae*. Based on the laboratory performance, the effective biocontrol agents were further evaluated in glasshouse and under field conditions. Among the various combination treatments, the mixture of bacterial and fungal biocontrol agents (Pf12 + Pf27+ TH3) significantly (85%) reduced the disease. From this experiment, it was concluded that biocontrol agents could serve as an alternative to chemical control measures in the management of onion basal rot leading concurrently to enhanced growth and yield of onion (Malathi 2015). In other study, *P. agglomerans* has been found as the most effective biocontrol agent against onion bacterial diseases caused by *P. marginalis*, *P. ananatis*, *P. viridiflava* and *X. retroflexus* with a per cent inhibition of 24.8, 25.6, 26.7 and 14.4%, respectively (Sadik et al. 2013).

9.5.2 Cucumber (*Cucumis sativus* L.)

Cucumber is yet another important vegetable belonging to family cucurbitaceae. It has remarkable economic and dietary value. The mature fruits are used as salad, and the immature fruits are used in pickles. It is soft, succulent with high water content and has sufficient amounts of vitamins such as vitamin A, C, K, B$_6$ and potassium. It also gives dietary fibres, pantothenic acid, magnesium, phosphorus, copper and manganese (Vimala et al. 1999). Since the fruits and seeds possess cooling properties, it is used as astringent and antipyretic vegetable (Gill et al. 2015). Moreover, it contains ascorbic acid and caffeic acid, both of which help to reduce the skin irritation and swelling.

Pathogen attacks cucumber plants throughout the world severely. Among pathogens, *F. oxysporum* f. sp. cucumerinum causes wilt disease and has been found as the most damaging pathogen (Zhou et al. 2008). However, no reliable, cost-effective and efficient method is available to manage the Fusarium wilt disease of cucumber. Despite these, some practices like the use of fungicide is being adopted to overcome the pathogen attack (Mishra et al. 2000). In contrast, in order to reduce the toxic impact of fungicides on crops, the efficacy of biocontrol agents, for example, *Bacillus subtilis*, *B. pumilus*, zinc oxide nanoparticles, castor and clove oils, and recommended rates of fungicide (famoxadone + cymoxanil) were applied during two growing seasons under greenhouse conditions against downy mildew of cucumber by Mohamed et al. (2016). The results revealed a significant increase in plant height, fruit number/plant and fruit yield following all treatments relative to control. Among bioagents, strain EMs1 of *B. pumilus* and *B. subtilis* showed significant reduction in the severity of downy mildew in cucumber plants. From this study, it was concluded that the combinations of treatments could safely be used to reduce the severity of downy mildew without any adverse effect on the cucumber plants. In a similar study, Raupach and Kloepper (2000) found that PGPR *B. pumilus* (strain INR7), *Curtobacterium flaccumfaciens* (strain ME1) and *B. subtilis* (strain GB03) significantly reduced the severity of foliar diseases of angular leaf spot of cucumber, caused by *Pseudomonas syringae* pv. lachrymans and a mixed infestation of angular leaf spot and anthracnose, caused by *Colletotrichum orbiculare*. Similar attempts have also been made by other workers against *P. cubensis*, and promising results have been obtained. For example, *Bacillus* strains Z-X-3 and Z-X-10, *P. fluorescens* and *T. harzianum* markedly reduced the pathogens affecting cucumber to the extent of 46% (Anand et al. 2007). Biocontrol potential of *Pseudomonas monteilii* in control of stem rot disease of cucumber has been established (Rakh et al. 2011). Biocontrol agents like *T. harzianum*, *T. viride* and *Gliocladium virens* have also been successfully exploited to control the pathogen to the extent of 68% and sclerotial production to 98%, respectively (Pant and Mukhopadhyay 2001; Dutta and Das 2002).

9.5.3   Lettuce (*Lactuca sativa*)

Lettuce, the world's most popular leafy salad vegetable, is a self-fertilizing diploid species belonging to the Compositae family. Over 79% of the world production (24.3 million tons) in 2011 (FAOSTAT 2013) originated from four countries: China, shared 55.2% of the total production (by weight), the United States (16%), India

(4.4%) and Spain (3.6%). Various types of lettuce for human consumption are cultivated in different countries. Lettuce is most often used for salads, although it is also seen in other kinds of food, such as soups, sandwiches and wraps, since it is a rich source of vitamin K and vitamin A, and it also gives a moderate source of folate and iron. Lettuce on the other hand can be contaminated by bacteria (*E. coli* and *Salmonella*), fungi, soil viruses and nematodes. And consequently, over 75 lettuce disorders of diverse causes and etiologies have been described (Raid 2004). On the one hand, some diseases are limited in their importance and distribution, while on the contrary, many diseases cause devastating yield and quality losses under favourable conditions. Some of the common diseases reported in lettuce include anthracnose (*Microdochium panattonianum*), Cercospora leaf spot, damping-off, downy mildew (*Bremia lactucae*), grey mould (*Botrytis cinerea*), Septoria leaf spot, Sclerotinia rot (lettuce drop) (*Sclerotinia sclerotiorum* and *S. minor*) and bottom rot (*Rhizoctonia solani*). However, there are no effective and efficient methods available for the management of fungal and bacterial diseases. In order to manage the fungal diseases of lettuce and to reduce the use of fungicides, the effectiveness of the biological control agents was assayed by Fiume and Fiume (2005). They used *Coniothyrium minitans*, an antagonist fungus that attacked and destroyed the sclerotia within the soil. In a similar study, *Streptomycetes* ($N = 94$) and non-streptomycete actinomycetes ($N = 35$) were examined in vitro to suppress the growth of *Sclerotinia minor*, a pathogen causing basal drop disease of lettuce (El-Tarabily et al. 2000). Among these cultures, three isolates (*Serratia marcescens*, *S. viridodiasticus* and *Micromonospora carbonacea*) were found most suppressive and showed significant antifungal activity and could effectively colonize the roots and rhizosphere of lettuce. They significantly reduced the growth of *S. minor* in vitro by producing high levels of chitinase and beta-1,3-glucanase. Some species of *Bacillus* are also reported to control the diseases caused by fungus in lettuce. For instance, *B. amyloliquefaciens* (strain FZB42) is reported to reduce the disease severity of bottom rot caused by soilborne pathogen *R. solani* on lettuce (Chowdhury et al. 2013). Considering the importance of biological preparation as an alternative to fungicides in the management of lettuce bottom rot, *Trichoderma*, among many biocontrol agents, has been found harmless to both humans and environments (Pinto et al. 2014). Hence, the commercial production of *Trichoderma* has considerably increased in recent times in most of the countries (Verma et al. 2007). It is well known that *Trichoderma* strains can produce a variety of secondary metabolites, like antibiotics, which can be inhibitory to microbial growth. The production of these substances is, however, strain-specific, and they can belong to a high variety of classes of volatile and non-volatile compounds such as, water-soluble compounds or peptaibols and linear oligopeptides (Vinale et al. 2008). Of these biomolecules, peptaibols inhibits the activity of the enzyme β-glucan synthase in the host fungus and acts synergistically with β-glucanases in preventing the reconstruction of the fungus cell wall, facilitating thus the disruptive action of β-glucanases (Vinale et al. 2008). Also, *T. viride* strains may produce antibiotics such as gliotoxin, gliovirin and viridiol, which have antagonistic activity against several plant pathogens (Vinale et al. 2008). In yet another investigation, the control of bottom rot with 0.5% weight/weight of one isolate of *T. harzianum*, with 1 kg/m² of *T. harzianum* in lettuce planted on beds under plastic tunnels, is reported (Maplestone et al. 1991).

9.5.4 Spinach (*Spinacia oleracea*)

Spinach is one of the most desirable dark green leafy vegetables belonging to family Amaranthaceae. World spinach production exceeded 20 million tons in 2010, of which about 90% was produced in China (FAOSTAT 2010). The United States was the second largest spinach producer, with about 2% of the production in the world. Spinach is an excellent source of vitamin K, vitamin A (in the form of carotenoids), b-carotene (provitamin A), manganese, folate, magnesium, iron, copper, vitamin B2, vitamin B6, vitamin E, calcium, potassium, vitamin C, lutein, folate, dietary fibre, phosphorus, vitamin B1, zinc, protein and choline. Additionally, spinach is also a good source of omega-3 fatty acids, niacin, pantothenic acid and selenium. Several pathogens are reported to attack spinach throughout the world. The chief among them has been the downy mildew. Other major diseases of spinach are white rust disease (*Albugo occidentalis*), Fusarium wilt (*F. oxysporum* f. sp. spinaciae) and root-knot nematode (*Meloidogyne* spp). Among foliar diseases, the most important are white rust and anthracnose (*Colletotrichum dematium*). These diseases can reduce the quality and commercial acceptance of spinach. However, such diseases are mostly controlled by systemic and biological fungicides. Increased concern over the impact of chemical fungicide on the environment has resulted in the increased interest in biocontrol strategies for the management of soilborne phytopathogenic fungi. The combination of biological and synthetic fungicides has been tested against many spinach diseases, for example, Fusarium wilts. Methyl benzimidazole carbamate (MBC) fungicides in combination with biological fungicides have been found very effective in reducing the incidence of Fusarium wilt of spinach (Elmer and McGovern 2004; Everts et al. 2014). Several demethylation inhibitor (DMI) fungicides (tebuconazole, difenconazole, prothioconazole and ipconazole) have also been shown to suppress Fusarium wilt in spinach and other leafy vegetables (Everts et al. 2014; Jones 2000). In Canada, the biological control agent Serenade ASO (*B. subtilis*, Bayer Crop Science, Bathesda, NC, USA) is registered for spinach for white rust suppression, as are the synthetic fungicides cyazofamid and fosetyl-AL (OMAFRA 2014).

Biological control of Fusarium wilt has also been reported using *Pseudomonas* spp., *Bacillus* spp., *Burkholderia* spp., *Penicillium oxalicum*, *Enterobacter cloacae*, *Streptomyces griseoviridis*, hypovirulent binucleate *Rhizoctonia*, *Trichoderma* spp., *Gliocladium catenulatum* and non-pathogenic *Fusarium* spp. (Bell et al. 1998; Benhamou and Cummings et al. 2009; Horinouchi et al. 2010; Larena et al. 2003; Larkin and Fravel 1998; Lemanceau et al. 1993; Minuto et al. 1995; Muslim et al. 2003; Nion and Toyota 2008; Rose et al. 2003). There are many commercial biological fungicides currently available in North America for the control of foliar and soilborne diseases in spinach and greenhouse, and field trials have been completed to determine the effectiveness of several organic products to control Fusarium wilt of spinach (Cummings et al. 2009; Cummings 2007). In a similar study, it is found that drench application of the biofungicide Prestop (*G. catenulatum*, Verdera Oy, Espoo, Finland) reduced the incidence of Fusarium wilt by 37–56% compared to the non-treated control plants. Although wilt symptoms were reduced, pre-emergence damping-off of spinach was increased by the soil drench (Cummings et al. 2009). This biological control treatment effectively reduced Fusarium wilt in greenhouse vegetables grown in soilless media (Rose et al. 2003).

9.5.5 Broccoli (*Brassica oleracea*)

Broccoli is an edible green plant in the cabbage family whose large flowering head is eaten as a vegetable in many countries. China is the top world producer of broccoli (9,596,000 tons) (FAO 2012), while India ranks second (7.9 million tons). Broccoli inflorescence provides some important health-promoting compounds such as glucosinolates, flavonoids, hydroxycinnamic acids and some other minor compounds. It has gastroprotective, antioxidant, antimicrobial, anticancer, hepato-protective, cardioprotective and anti-inflammatory activities (Owis 2015). Broccoli plants are infected by many disease causing biotic agents and hence, serious losses in the broccoli yield is reported (El-Mohamedy 2012). Among these, *Pythium* root rot is the most serious disease (Abdelzaher 2003; Tanina et al. 2004). Although, fungicides have been used to suppress Pythium diseases, but the use of fungicides also adversely affect the environment. Currently, attention has been shifted towards the employment of biological control approach involving the use of antagonistic microorganisms such as fungi and bacteria as an alternative to fungicides (Gravel et al. 2004). In this context, *P. fluorescens* was used as biocontrol agent for controlling broccoli root rot disease caused by *P. ultimum* and *Rhizoctonia* pathogens. *P. fluorescens* 2–79 has been shown to produce PCA (Phenazine-1-carboxylic acid) and AAP (2-acetamidophenol) (Slininger et al. 2000). *Pseudomonas* sp. DF41 is reported to be very effective inhibitors of *S. sclerotiorum* mycelial growth and suppresses germination of sclerotia and ascospores (Savchuk 2002). Presence of phenazine biosynthetic genes and PCA production by PA-23 accounted for the inhibition of mycelial growth of *S. sclerotiorum* (Fernando et al. 2007). Georgakopoulos et al. (2002) found that better biocontrol in broccoli was achieved when *B. subtilis* and *P. fluorescens* were applied by drenching or by coating seed in peat carrier. In other study, the efficacy of fungal and bacterial isolates against Pythium root rot diseases was assessed in a pot experiment using soil artificially infested under greenhouse conditions (El-Mohamedy 2012). For this, two isolates each of *T. harzianum* (G1 and B1), *T. viride* (G1 and G3) and one isolate each of *B. subtilis* (B1) and *P. fluorescens* (B3) were tested. The results of this study clearly revealed complete reduction in the percentages of Pythium root rot incidence of broccoli plants following microbial inoculations. Applying biocontrol agents as a combination of soil mixing plus root dipping method was generally most effective than each individual treatments for suppressing Pythium root rot incidence followed by soil mixing and root dipping methods. A high reduction in Pythium root rot incidence was observed due to application of *T. harzianum* isolate G1 and B1 (88 and 84.4%, respectively), *T. viride* isolate G1 and G3 (80.2 and 77.2%, respectively), *P. fluorescens* isolate B3 and *B. subtilis* isolate B1 (80 and 75%, respectively) when applied in soil mixing plus root dipping methods. The inhibitory effects of *T. harzianum* (isolates G1 and B1) and *T. viride* (isolates G1 and G3) were higher than other isolates of *T. harzianum* and *T. viride*, as they completely reduced the growth of the pathogen. The antagonistic bacteria also showed significant reduction in the growth of *P. ultimum*. The higher inhibitory effect was recorded for *B. subtilis* isolate B3 and *P. fluorescens* isolate B3. Based on these findings, it was concluded that the antagonistic fungi in general had a greater inhibitory effect on the growth compared with the bacterial agents.

The use of the volatile compound-producing fungus *Muscodor albus* for the biological control of soilborne diseases in greenhouse soilless growing mix was investigated by Julien and Manker (2004). In this study, fresh rye grain culture of *M. albus* incorporated into *R. solani*-infested growing mix at a rate of 15 g/L or greater provided complete control of damping-off of broccoli seedlings, restoring seedling emergence to levels similar to the non-infested control without deleterious effect to plant growth. The ability of *M. albus* to control damping-off declined rapidly after its incorporation to the growing mix, suggesting that its activity occurs in the initial hours of treatment. In treated mix, damping-off remained under control regardless of planting time after treatment, suggesting that a biological fumigation had killed *R. solani*. *M. albus* also completely controlled root rot of bell pepper caused by *Phytophthora capsici*.

Conclusion

Vegetables are one of the most important components of common food habit because they provide essential micro- and macronutrients, proteins, antioxidants and vitamins to the human body. Application of synthetic pesticides has been the traditional strategy for the management of phytopathogenic diseases of vegetables. The cost of synthetic pesticides, associated human health problems, and environmental pollution have, however, necessitated the development of alternative strategies for the control of diseases of vegetables. One of the emerging but promising strategies for managing diseases of vegetable crops is the use of microbial biocontrol agents which are both inexpensive and environmentally safe. The use of such microbial preparations in disease management has indeed been found very effective and affordable by the growers, but still it requires greater understanding of the mechanistic basis of disease suppression. Also, there is a need to generate awareness among growers so that its use under different agronomic regions is increased, and the benefit of this low cost technology is achieved by vegetable growers. However, the reports/results presented in this chapter on use of biocontrol agents in vegetable production are likely to benefit vegetable growers in a big way while reducing the dependence on chemicals use in vegetable cultivation. Therefore, commercialization of some of these antagonists to control loss of vegetables appears to be feasible and may present an alternative to synthetic pesticides.

References

Abdelzaher HM (2003) Biological control of root rot of cauliflower (caused by *Pythium ultimum* var. *ultimum*) using selected antagonistic rhizospheric strains of *Bacillus subtilis*. New Zeal J Crop Hort Sci 31:209–220

Abeer HM, Rehab A (2015) Biological and nanocomposite control of Fusarium wilt of potato caused by *Fusarium oxysporum* f sp. Tuberosi. Glob J Biol Agric Health Sci 4:151–163

Agnihotri NP (1999) Pesticide: safety evaluation and monitoring. All India Co-ordinated Project (AICRP) on Pesticide Residues, Division of Agricultural Chemicals, Indian Agricultural Research Institute (IARI), New Delhi, pp 132–142

Ahmed AQ, Javed N, Khan SA (2016) Efficacy of rhizospheric organism *Rhizobium leguminosarum* against *Meloidogyne incognita* in soybean. Pak J Agric Sci 53:377–381

Aktar W, Sengupta D, Chowdhury A (2009) Impact of pesticides use in agriculture: their benefits and hazards. Interdiscip Toxicol 2:1–12

Al Mughrabi KI, Vikram A, Poirier R et al (2016) Management of common scab of potato in the field using biopesticides, fungicides, soil additives, or soil fumigants. Biocontrol Sci Technol 26:125–135

Anand P, Kunnumakkara AB, Newman RA et al (2007) Bioavailability of curcumin: problems and promises. Mol Pharmacol 4:807–818

Anonymous (2011) IIHR annual report-2010-11; Indian Institute of Horticultural Research, Bangalore

Anonymous (2011a) Indian horticulture database. http://www.des.kar.nic.in

Anonymous (2011b). Vision 2030. Indian Institute of Vegetable Research, Varanasi, UP

Behrani GQ, Syed RN, Abro MA et al (2015) Pathogenicity and chemical control of basal rot of onion caused by *Fusarium oxysporum* f. sp. *cepae*. Pak J Agric Agril Engg Vet Sci 31:60–70

Bell AA, Hubbard JC, Liu L et al (1998) Effects of chitin and chitosan on the incidence and severity of Fusarium yellows of celery. Plant Dis 82:322–328

Beneduzi A, Ambrosini A, Passaglia LM (2012) Plant growth-promoting rhizobacteria (PGPR): their potential as antagonists and biocontrol agents. Genet Mol Biol 35:1044–1051

Beura SK, Mahanta IC, Mohapatra KB (2008) Economics and chemical control of Phomopsis twig blight and fruit rot of brinjal. J Mycopathol Res 46:73–76

Bjelic DD, Marinkovic JB, Tintor BB (2015) Screening of *Azotobacter* isolates for PGP properties and antifungal activity. MS Nat Sci 2015(129):65–72

Bottiglieri M, Keel C (2006) Characterization of PhlG, a hydrolase that specifically degrades the antifungal compound 2,4-diacetylphloroglucinol in the biocontrol agent *Pseudomonas fluorescens* CHAO. Appl Environ Microbiol 72:418–427

Brewer JH, Thrasher JD, Hooper D (2013) Chronic illness associated with mold and mycotoxins: is naso-sinus fungal biofilm the culprit? Toxins (Basel) 6:66–80

Chen PS, Peng YH, Chung WC (2016) Inhibition of *Penicillium digitatum* and Citrus green mold by volatile compounds produced by *Enterobacter cloacae*. J Plant Pathol Microbiol 7:339

Chethana BS, Ganesh G, Manjunath B (2011) Screening of genotypes and effect of fungicides against purple blotch of onion. J Agric Technol 7:1369–1374

Chowdhury SP, Dietel K, Rändler M et al (2013) Effects of *Bacillus amyloliquefaciens* FZB42 on lettuce growth and health under pathogen pressure and its impact on the rhizosphere bacterial community. PLoS One 8(7):e68818. doi:10.1371/journal.pone.0068818

Cummings, JA (2007) Evaluation of seed and drench treatments for management of damping-off and seedling blight pathogens of spinach for organic production. Doctoral dissertation, Washington State University

Cummings JA, Miles CA, Toit LJ (2009) Greenhouse evaluation of seed and drench treatments for organic management of soilborne pathogens of spinach. Plant Dis 93:1281–1292

Damalas CA, Eleftherohorinos IG (2011) Pesticide exposure, safety issues, and risk assessment indicators. Int J Environ Res Public Health 8:1402–1419

Das SR, Mohanty KC (1988) Management of die back/twig blight of chilly with fungicides. Indian J Plant Protect 16:109–111

Deshwal VK, Pandey P, Kang SC (2003) Rhizobia as a biological control agent against soil borne plant pathogenic fungi. Indian J Exp Biol 41:1160–1164

Dhanasekaran P, Sivamani A, Panneerselvam N et al (2005) Biological control of tomato seedling damping off with *Streptomyces* sp. Plant Pathol J 4:91–95

Dubey RC, Khare S, Kumar P (2014) Combined effect of chemical fertilisers and rhizosphere-competent *Bacillus subtilis* BSK17 on yield of *Cicer arietinum*. Arch Phytopathol Plant Protect 47:2305–2318

Dutta P, Das BC (2002) Management of collar rot of tomato by *Trichoderma* spp. and chemicals. Indian Phytopathol 55:235–237

Egel DS, Martyn RD (2007) Fusarium wilt of watermelon and other cucurbits. Plant Health Instruct. doi:10.1094/PHI-I-2007-0122-01

El-Katatny MH, Somitsch W, Robra KH et al (2000) Production of chitinase and β-1, 3-glucanase by Trichoderma harzianum for control of the phytopathogenic fungus Sclerotium rolfsii. Food Technol Biotechnol 38:173–180

Elmer WH, McGovern RJ (2004) Efficacy of integrating biologicals with fungicides for the suppression of Fusarium wilt of cyclamen. Crop Protect 23:909–914

El-Mohamedy RSR (2012) Biological control of Pythium root rot of broccoli plants under greenhouse conditions. J Agric Technol 8:1017–1028

El-Tarabily KA, Nassar AH, Hardy GSJ (2009) Plant growth promotion and biological control of *Pythium aphanidermatum*, a pathogen of cucumber, by endophytic actinomycetes. J Appl Microbiol 106:13–26

El-Tarabily KA, Soliman MH, Nassar AH et al (2000) Biological control of *Sclerotinia minor* using a chitinolytic bacterium and actinomycetes. Plant Pathol 49:573–583

Everts KL, Egel DS, Langston D et al (2014) Chemical management of Fusarium wilt of watermelon. Crop Protect 66:114–119

FAO statistics (2012) Production year book. Food and Agriculture Organization

FAOSTAT (2010) Production, food and agricultural organization of the United Nations. http://faostat.fao.org/site/339/default.aspx

FAOSTAT (2013) Crops data for 2011. Food and Agriculture Organization of the United Nations. http://faostat3.fao.org (accessed 21 Feb 2013)

Fatima N, Batool H, Sultana V et al (2009) Prevalence of post-harvest rot of vegetables and fruits in Karachi, Pakistan. Pak J Bot 41:3185–3190

Fernando WGD, Nakkeeran S, Zhang Y et al (2007) Biological control of *Sclerotinia sclerotiorum* (Lib.) de Bary by *Pseudomonas* and *Bacillus* species on canola petals. Crop Protect 26:100–107

Figueroa-López AM, Cordero-Ramírez JD, Martínez-Álvarez JC et al (2016) Rhizospheric bacteria of maize with potential for biocontrol of *Fusarium verticillioides*. SpringerPlus 5:1

Fiume F, Fiume G (2005) Biological control of Botrytis gray mould and Sclerotinia drop in lettuce. Commun Agric Appl Biol Sci 70:157

Frankowski J, Lorito M, Scala F et al (2001) Purification and properties of two chitinolytic enzymes of *Serratia plymuthica* HRO-C48. Arch Microbiol 176:421–426

Gade RM, Zote KK, Mayee CD (2007) Integrated management of pigeon pea wilt using fungicides and bioagents. Indian Phytopathol 60:24–30

Georgakopoulos DG, Fiddaman P, Leifert C, Malathrakis NE (2002) Biological control of cucumber and sugar beet damping-off caused by *Pythium ultimum* with bacterial and fungal antagonists. J Appl Microbiol 92:1078–1086

Gilden RC, Huffling K, Sattler B (2010) Pesticides and health risks. J Obstet Gynecol Neonatal Nurs 39:103–110

Gill NS, Sharma G, Arora R (2015) Cucumis trigonus roxb: a review. Int J Recent Adv Pharma Res 5:45–50

Gomes RC, Semedo LTAS, Soares RMA (2001) Purification of a thermostable endochitinase from *Streptomyces* RC1071 isolated from a cerrado soil and its antagonism against phytopathogenic fungi. J Appl Microbiol 90:653–661

Gravel V, Martinez C, Antoun H et al (2004) Evaluation of antagonistic microorganisms as bio control agents of root rot (*Pythium ultimum*) of greenhouse tomatoes in rock wool. Can J Plant Pathol 26:152

Gunes A, Karagoz K, Turan M et al (2015) Fertilizer efficiency of some plant growth promoting rhizobacteria for plant growth. Res J Soil Biol 7:28–45

Guo JH, Qi HY, Guo YH et al (2004) Biocontrol of tomato wilt by plant growth-promoting rhizobacteria. Biol Control 29:66–72

Gupta G, Parihar SS, Ahirwar NK, Snehi SK et al (2015) Plant growth promoting rhizobacteria (PGPR): current and future prospects for development of sustainable agriculture. J Microbiol Biochem Tech 7:96–102

Habib A, Talib S, Sahi MU (2007) Evaluation of some fungicides against seed born mycoflora of eggplant and their comparative efficacy regarding seed germination. Int J Agric Biol 9:3

Hammer PE, Hill DS, Lam ST (1997) Four genes from *Pseudomonas fluorescence* that present the synthesis of pyrrolnitrin. Appl Environ Microbiol 63:2147–2152

Hassen AI, Bopape FL, Sanger LK (2016) Microbial inoculants as agents of growth promotion and abiotic stress tolerance in plants. In: Singh DP, Singh HB, Prabha R (eds) Microbial inoculants in sustainable agricultural productivity: vol 1 of research perspective. Springer, India, pp 23–36

Herr I, Buchler MW (2010) Dietary constituents of broccoli and other cruciferous vegetables implications for prevention and therapy of cancer. Cancer Treat Rev 36:377–383

Hobbelen PH, Paveley ND, Van DBF (2014) The emergence of resistance to fungicides. PLoS One 9:91910

Horinouchi H, Muslim A, Hyakumachi M (2010) Biocontrol of Fusarium wilt of spinach by the plant growth promoting fungus *Fusarium equiseti* GF183. J Plant Pathol:249–254

Hossain MI, Islam MR, Uddin MN et al (2013) Control of Phomopsis blight of egg plant through fertilizer and fungicide management. Int J Agric Res Innov Technol 3:66–72

Huang CJ, Wang TK, Chung SC (2005) Identification of an antifungal chitinase from a potential biocontrol agent, *Bacillus cereus* 28-9. BMB Rep 38:82–88

Iqbal MA, Khalid M, Shahzad SM (2012) Integrated use of *Rhizobium leguminosarum*, plant growth promoting rhizobacteria and enriched compost for improving growth, nodulation and yield of lentil (*Lens culinaris* Medik.) Chil. J Agric Res 72:104–110

Islam MR, Meah MB (2011) Association of *Phomopsis vexans* with eggplant (*Solanum melongena*) seeds, seedlings and its management. Agriculturists 9:8–17

Jones RK (2000) Assessments of Fusarium head blight of wheat and barley in response to fungicide treatment. Plant Dis 84:1021–1030

Joseph B, Patra RR, Lawrence R (2007) Characterization of plant growth promoting rhizobacteria associated with chickpea (*Cicer arietinum* L.) Int J Plant Prod 1:141–152

Julien M, Manker DC (2004) Biocontrol of soil-borne diseases and plant growth enhancement in greenhouse soilless mix by the volatile-producing fungus *Muscodor albus*. Crop Protect 24:355–362

Kang YJ, Shen M, Wang HL et al (2012) Biological control of tomato bacterial wilt caused by *Ralstonia solanacearum* with *Erwinia persicinus* RA2 and *Bacillus pumilus* WP8. Chin J Biol Cont 28:255–261

Kanjanamaneesathian M (2015) Biological control of diseases of vegetables grown hydroponically in Thailand: challenge and opportunity. Recent Pat Biotechnol 9(3):214–222

Kemmitt G (2002) Early blight of potato and tomato. Plant Health Instruct. doi:10.1094/PHI-I-2002-0809-01

Kestwal RM, Lin JC, Bagal-Kestwal D et al (2011) Glucosinolates fortification of cruciferous sprouts by sulphur supplementation during cultivation to enhance anti-cancer activity. Food Chem 126:1164–1171

Khokhar I, Haider MS, Mukhtar I, Mushtaq S (2013) Biological control of Aspergillus niger the cause of Black-rot disease of (Allium cepa L). Onion by Penicillium species. J Agrobiology 29:23–28

Kim MK, Choi GJ, Lee HS (2003) Fungicidal property of *Curcuma longa* L rhizome-derived curcumin against phytopathogenic fungi in a greenhouse. J Agric Food Chem 51:1578–1581

Kloepper JW, Schroth MN (1978) Plant growth-promoting rhizobacteria on radishes. In: Proceedings of the 4th international conference on plant pathogenic bacteria, Gibert-Clarey, Tours, France vol 2, pp 879–888

Kokalis BN (2015) *Pasteuria penetrans* for control of *Meloidogyne incognita* on tomato and cucumber, and *M. arenaria* on snapdragon. J Nematol 47:207

Kotan G, Kardaş F, Yokuş OA et al (2016) A novel determination of curcumin via Ru@ Au nanoparticle decorated nitrogen and sulfur-functionalized reduced graphene oxide nanomaterials. Anal Methods 8:401–408

Kumar A, Prakash A, Johri BN (2011) Bacillus as PGPR in Crop Ecosystem. In: Maheshwari, Dinesh K. (ed) Bacteria in Agrobiology: Crop Ecosystems, Springer, New York, p 37–60

Kumar M, Das B, Prasad KK et al (2013) Effect of integrated nutrient management on growth and yield of broccoli (*Brassica oleracea* var. *italica*) under Jharkhand conditions. Veg Sci 40:117–120

Kumar GP, Desai S, Amalraj ELD (2015) Isolation of Fluorescent *Pseudomonas* spp. from diverse agro-ecosystems of India and characterization of their PGPR traits. Bacteriol J 5:3–24

Kumar P, Dubey RC, Maheshwari DK (2012) *Bacillus strains* isolated from rhizosphere showed plant growth promoting and antagonistic activity against phytopathogens. Microbiol Res 167:493–499

Lanteigne C, Gadkar VJ, Wallon T (2012) Production of DAPG and HCN by *Pseudomonas* sp. LBUM300 contributes to the biological control of bacterial canker of tomato. Phytopathol 102:967–973

Larena I, Sabuquillo P, Melgarejo P et al (2003) Biocontrol of Fusarium and Verticillium wilt of tomato by *Penicillium oxalicum* under greenhouse and field conditions. J Phytopathol 151:507–512

Larkin RP, Fravel DR (1998) Efficacy of various fungal and bacterial biocontrol organisms for control of Fusarium wilt of tomato. Plant Dis 82:1022–1028

Lemanceau P, Bakker PA, De Kogel WJ et al (1993) Antagonistic effect of nonpathogenic *Fusarium oxysporum* Fo47 and pseudobactin 358 upon pathogenic *Fusarium oxysporum* f. sp. dianthi. Appl Environ Microbiol 59:74–82

Madhavi M, Kavitha A, Vijayalakshmi M (2012) Studies on *Alternaria porri* (Ellis) ciferri pathogenic to onion (*Allium cepa* L.) Arch Appl Sci Res 4:1–9

Makun HA (2013). Mycotoxin and food safety in developing countries. InTech, p. 280. ISBN: 978-953-51-1096-5 doi:10.5772/3414

Malathi S (2013) Production of cell wall degrading enzymes by *Fusarium oxysporum* f. sp. *cepae* causing basal rot of onion and its histopathological changes. Int J Plant Protect 6:412–418

Malathi S (2015) Biological control of onion basal rot caused by *Fusarium oxysporum* f. sp. *cepae*. Asian J Biol Sci 10:1

Mamgain A, Roychowdhury R, Tah J (2013) Alternaria pathogenicity and its strategic controls. Res J Biol 1:1–9

Manukinda A, Narasegowda NC, Premalatha Y et al (2016) Effect of Taegro as bio-fungicide on yield and control of early and late blight diseases in tomato (*Solanum lycopersicum* L.) IJSN 7:89–93

Maplestone PA, Whipps JM, Lynch JM (1991) Effect of peat-bran inoculum of *Trichoderma species* on biological control of *Rhizoctonia solani* in lettuce. Plant Soil 136:257–263

Mari M, Guizzardi M (1998) The postharvest phase: emerging technologies for the control of fungal diseases. Phytoparasitica 26:59–66

Martinez C, Avis TJ, Simard JN (2006) The role of antibiosis in the antagonism of different bacteria towards *Helminthosporium solani*, the causal agent of potato silver scurf. Phytoprotection 87:69–75

Massomo SMS, Mortensen CN, Mabagala RB (2004) Biological control of black rot (*Xanthomonas campestris* pv. *campestris*) of cabbage in Tanzania with *Bacillus strains*. J Phytopathol 152:98–105

Mavrodi DV, Blankenfeldt W, Thomashow LS (2006) Phenazine compounds in fluorescent *Pseudomonas* spp. biosynthesis and regulation. Annu Rev Phytopathol 44:417–445

Mehrabi Z, McMillon EV, Clark IM (2016) *Pseudomonas* spp. diversity is negatively associated with suppression of the wheat take-all pathogen. Sci Rep 6:29905. doi:10.1038/srep29905

Minuto A, Migheli Q, Garibaldi A (1995) Evaluation of antagonistic strains of *Fusarium* spp. in the biological and integrated control of Fusarium wilt of cyclamen. Crop Protect 14:221–226

Mishra RK, Gupta RP (2012) In vitro evaluation of plant extracts, bio-agents and fungicides against purple blotch and Stemphylium blight of onion. J Med Plant Res 6:5840–5843

Mishra AK, Liu CH, He B et al (2000) Efficacy of isoalantolactone, a sesquiterpene lactone from Inularacemosa, as herbal fungitoxicant to control 'take-all' disease of wheat. Int Pest Control 42:131–133

Mohamed A, Hamza A, Derbalah A (2016) Recent approaches for controlling downy mildew of cucumber under greenhouse conditions. Plant Protect Sci 52:1

Muneeshwar S, Razdan VK, Mohd R (2012) *In vitro* evaluation of fungicides and biocontrol agents against brinjal leaf blight and fruit rot pathogen *Phomopsis vexans*. Q J Life Sci 9:327–332

Muslim A, Horinouchi H, Hyakumachi M (2003) Suppression of Fusarium wilt of spinach with hypovirulent binucleate *Rhizoctonia*. J Gen Plant Pathol 69:143–150

Naguleswaran V, Pakeerathan K, Mikunthan G (2014) Biological control: a promising tool for bulb-rot and leaf twisting fungal diseases in red onion (*Allium cepa* L.) in Jaffna district. World Appl Sci J 31:1090–1109

Ngowi AVF, Mbise TJ, Ijani ASM, London L, Ajayi OC (2007) Pesticides use by smallholder farmers in vegetable production in northern Tanzania. Crop Protect 26:1617–1624

Nion YA, Toyota K (2008) Suppression of bacterial wilt and Fusarium wilt by a *Burkholderia nodosa* strain isolated from Kalimantan soils, Indonesia. Microbes Environ 23:134–141

Noyan Ashraf MH, Wu L, Wang R et al (2006) Dietary approaches to positively influence fetal determinants of adult health. FASEB J 20:371–373

Ogundana SK, Dennis C (1981) Assessment of fungicides for the prevention of storage rot of yam tubers. Pest Sci 12:491–494

Okoi AI, Okon EA, Ataga AE (2015) Evaluation of fungicide efficacy for the control of leaf spot disease on okra (*abelmoschus esculentus*) caused by *curvularia lunata* (Wakker) Boedijn in port Harcourt. Eur J Pharm Med 2:93–102

Omafra (2014) Spinach and Swiss chard. In: Vegetable crop production guide: publication, vol 838. Ontario Ministry of Agric Food and Rural Affairs, Guelph, ON, pp 182–183

Owis AI (2015) Broccoli; the green beauty: a review. J Pharm Sci Res 7:696–703

Pan S, Acharya S (1995) Studies on the seed borne nature of *Phomopsis vexans*. Indian Agric 39:193–198

Pani BK, Singh DV, Nanda SS (2013) Chemical control and economics of phomopsis blight and fruit rot of brinjal in the eastern Ghat highland zone of Odisha. Int J Agric Environ Biotechnol 6:581–584

Panpatte DG, Jhala YK, Shelat HN (2016) *Pseudomonas fluorescens*: a promising biocontrol agent and PGPR for sustainable agriculture. In: Singh DP, Singh HB, Prabha R (eds) Microbial inoculants in sustainable agricultural productivity. Springer, India, pp 257–270

Pant R, Mukhopadhyay AN (2001) Integrated management of seed and seedling rot complex of soybean. Indian Phytopath 54:346–350

Pereg L, McMillan M (2015) Scoping the potential uses of beneficial microorganisms for increasing productivity in cotton cropping systems. Soil Biol Biochem 80:349–358

Pertot I, Alabouvette C, Esteve EH (2015) The use of microbial biocontrol agents against soil-borne diseases. EIP-AGRI FOCUS GROUP SOIL-BORNE DISEASES, MINI PAPER-MICROBIAL BIOCONTROL AGENTS, http://ec.europa.eu/eip/agriculture/sites/agrieip/files/8_eip_sbd_mp_biocontrol_final.pdf

PHDEB (2006) Pakistan horticulture development and export board. http://www.phdeb.org.pk

Pinto ZV, Cipriano MAP, Santos AS, Pfenning LH, Patrício FRA (2014) Control of lettuce bottom rot by isolates of Trichoderma spp. Summa Phytopathologica 40:141–146

Poly S, Srikanta D (2012) Assessment of yield loss due to early blight (*Alternaria solani*) in tomato. Indian J Plant Protect 40:195–198

Priya RU, Sataraddi A, Darshan S (2015) Efficacy of non-systemic and systemic fungicides against purple blotch of onion (*Allium cepa* L) caused by *Alternaria porri* (Ellis) Cif. Int J Recent Sci Res 9:6519–6521

Pujeri US, Pujar AS, Hiremath SC et al (2015) Analysis of pesticide residues in vegetables in Vijayapur, Karnataka India. World J Pharm Pharm Sci 4:1743–1750

Quarles W (2004) Least-toxic controls of plant diseases. BBG, Brooklyn NY

Raaijmakers JM, Bonsall RF, Weller DM (2002) Effect of population density of *Pseudomonas fluorescens* on production of 2,4-diacetylphloroglucinol in the rhizosphere of wheat. Phytopathology 89:470–475

Rahman MA, Bhattiprolu SL (2005) Efficacy of fungicides and mycorrhizal fungi for the control of damping-off disease in nurseries of tomato, chilli and brinjal crops. Karnataka J Agric Sci 18:2

Rahman ML, Ahmed HU, Mian IH (1988) Efficacy of fungicides in controlling purple blotch of onion. Plant Pathol 4:71–76

Rahman MME, Hossain DM, Suzuki K (2016) Suppressive effects of *Bacillus* spp. on mycelia, apothecia and sclerotia formation of *Sclerotinia sclerotiorum* and potential as biological control of white mold on mustard. Aust Plant Pathol 45:103–117

Raid RN (2004) Lettuce diseases and their management. In: Naqvi SAMH (ed) Diseases of fruits and vegetables: volume II. Springer, Netherlands, pp 121–147

Rakh RR, Raut LS, Dalvi SM (2011) Biological control of *Sclerotium rolfsii*, causing stem rot of groundnut by *Pseudomonas* cf. *monteilii*. Recent Res Sci Technol 3:3

Ramyabharathi SA, Meena B, Raguchander T (2012) Induction of chitinase and β-1, 3-glucanase PR proteins in tomato through liquid formulated *Bacillus subtilis* EPCO 16 against Fusarium wilt. J Today's Biol Sci Res Rev 1:50–60

Raupach GS, Kloepper JW (2000) Biocontrol of cucumber diseases in the field by plant growth-promoting rhizobacteria with and without methyl bromide fumigation. Plant Dis 84:1073–1075

Reddy PP (2014) Plant growth promoting rhizobacteria for horticultural crop protection. Springer, India

Reddy BV, Ramesh S, Reddy PS et al (2009) Genetic enhancement for drought tolerance in sorghum. Plant Breed Rev 31:189

Rohini, Gowtham HG, Niranjana SR (2015) Evaluation of efficacy of fungicides against phomopsis leaf blight of brinjal (*Solanum melongena* L.) Int J Agric Sci Res 5:45–50

Rose S, Parker M, Punja ZK (2003) Efficacy of biological and chemical treatments for control of Fusarium root and stem rot on greenhouse cucumber. Plant Dis 87:1462–1470

Sabalpara AN, Patel DU, Pandya JR et al (2009) Evaluation of fungicides against *Phomopsis vexans*. J Mycol Plant Pathol 39:55

Sadik S, Mazouz H, Bouaichi A et al (2013) Biological control of bacterial onion diseases using a bacterium, *Pantoea agglomerans* 2066-7. Int J Sci Res:2319–7064

Sahar P, Sahi ST, Jabbar A et al (2013) Chemical and biological management of *Fusarium oxysporum* f. sp. *melongenae*. Pak Phytopathol 25:155–159

Salau IA, Shehu K, Kasarawa AB (2015) A fungi associated with post-harvest rot of commonly consumed fruits in Sokoto metropolis, Nigeria. J Adv Bot Zool 3:3

Sang MK, Kim KD (2012) The volatile-producing *Flavobacterium johnsoniae* strain GSE09 shows biocontrol activity against *Phytophthora capsici* in pepper. J Appl Microbiol 113:383–398

Satish S, Mohana DC, Ranhavendra MP et al (2007) Antifungal activity of some plant extracts against important seed borne pathogens of *Aspergillus* sp. Int J Agric Technol 3:109–119

Savchuk SC (2002) Evaluation of biological control control of *Sclerotinia sclerotiorum* on Canola (*Brassica napus*) in the lab, in the greenhouse, and in the field. Doctoral dissertation, M.Sc. thesis, University of Manitoba, Manitoba, Canada

Sayyed RZ, Patel PR (2011) Biocontrol potential of siderophore producing heavy metal resistant *Alcaligenes* sp and *Acinetobacter* sp vis-a vis organophosphorus fungicides. Indian J Microbiol 3:266–273

Shafique HA, Noreen R, Sultana V et al (2015) Effect of endophytic *Pseudomonas aeruginosa* and *Trichoderma harzianum* on soil-borne diseases, mycorrhizae and induction of systemic resistance in okra grown in soil amended with *Vernonia anthelmintica* (L.) seed's powder. Pak J Bot 47:2421–2426

Shankar R, Harsha S, Bhandary R (2014) A practical guide to identification and control of cucumber diseases.http://www.tropicaseeds.com

Shanmugaiah V, Nithya K, Harikrishnan H (2015) Biocontrol mechanisms of siderophores against bacterial plant pathogens. Sustainable approaches to controlling plant pathogenic bacteria. CRC Press, Boca Raton, FL, pp 167–190

Sharma SR, Sohi HS (1981) Effect of different fungicides against *Rhizoctonia* root rot of French bean (*Phaseolus vulgaris*). Indian J Mycol Plant Pathol 11:216

Sharma N, Tripathi A (2006) Fungitoxicity of the essential oil of *Citrus sinensis* on post-harvest pathogens. World J Microbiol Biotechnol 22:587–593

Shrivastava P, Kumar R, Yandigeri MS (2015) In vitro biocontrol activity of halotolerant *Streptomyces aureofaciens* K20: a potent antagonist against *Macrophomina phaseolina* (Tassi) Goid. Saudi J Biol Sci. doi:10.1016/j.sjbs

Singh AK, Agrawal KC (1999) Fungicidal control of phomopsis fruit rot of brinjal. Veg Sci 26(2):192

Singh D, Yadav DK, Chaudhary G et al (2015) Potential of *Bacillus amyloliquefaciens* for biocontrol of bacterial wilt of tomato incited by *Ralstonia solanacearum*. J Plant Pathol Microbiol 7:1–6

Slininger PJ, Burkhead KD, Schisler DA et al (2000) Isolation, identification, and accumulation of 2-acetamidophenol in liquid cultures of the wheat take-all biocontrol agent *Pseudomonas fluorescens* 2-79. Appl Microbiol Biotechnol 54:376–381

Solecki R, Davies L, Dellarco V et al (2005) Guidance on setting of acute reference dose (ARfD) for pesticides. Food Chem Toxicol 43:1569–1593

Srivastava M, Gupta SK, Saxena AP (2011) A review of occurrence of fungal pathogens on significant brassicaceous vegetable crops and their control measures. Asian J Agric Sci 3:70–79

Subhani MN, Sahi ST, Ali L et al (2015) Genotypic variations in potassium contents of potato leaves infested with late blight of potato incited by *Phytophthora infestans* (Mont.) de Bary. J Environ Agric Sci 2:6

Sudhasha S, Usharani S, Ravimycin T (2008) Surveillance of onion basal rot disease incidence caused by *Fusarium oxysporum* f. sp. *cepae* and varietal reaction under field condition. Asian J Biol Sci 3:369–371

Suryawanshi AV, Deokar CD (2001) Effect of fungicides on growth and sporulation of fungal pathogens causing fruit rot of chilli. Madras Agric J 88:181–182

Taiga A, Suleiman MN, Sule W et al (2008) Comparative in vitro inhibitory effects of cold extracts of some fungicidal plants on *Fusarium oxysporum* mycelium. Afr Biotechnol 7(18):3306–3308

Tanina K, Tojo M, Date H et al (2004) Pythium rot of chingensai (*Brassica campestris* L. chinensis group) caused by *Pythium ultimum* var. *ultimum* and *Pythium aphanidermatum*. J Gen Plant Pathol 70:188–191

Thabet WM, Shendy AH, Gad Alla SA et al (2016) Chronic exposure of insecticide and fungicide as indicator of health impact in some commonly consumed leafy vegetables: case study. Cogent Food Agric. doi:10.1080/23311932.2016.1193926

Thakur N, Tripathi A (2015) Biological management of damping-off, buckeye rot and fusarial wilt of tomato (cv. Solan Lalima) under Mid-Hill conditions of Himachal Pradesh. Agric Sci 6:535

Thippeswamy B, Krishnappa M, Chakravarthy CN (2006) Pathogenicity and management of phomopsis blight and leaf spot in brinjal caused by *Phomopsis vexans* and *Alternaria solani*. Indian Phytopathol 59:475–481

Tripathi P, Dubey NK, Shukla AK (2008) Use of some essential oils as post-harvest botanical fungicides in the management of grey mould of grapes caused by *Botrytis cinerea*. World Microbiol Biotechnol 24:39–46

Twisha P, Desai PB (2014) Study on rhizospheric microflora of wild and transgenic varieties of *Gossypium* species in monsoon. Res J Recent Sci 3:42–51

Ulloa-Ogaz AL, Muñoz-Castellanos LN, Nevárez-Moorillón GV (2015) Biocontrol of phyto-pathogens: antibiotic production as mechanism of control, In: Mendez-Vilas (ed) The battle against microbial pathogens: basic science, technological advances and educational programes. FORMATEX RESEARCH CENTER, Spain, pp 305–309

Vaish DK, Sinha AP (2003) Determination of tolerance in *Rhizoctonia solani*, *Trichoderma virens* and *Trichoderma* sp (isolate 20) to systemic fungicides. Indian J Plant Pathol 21:48–50

Verma M, Brar SK, Tyagi RD et al (2007) Industrial wastewaters and dewatered sludge: rich nutri-ent source for production and formulation of biocontrol agent, *Trichoderma viride*. World J Microbiol Biotechnol 23:1695–1703

Vimala P, Ting CC, Salbiah H et al (1999) Biomass production and nutrient yields of four green manures and their effect on the yield of cucumber. J Trop Agrifood Sci 27:47–56

Vinale F, Sivasithamparam K, Ghisalberti EL et al (2008) *Trichoderma*–plant–pathogen interac-tions. Soil Biol Biochem 40:1–10

Whipps JM (2001) Microbial interactions and biocontrol in the rhizosphere. J Exp Bot 52:487–511

Wu L, Ashraf MHN, Facci M et al (2004) Dietary approach to attenuate oxidative stress, hyperten-sion, and inflammation in the cardiovascular system. Proc Natl Acad Sci U S A 101:7094–7099

Xiao L, Xie CC, Cai J et al (2009) Identification and characterization of a chitinase-produced *Bacillus* showing significant antifungal activity. Curr Microbiol 58:528–533

Yan A, Wang X, Zhang X et al (2007) LysR family factor PltR positively regulates pyoluteorin production in a pathway-specific manner in *Pseudomonas* sp. M18. Sci China Ser C Life Sci 50:518–524

Yu X, Ai C, Xin L et al (2011) The siderophore-producing bacterium, *Bacillus subtilis* CAS15, has a biocontrol effect on Fusarium wilt and promotes the growth of pepper. Eur J Soil Biol 47:138–145

Zaidi A, Ahmad E, Khan MS (2015) Role of plant growth promoting rhizobacteria in sustainable production of vegetables: current perspective. Sci Hortic 193:231–239

Zamoum M, Goudjal Y, Sabaou N (2015) Biocontrol capacities and plant growth-promoting traits of endophytic actinobacteria isolated from native plants of Algerian Sahara. J Plant Dis Protect 122:215–223

Zhou XG, Wu FZ, Wang XZ et al (2008) Progresses in the mechanism of resistance to *Fusarium* wilt in cucumber (*Cucumis sativus* L.) J Northeast Agric Univ 15:1–6

Printed by Printforce, the Netherlands